the complete
A-Z
CHEMISTRY
handbook
SECOND EDITION
Andrew Hunt

Hodder & Stoughton

A MEMBER OF THE HODDER HEADLINE GROUP

British Library Cataloguing in Publication Data
A catalogue entry for this title is available from the British Library

ISBN 0–340–77218–2

First published 1998
Second edition published 2000

Impression number   10 9 8 7 6 5 4 3 2
Year   2004 2003 2002 2001 2000

Typeset by GreenGate Publishing Services, Tonbridge, Kent.
Printed and bound in Great Britain for Hodder and Stoughton Educational,
a division of Hodder Headline plc, 338 Euston Road, London NW1 3BH,
by Redwood Books, Trowbridge, Wilts.

HOW TO USE THIS BOOK

The Complete A–Z Chemistry Handbook is an alphabetical textbook designed to help you at every stage of an advanced level chemistry course. Entries begin with one sentence definitions which explain why the terms are important. These are followed by fuller explanations and examples, together with diagrams, tables and equations. Words in italics are cross-references to other entries in the book.

You can use the book as a course companion and revision aid.

Course companion

The book covers all the main ideas in advanced level chemistry specifications. This concise, alphabetical book makes it easy for you to find the explanations, information and examples that you need when starting to study a new topic. Worked examples show you how to tackle the calculations.

Entries include many aspects of practical and applied chemistry, including both analytical chemistry and organic preparations. Furthermore, the book explains some of the terms in more specialist options so it will help you to get started on topics such as spectroscopy, environmental chemistry and biochemistry.

Revision aid

The appendices will help your revision. Appendix 1 includes revision lists for all the important topics. Appendices 2 and 3 offer hints for exam success, including advice on synoptic assessment. Appendix 4 shows which key terms you need to revise for your course. This appendix covers the modules in each of the five course specifications. Finally, Appendix 5 explains the terms that examiners use when setting questions.

Andrew Hunt

ACKNOWLEDGEMENTS

Firstly I thank Geoffrey Barraclough who both read and checked the whole book and contributed to the appendices. His experience of chemistry teaching and detailed understanding of examining at advanced level have made sure that the key term entries in this book will help students to improve the quality of their examination answers and avoid common errors.

Secondly I thank Ian Marcousé, the series editor and three experienced chemistry teachers: Rod Clough, Jeffrey Hancock, and Ted Lister, who read early drafts. Their wise comments helped me to develop the content and style of this book.

Thirdly I thank Tim Gregson-Williams of Hodder & Stoughton, David Mackin of GreenGate Publishing Services and all the people who work with them.

Finally I thank all the people I have worked with while contributing to the Nuffield Advanced Chemistry, SATIS 16–19, Salters Advanced Chemistry and other Nuffield Science projects. Writing and editing for these projects has clarified my understanding of chemical ideas and taught me much about the importance of chemistry in our lives.

Andrew Hunt

The publishers would like to thank the following for permission to reproduce illustrations in this book:

(page 76) from *Science for Understanding Tomorrow's World: Global Change for Education* with permission of the International Council for Science, 51, Bd. de Montmorency, 75016 Paris, France.

(page 82) from *Salters' Advanced Chemistry, Chemical Storylines* by Salters, with permission of Heinemann Educational Publishers, a division of Reed Educational & Professional Publishing Ltd.

While every effort has been made to contact copyright holders, any omissions brought to our attention will be remedied in future printings.

A$_r$ is the symbol for *relative atomic mass.*

absolute zero (0 K) is the temperature at which atoms and molecules in crystals are effectively motionless. It is the lowest temperature on the absolute or *Kelvin* temperature scale. A plot of the *volume* of a sample of gas against temperature at constant pressure is a straight line which on extrapolation cuts the temperature axis at −273.15°C.

Gases appear to behave as if they would have zero volume at absolute zero. This would be true of a non-existent *ideal gas* but in practice real gases turn to liquids and solids and occupy a definite volume before absolute zero is reached.

Absolute zero is now defined precisely by setting the *triple point* of water at 273.16 K.

absorption of liquids and gases happens when a *fluid* soaks into the pores of a material, like water soaking into a sponge. Absorption should be carefully distinguished from *adsorption.*

absorption of radiation: when *electromagnetic radiation* passes through a material some or all of the wavelengths of the radiation may be absorbed. When white light passes through a red filter, for example, all the wavelengths are absorbed except for red which passes through – so it looks red.

The gases in the *atmosphere* absorb most of the radiation reaching the Earth from the Sun.

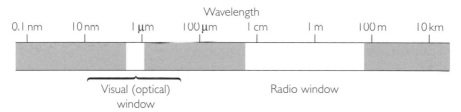

The bands of wavelengths in the electromagnetic spectrum of sunlight which can pass through the Earth's atmosphere to reach the surface of the Earth It is as if there arc two 'windows' letting through bands of radiation while all other wavelengths are shut out (by absorption).

Ozone in the upper atmosphere helps to protect living things by absorbing harmful *ultraviolet radiation.* Chemists use *spectroscopy* to study the absorption of radiation. From an *absorption spectrum* they can make deductions about the composition and structure of the sample.

absorption spectrum: a plot showing how strongly a sample absorbs radiation over a range of frequencies. Absorption spectra from *infra-red spectroscopy* and *ultraviolet spectroscopy* give chemists valuable information about the composition and structure of chemicals.

accuracy of data is determined by the agreement between a measured quantity and the correct value. In chemical analysis the correct value is often not known and so chemists need to estimate the *uncertainty of the measurements* (see also *errors of measurement*).

acids are compounds with the following characteristic properties:

- in water they form solutions with a *pH* below 7
- they change the colours of *acid–base indicators*
- they react with *metals* such as magnesium to produce hydrogen gas;

$$Mg(s) + 2HCl(aq) \longrightarrow MgCl_2(aq) + H_2(g)$$

- they react with carbonates such as calcium carbonate to form carbon dioxide gas;

$$CaCO_3(s) + 2HCl(aq) \longrightarrow CaCl_2(aq) + CO_2(g) + H_2O(l)$$

- they react with *basic oxides* to form salts and water;

$$CuO(s) + H_2SO_4(aq) \longrightarrow CuSO_4(aq) + H_2O(l)$$

Pure acids may be solids (such as citric and tartaric acids), liquids (such as *sulfuric, nitric* and *ethanoic* acids) or gases (such as hydrogen chloride which becomes *hydrochloric acid* when it dissolves in water).

Definitions of acids are based on the theories used to explain why they have similar properties. Chemists use different theories in different contexts. When considering acids in solution in water they still sometimes refer to the theory which the Swedish chemist Svante Arrhenius (1859–1927) suggested. He explained the behaviour of acids in terms of hydrogen ions. What acids have in common, according to this theory, is that they produce hydrogen ions when they dissolve in water.

$$HCl(g) \longrightarrow H^+(aq) + Cl^-(aq)$$

The definition of an acid generally used today is based on the *Brønsted–Lowry theory.* Another definition is based on the *Lewis acid/base theory.*

acid anhydrides are related to *carboxylic acids.* The *functional group* in an anhydride is formed by eliminating a molecule of water from two carboxylic acid groups. Sometimes this happens simply on heating the acid but generally anhydrides are made in other ways.

With water, alcohols and amines, acid anhydrides react in a similar way to *acyl chlorides.* Chemists call this type of reaction *acylation.*

Reactions of ethanoic anhydride

Anhydrides are less reactive than acyl chlorides and so are often preferred for laboratory and industrial syntheses. Their reactions may require heating in a flask fitted with a *reflux condenser*.

acid–base equilibria are equilibrium systems involving *acid–base reactions*. Acid–base reactions are *reversible*. This is illustrated by a solution of an *ammonium salt* in water.

$$NH_4^+(aq) + H_2O(l) \rightleftharpoons NH_3(aq) + H_3O^+(aq)$$

acid 1   base 2   base 1   acid 2

There is competition for *protons* (hydrogen ions) between ammonia molecules and water molecules. On the left-hand side of the equation the protons are held by *lone pairs of electrons* on the ammonia molecules. On the right-hand side they are held by lone pairs on water molecules.

The *equilibrium law* applies. The position of equilibrium is determined by the value of the equilibrium of constant for the reaction. For the example in the equation, the relevant equilibrium constant is the *acid dissociation constant* for the ammonium ion.

The equilibrium involves two conjugate acid–base pairs:

- NH_4^+ and NH_3
- H_3O^+ and H_2O.

An acid turns into its conjugate base when it loses a proton. A base turns into its conjugate acid when it gains a proton.

acid–base indicators are used to show up changes in pH of solutions. Indicators are *weak acids* or *bases* which change colour when they lose or gain hydrogen ions. When added to a solution an indicator gains or loses protons depending on the *pH* of the solution. It is conventional to represent a weak acid indicator as HIn where In is a shorthand for the rest of the molecule. In water:

$$HIn(aq) + H_2O(l) \rightleftharpoons H_3O^+(aq) + In^-(aq)$$

un-ionised   indicator after
indicator   losing a proton
colour 1   colour 2

The structures of methyl orange in acid and alkaline solutions. In acid solution the added hydrogen ion (proton) localises two electrons to form a covalent bond. In alkaline solution the removal of the hydrogen ion allows the two electrons to join the other delocalised electrons. The change in the number of delocalised electrons causes a shift in the peak of the wavelengths of light absorbed, so the colour changes and the molecule acts as an indicator.

Indicators such as phenolphthalein, methyl orange and bromothymol blue change colour over a range of pH values and are used to detect the *end-point* in *acid–base titrations*.

The pH range over which an indicator changes colour is determined by its strength as an acid (or base). Typically the range is given roughly by $pK_a \pm 1$ (see the *Henderson–Hasselbalch equation* for an explanation).

Indicator	pK_a	colour change HIn/In⁻	pH range over which colour change occurs
methyl orange	3.6	red/yellow	3.2–4.2
methyl red	5.0	yellow/red	4.2–6.3
bromothymol blue	7.1	yellow/blue	6.0–7.6
phenolphthalein	9.4	colourless/red	8.2–10.0

acid–base reactions, according to the *Brønsted–Lowry theory*, are reactions involving the transfer of protons from an acid to a base.

acid–base titration: a practical technique used to determine the *concentration* of an *acid* or an *alkali*. A *titration* measures the volume of a *standard solution* of alkali or acid needed to react exactly with a measured volume of the unknown solution. The procedure for an acid–base titration is the same as for any other titration but the method for finding the end-point is distinctive: the analyst either adds an *acid–base indicator* or uses a *pH meter*.

Worked example:

Limewater is a saturated solution of calcium hydroxide in water. 25.0 cm³ of 0.04 mol dm⁻³ hydrochloric acid from a burette neutralised 20.0 cm³ of limewater. What was the concentration of the limewater?

Notes on the method
Always start by writing the equation for the reaction. See *titration* for a general method for the calculations.

Remember to convert volumes in cm³ to volumes in dm³ by dividing by 1000.

In any titration there is one unknown – in this case the concentration of the limewater, c_A.

Answer
The equation for the reaction is:

$$Ca(OH)_2(aq) + 2HCl(aq) \longrightarrow CaCl_2(aq) + 2H_2O(l)$$

The volume of calcium hydroxide in the flask, $V_A = \dfrac{20.0}{1000}$ dm³

Let the concentration of calcium hydroxide be c_A.

The volume of hydrochloric acid added from the burette, $V_B = \dfrac{25.0}{1000}$ dm³

The concentration of hydrochloric acid, $c_B = 0.04$ mol dm⁻³

$$\frac{V_A \times c_A}{V_B \times c_B} = \frac{n_A}{n_B}$$

$$\frac{20.0/000 \times c_A}{25.0/000 \times c_B} = \frac{1}{2}$$

Therefore $c_A = \dfrac{25.0 \times 0.04}{2 \times 20.0} = 0.025 \text{ mol dm}^{-3}$

The concentration of the limewater was 0.025 mol dm^{-3}.

acid catalysis: any reaction speeded up by an acid *catalyst*. One example is the acid-catalysed *hydrolysis* of an *ester* to give a *carboxylic acid* and an *alcohol*. All the chemicals are in the same solution so this is an example of *homogeneous catalysis*.

acid chlorides: see *acyl chlorides*.

acid dissociation constants measure the strength of acids. They show the extent to which acids dissociate into ions in solution. Acid dissociation constants are used to compare the strengths of relatively *weak acids* which are only partly ionised in solution.

For a weak acid represented by the formula HA:

$$HA(aq) + H_2O(l) \rightleftharpoons H_3O^+(aq) + A^-(aq)$$

According to the *equilibrium law*, the equilibrium constant,

$$K_c = \frac{[H_3O^+(aq)][A^-(aq)]}{[HA(aq)][H_2O(l)]}$$

In dilute solution the concentration of water is effectively constant, so the expression can be written in this form:

$$K_a = \frac{[H_3O^+(aq)][A^-(aq)]}{[HA(aq)]}, \text{ where } K_a \text{ is the acid dissociation constant}$$

Acid	$K_a/\text{mol dm}^{-3}$	
nitric(III) acid (nitrous acid), HNO$_2$	4.7×10^{-4}	stronger
methanoic acid, HCO$_2$H	1.6×10^{-4}	
ethanoic acid, CH$_3$CO$_2$H	1.7×10^{-5}	
chloric(I) acid, HOCl	3.7×10^{-8}	
phenol, C$_6$H$_5$OH	1.3×10^{-10}	weaker

Worked example:

Calculate the hydrogen ion concentration and the pH of a 0.01 mol dm^{-3} solution of ethanoic acid. K_a for the acid is 1.7×10^{-5} mol dm^{-3}.

Notes on the method

Two approximations simplify the calculation.

1 The first assumption is that $[H_3O^+(aq)] = [A^-(aq)]$. In this example A$^-$ is the ethanoate ion CH$_3$CO$_2^-$. This assumption seems obvious from the equation

for the ionisation of a weak acid but it ignores the hydrogen ions from the ionisation of water. Water produces far fewer hydrogen ions than most weak acids so its ionisation can be ignored.

2 The second assumption is that so little of the ethanoic acid ionises in water that $[HA(aq)] \approx 0.01$ mol dm^{-3}. Here HA represents ethanoic acid. This is a riskier assumption which has to be checked because in very dilute solutions the degree of ionisation may become quite large relative to the amount of acid in the solution.

Answer

$$K_a = \frac{[H_3O^+(aq)][A^-(aq)]}{[HA(aq)]} = \frac{[H_3O^+(aq)]^2}{0.01 \text{ mol dm}^{-3}} = 1.7 \times 10^{-5} \text{ mol dm}^{-3}$$

Therefore $[H_3O^+(aq)]^2 = 1.7 \times 10^{-7}$ mol^2 dm^{-6}

So $[H_3O^+(aq)] = 4.12 \times 10^{-4}$ mol dm^{-3}

pH $= -\lg [H_3O^+(aq)] = -\lg [4.12 \times 10^{-4}] = 3.39$

Check the second assumption: in this case about 0.0004 mol dm^{-3} of the 0.0100 mol dm^{-3} of acid (4%) has ionised. In this instance the degree of ionisation is just about small enough to justify the assumption that $[HA(aq)] \approx$ the concentration of un-ionised acid.

---

acidic oxide: an oxide of a *non-metal* which reacts with water to form an *acid*. Some acidic oxides are insoluble but they can be recognised because they react directly with *basic oxides* to form *salts*. Note that acidic oxides are not themselves *acids* as defined by the Brønsted–Lowry theory because they do not contain ionisable hydrogen atoms, so they cannot act as proton donors.

Oxide	Acid formed with water
carbon dioxide, CO_2	carbonic acid, H_2CO_3
sulfur dioxide, SO_2	sulfurous acid, H_2SO_3 [sulfuric(IV) acid]
sulfur trioxide, SO_3	sulfuric acid, H_2SO_4 [sulfuric(VI) acid]
phosphorus(V) oxide, P_2O_5	phosphoric acid, H_3PO_4

Silica or silicon dioxide (SiO_2) is an acidic oxide which is insoluble in water. Silica reacts with basic metal oxides at high temperatures in furnaces during glass making and in steelmaking. Calcium oxide (quicklime) is added to a *blast furnace* to remove silica and other impurities. The calcium silicate is a liquid at the temperature of the furnace; it runs towards the bottom where it floats on top of the molten iron as a slag which can be tapped off separately.

$$CaO(s) + SiO_2(s) \longrightarrow CaSiO_3(l)$$

acid rain is a type of pollution produced by burning fuels, wastes or when industrial processes release *acidic oxides* into the air. Sulfur dioxide (SO_2) and *nitrogen oxides*, (NO_x) form during *combustion* in engines, furnaces, incinerators and power stations. These primary pollutants are converted to secondary pollutants by chemical reactions in the air. Among the secondary pollutants are sulfuric acid, nitric acid and ammonium sulfate The pollutants cause acidification by being deposited in the environment as gases or particles (dry deposition) or in the form of rain or mist (wet deposition).

acid salt: a compound formed by partly neutralising *acids* such as *sulfuric acid*, H_2SO_4, or *phosphoric acid*, H_3PO_4. These are acids with two or three ionisable hydrogen atoms. Sodium hydrogensulfate, $NaHSO_4$, and sodium dihydrogenphosphate, NaH_2PO_4 are examples of acid salts.

The negative ion in an acid salt can act either as an acid by giving away a further proton, or as a *base* by accepting a proton. The pH of an *aqueous solution* of an acid salt depends on the *acid strength* of the parent acid.

A solution of sodium hydrogensulfate, a salt of a *strong acid*, is acidic with a pH below 7 because the hydrogensulfate ion is also an acid, giving protons to water molecules:

$$HSO_4^- + H_2O \rightleftharpoons SO_4^{2-} + H_3O^+$$

A solution of sodium hydrogencarbonate, a salt of the *weak acid* called carbonic acid, is alkaline with a pH above 7 because the hydrogencarbonate ion is a base, taking protons (H^+) from water molecules:

$$HCO_3^- + H_2O \rightleftharpoons H_2CO_3 + OH^-$$

acid strength: see *strong acid, weak acid* and *acid dissociation constant*.

acid strength of organic hydroxy compounds: the order of acid strength for organic compounds with — OH groups is:

carboxylic acids > phenols > water > alcohols.

Carboxylic acids are *weak acids* which ionise to a significant extent by giving protons to water forming a solution with pH 3–4. *Phenol* is acid enough to lower the pH of a solution in water to 5–6 but it ionises significantly only when mixed with a stronger base such as the hydroxide ions in a solution of sodium hydroxide. Ethanol and other *alcohols* are such weak acids that they do not ionise to a significant extent in water or in a solution of a strong alkali.

actinides (or actinoids) are the *f-block elements* in period 7 of the *periodic table*. They are the 14 elements from thorium to lawrencium which lie between actinium (element 89) and element 104, rutherfordium.

activated complex: a combination of reacting atoms, molecules or ions at that point during a chemical reaction when they are at the top of the *activation energy* barrier

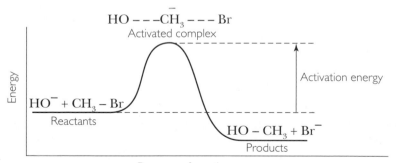

Reaction profile to show the activated complex formed in the course of a *nucleophilic substitution reaction*

between reactants and products. An activated complex is a combination of reacting particles at a higher energy because chemical bonds are stretched. The activated complex is the *transition state* for the reaction step.

activation energy: the minimum energy needed in a collision between molecules if they are to react. The activation energy is the height of the energy barrier separating reactants and products during a chemical reaction (see *activated complex*).

Activation energy is an important idea in the *collision theory* of reaction rates. It accounts for the fact that reactions go much more slowly than would be expected if every collision between atoms and molecules led to a reaction. Only a very small proportion of collisions bring about chemical change. Molecules react only if they collide with enough energy between them to overcome the energy barrier. At around room temperature only a small proportion of molecules have enough energy to react.

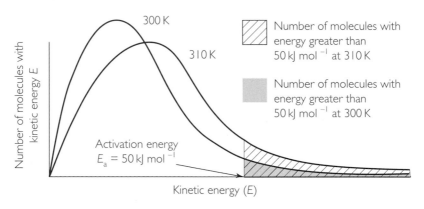

Maxwell–Boltzmann distribution curve for the kinetic energies of molecules at about room temperature, 300 K and 310 K. The shaded area shows the proportion of molecules having at least the activation energy for a reaction. This area is bigger at a higher temperature.

Activation energies account for the way in which reaction rates vary with temperature, *T*. At a higher temperature there are more molecules with enough energy to react when they collide.

The activation energies for many reactions in biochemical systems are around 50 kJ mol^{-1} and for these reactions the rate of reaction doubles for each 10 K rise in temperature.

It is possible to determine the activation energy for a reaction by measuring its rate over a range of temperatures and then using the *Arrhenius equation*.

active site: the part of a *catalyst* which is responsible for its catalytic activity. *Zeolites* are examples of inorganic catalysts, for example, which owe their catalytic activity to active sites in the crystal structure which can adsorb specific molecules so that they can react to produce the required products.

Enzymes are biological catalysts. An enzyme molecule consists of a coiled *protein* chain. The coiling gives rise to an active site with a precise three dimensional shape. The enzyme can act only on molecules with the right shape to fit into the active site. This accounts for the specificity of enzymes.

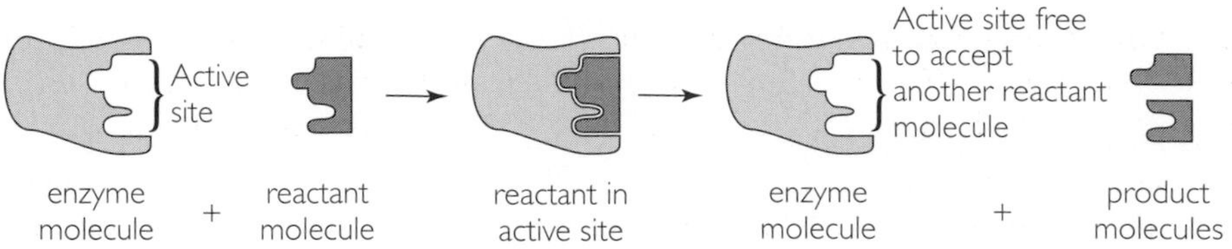

| enzyme molecule | + | reactant molecule | reactant in active site | enzyme molecule | + | product molecules |

A model illustrating an enzyme with an active site splitting a molecule into two smaller molecules

acyl chlorides are made from *carboxylic acids* by replacing the hydroxyl (— OH) group with a chlorine atom.

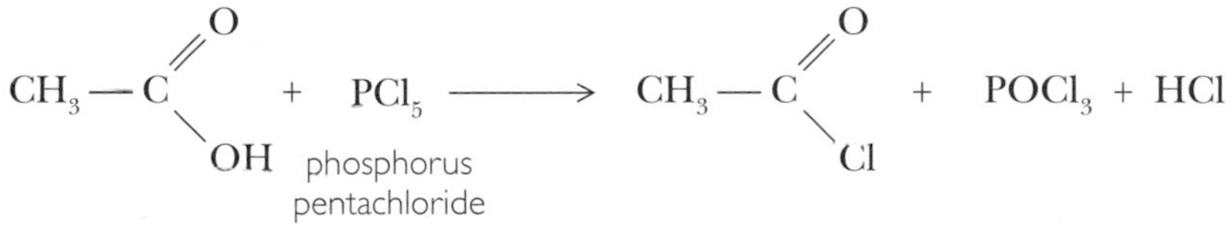

$$CH_3 - C\overset{O}{\underset{OH}{<}} \quad + \quad PCl_5 \quad \longrightarrow \quad CH_3 - C\overset{O}{\underset{Cl}{<}} \quad + \quad POCl_3 + HCl$$

phosphorus pentachloride

Formation of ethanoyl chloride from ethanoic acid

Acyl chlorides are very reactive; they are powerful acylating agents. Most are quickly *hydrolysed* back to the acid by cold water. Acyl chlorides react rapidly with *alcohols* and *phenols* to form *esters*. These reactions are all examples of *acylation*.

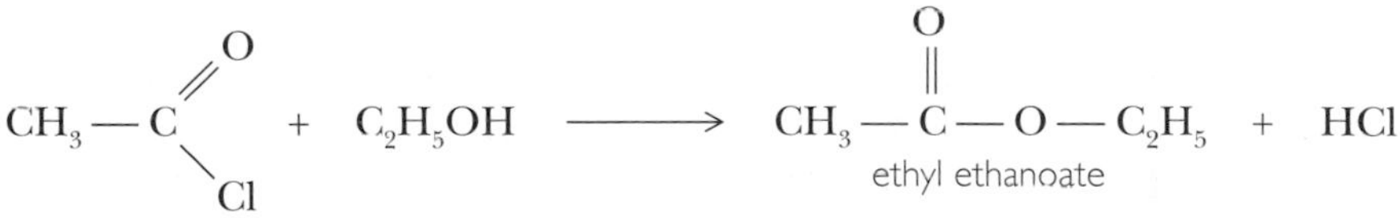

$$CH_3 - C\overset{O}{\underset{Cl}{<}} \quad + \quad C_2H_5OH \quad \longrightarrow \quad CH_3 - \overset{O}{\overset{\|}{C}} - O - C_2H_5 \quad + \quad HCl$$

ethyl ethanoate

Formation of an ester

They react with *ammonia* and *amines* to form *amides*.

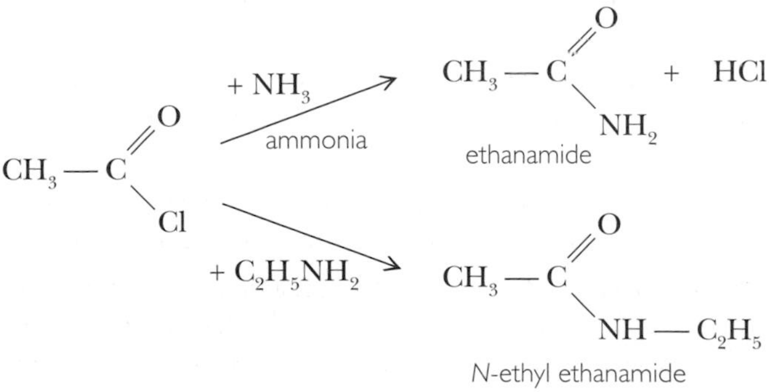

$$CH_3 - C\overset{O}{\underset{Cl}{<}}$$

$+ NH_3$ (ammonia) $\nearrow CH_3 - C\overset{O}{\underset{NH_2}{<}} \quad + \quad HCl$ (ethanamide)

$+ C_2H_5NH_2 \searrow CH_3 - C\overset{O}{\underset{NH - C_2H_5}{<}}$ (N-ethyl ethanamide)

Formation of an amide and an *N*-substituted amide

Acyl chlorides can also acylate *benzene* by the *Friedel–Crafts* reaction.

acylation is a reaction which substitutes an acyl group for a hydrogen atom. The H atom may be part of an — OH group, an — NH_2 group or a *benzene* ring. Acylating agents are either *acyl chlorides* or *acid anhydrides*.

$$CH_3 - C \overset{O}{\underset{\diagdown}{\parallel}}$$

The ethanoyl group – an acyl group

Acylation is the reaction which converts 2-hydroxybenzoic acid to *aspirin*.

Formation of aspirin by acylation of 2-hydroxybenzoic acid with ethanoic anhydride

Acylation of benzene is an example of the *Friedel–Crafts reaction.*

addition–elimination reactions take place when two molecules first add together and then immediately split off a small molecule such as water or hydrogen chloride. They are sometimes called *condensation reactions.*

The reactions take place in three stages:

- nucleophilic addition
- gain and loss of hydrogen ions (protons)
- elimination of water.

acyl halides – the reactions of *acyl chlorides* with the *nucleophiles* water, alcohols, ammonia and amines are all addition–elimination reactions.

Mechanism of the reaction of an alcohol (ethanol) with an acyl halide (ethanoyl chloride) to form an ester (ethyl ethanoate)

carbonyl compounds – the addition–elimination reactions of *carbonyl compounds* take place with compounds of the form $X - NH_2$.

$$X - \overset{\overset{\displaystyle H}{|}}{\underset{\underset{\displaystyle H}{|}}{N\!:}} \overset{R}{\underset{R'}{\diagdown}} C = O \xrightarrow{\text{addition}} X - \overset{\overset{\displaystyle H}{|}}{\underset{\underset{\displaystyle H}{|}}{\overset{+}{N}}} - \overset{\overset{\displaystyle R}{|}}{\underset{\underset{\displaystyle R'}{|}}{C}} - O^-$$

gain and loss of protons

$$H_2O \; + \; X - N = C \overset{\diagup R}{\diagdown R'} \xleftarrow{\text{elimination}} X - \overset{}{\underset{\underset{\displaystyle H}{|}}{N}} - \overset{\overset{\displaystyle R}{|}}{\underset{\underset{\displaystyle R'}{|}}{C}} - OH$$

Generalised reaction of a carbonyl compounds with $X - NH_2$. R and R' are both alkyl groups in ketones. In an aldehyde R = H while R' is an alkyl group.

The 2,4-dinitrophenylhydrazine in *Brady's reagent* reacts with carbonyl compounds in this way to form 2,4-dinitrophenylhydrazone *derivatives*.

addition polymerisation is a process for making *polymers* from compounds containing *double bonds*. Many molecules of the *monomer* add together to form a long chain polymer. Ethene, for example, polymerises to form poly(ethene).

$$nCH_2 = CH_2 \longrightarrow \left[CH_2 - CH_2 \right]_n$$

Monomer	Polymer	Notes
ethene	poly(ethene) or polythene	A high-pressure, high-temperature process in the presence of a peroxide initiator produces low-density poly(ethene) with branched chains. The mechanism involves a *free-radical chain reaction.*
$\overset{H}{\underset{H}{\diagdown}} C = C \overset{\diagup H}{\underset{\diagdown H}{}}$	$\left[\overset{\overset{\displaystyle H}{|}}{\underset{\underset{\displaystyle H}{|}}{C}} - \overset{\overset{\displaystyle H}{|}}{\underset{\underset{\displaystyle H}{|}}{C}} \right]_n$	A low-pressure, low-temperature process with a *Ziegler–Natta catalyst* produces high-density poly(ethene) in which the polymer chains pack closer because they have no side branches.

(Cont'd)

propene 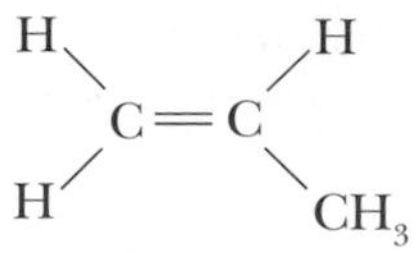	poly(propene) or polypropylene 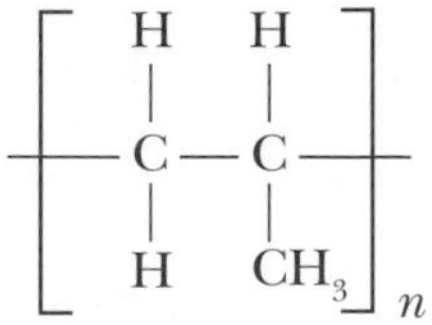	*Isotactic* poly(propene) is made using a Ziegler–Natta catalyst. This strong polymer is used to make pipes, wrapping films, carpet fibres and ropes.

$$\underset{H}{\overset{H}{\diagup}}C=C\underset{CH_3}{\overset{H}{\diagdown}}$$

$$\left[\begin{array}{c} H \ \ H \\ | \ \ \ | \\ -C-C- \\ | \ \ \ | \\ H \ \ CH_3 \end{array}\right]_n$$

phenylethene 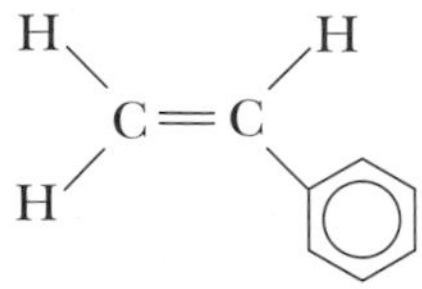	poly(phenylethene) or polystyrene 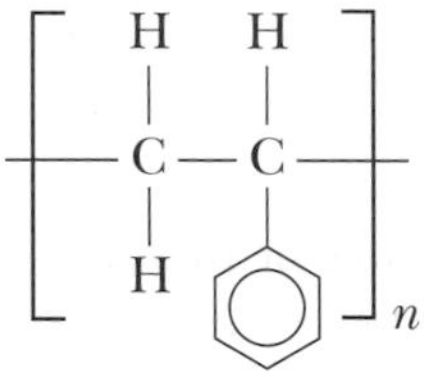	Expanded polystyrene has low density and is an excellent thermal insulator. It is used for packaging because it absorbs shocks.

chloroethene 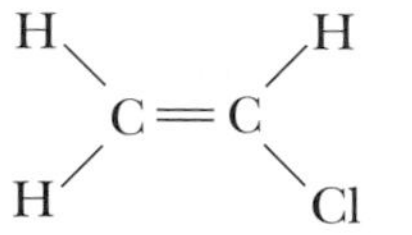	poly(chloroethene)or polyvinylchloride (PVC) 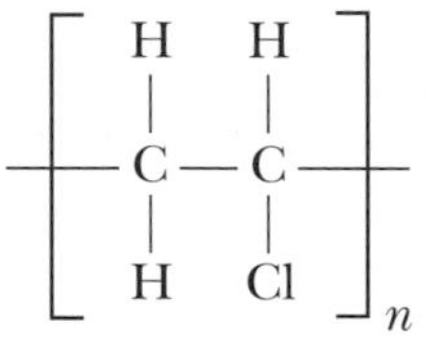	Unplasticised uPVC is a rigid polymer suitable for guttering and window frames. PVC with a *plasticiser* is flexible and used for packaging, flooring and cable insulation.

tetrafluorethene 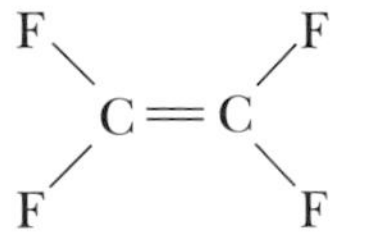	poly(tetrafluorethene) or ptfe 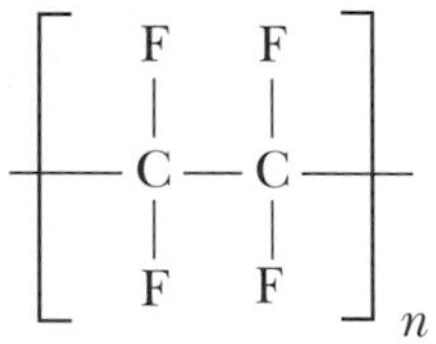	Engineers use ptfe to provide low-friction surfaces to allow bridges and other engineering structures to move slightly as the metals expand and contract with temperature changes. It is the polymer used to coat non-stick pans.

addition reaction: a reaction in which two molecules combine to form a single product. Bromine, for example, adds to ethene to form the addition product 1,2-dibromoethane.

Structure diagram to show addition of bromine to ethene

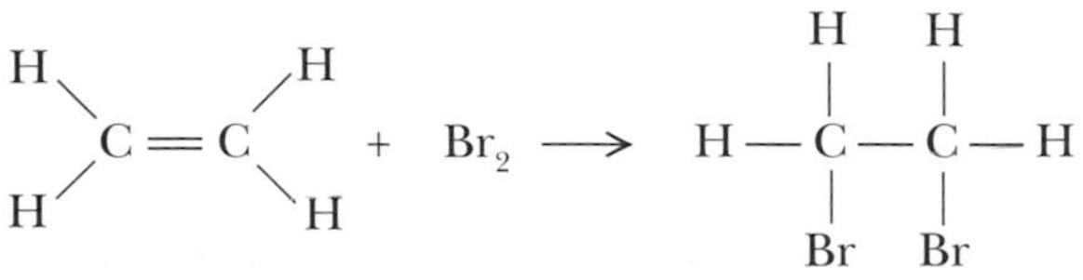

$$\underset{H}{\overset{H}{\diagup}}C=C\underset{H}{\overset{H}{\diagdown}} \ + \ Br_2 \longrightarrow H-\underset{Br}{\overset{H}{C}}-\underset{Br}{\overset{H}{C}}-H$$

1,2-dibromoethane

Addition reactions are characteristic of *unsaturated compounds* such as the *alkenes* and *carbonyl compounds*.

adsorption is a process in which atoms, molecules or ions are held on the surface of a solid. Adsorption processes are important in *heterogeneous catalysis* and in some types of *chromatography*. Adsorption should be carefully distinguished from *absorption*.

aerobic respiration: see *respiration*.

aerosols are *colloids* in which particles of a solid, or droplets of a liquid are finely dispersed in a gas. Smoke is an aerosol in which the dispersed particles are solid. Examples of aerosols with liquid droplets are mist, clouds and insecticide or paint sprays.

agrochemicals include *pesticides* which destroy the organisms that damage crops such as insects (insecticides) and weeds (*herbicides*). Other agrochemicals are the growth regulators which can stimulate or inhibit plant growth. The chemical industry distinguishes agrochemicals from *fertilisers*, which provide the chemicals needed for healthy plant growth.

air is a mixture of gases as shown in the table below. The proportion of water *vapour* in the air varies widely according to the weather conditions. Air pollution adds other gases such as *nitrogen oxides* , sulfur dioxide, *ozone* and *hydrocarbons* (see also *acid rain* and *photochemical smog*).The gases are separated on a large scale by *fractional distillation* of liquid air. UK production of oxygen is about two million tonnes per year.

gas	percentage by volume in dry air	boiling point/K
nitrogen	78.0	77
oxygen	21.0	90
argon	0.9	87
carbon dioxide (about 0.04%) and small traces of neon, helium and krypton	0.1	

air condenser: a glass tube without a water jacket used for condensing *vapours* during *distillation* or refluxing (see *reflux condenser*). An air condenser is fitted when the temperature of the vapour is so high that the surrounding air is cool enough to condense the liquid. Normally this is when the boiling point of the liquid being distilled is above about 150°C.

alcohols are compounds with the formula R — OH where R represents an *alkyl group*. The hydroxy group — OH is the *functional group* which gives the compounds their characteristic reactions.

The chemical industry makes alcohols by the *hydration* of *alkenes* in the presence of an acid *catalyst*.

$$H_2C = CH_2 + H_2O \xrightarrow[\text{300°C, 60 atm}]{\substack{\text{phosphoric acid} \\ \text{catalyst}}} H_3C - CH_2 - OH$$

Equation for the industrial production of ethanol. Ethanol is also produced by fermentation.

Alcohols are named by changing the ending of the corresponding *alkane* to –ol. So ethane becomes ethanol.

methanol   ethanol

$$CH_3 - CH_2 - CH_2 - CH_2 - OH$$ butan-1-ol, a primary alcohol

$$CH_3 - CH_2 - CH - CH_3$$
$$|$$
$$OH$$ butan-2-ol, a secondary alcohol

$$CH_3 - \underset{\underset{OH}{|}}{\overset{\overset{CH_3}{|}}{C}} - CH_3$$ 2-methylpropan-2-ol, a tertiary alcohol

Names and structures of alcohols

Even the simplest alcohols such as methanol and ethanol are liquids at room temperature because of *hydrogen bonding* between the hydroxy groups. For the same reason alcohols with relatively short hydrocarbon chains mix freely with water.

The reactions of ethanol are typical of the alcohols in general.

Reactions of ethanol

Oxidation reactions distinguish primary alcohols, secondary alcohols and tertiary alcohols (see *primary, seconday and tertiary organic compounds*).

$$\text{primary alcohol} \xrightarrow[\text{warm}]{Cr_2O_7^{2-}, H^+(aq)} \text{aldehyde} \xrightarrow[\text{reflux}]{\text{excess } Cr_2O_7^{2-}, H^+(aq)} \text{carboxylic acid}$$

$$\text{secondary alcohol} \xrightarrow[\text{reflux}]{Cr_2O_7^{2-}, H^+(aq)} \text{ketone}$$

tertiary alcohol   no reaction

Use of oxidation with an acidified solution of dichromate(VI) ions to distinguish primary, secondary and tertiary alcohols

aldehydes are *carbonyl compounds* in which the carbonyl group is attached to two hydrogen atoms or an *alkyl group* and a hydrogen atom. So the carbonyl group is at the end of a carbon chain. The — CHO group is the *functional group* which gives aldehydes their characteristic reactions. They are named after the alkane with the same carbon skeleton by changing the ending 'e' to 'al' (so, for example, ethane becomes ethanal).

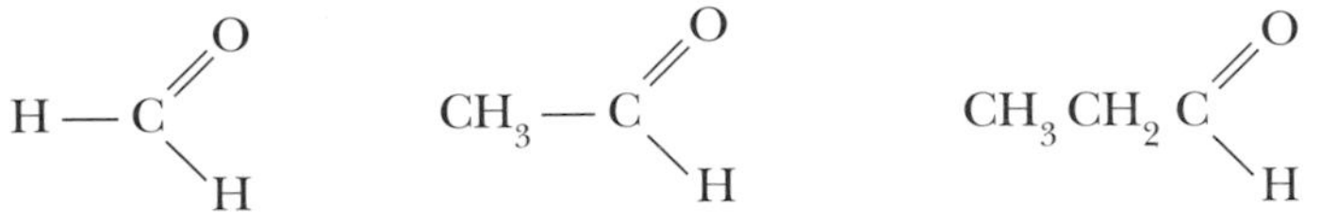

methanal   ethanal   propanal

Structures and names of aldehydes

Methanal is a gas at room temperature. Ethanal boils at $21°C$ so it may be a liquid or gas at room temperature depending on the conditions.

Aldehydes are formed by *oxidation* of primary *alcohols* on heating with a mixture of dilute sulfuric acid and potassium dichromate(VI) under conditions which allow the aldehyde to distil off as it forms. Unlike *ketones*, aldehydes can easily be oxidised further to *carboxylic acids* by longer heating with an excess of the reagent.

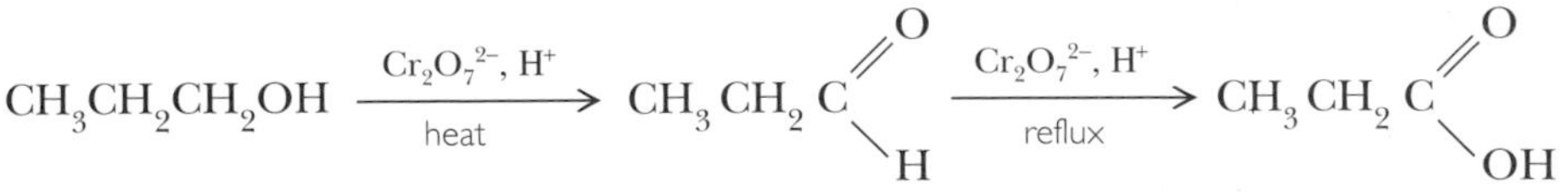

$$CH_3CH_2CH_2OH \xrightarrow[\text{heat}]{Cr_2O_7^{2-}, H^+} CH_3CH_2C{\overset{O}{\underset{H}{}}} \xrightarrow[\text{reflux}]{Cr_2O_7^{2-}, H^+} CH_3CH_2C{\overset{O}{\underset{OH}{}}}$$

propan-1-ol   propanal   propanoic acid

Two stage oxidation of propan-1-ol

Fehling's solution and *Tollen's reagent* (ammoniacal silver nitrate) are mild oxidising agents, used to distinguish aldehydes from ketones. Preparing a crystalline derivative with *Brady's reagent* (2,4-dinitrophenylhydrazine) makes it possible to identify aldehydes. Ethanal is the only aldehyde to undergo the *triiodomethane reaction*.

Sodium tetrahydridoborate(III) (NaBH$_4$) reduces aldehydes to primary alcohols. An alternative reducing agent is *lithium tetrahydridoaluminate(III)* (LiAlH$_4$) but it is not so easy to use.

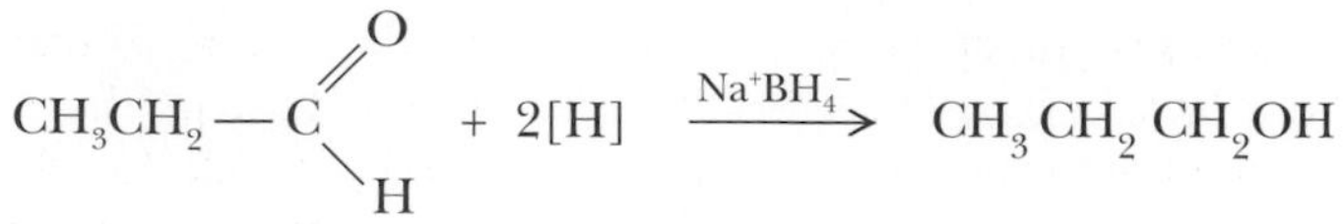

$$CH_3CH_2 - \overset{\displaystyle O}{\underset{\displaystyle H}{C}} \quad + \ 2[H] \quad \xrightarrow{Na^+BH_4^-} \quad CH_3CH_2CH_2OH$$

Reduction of propanal to propan-1-ol. The 2[H] comes from the reducing agent. This is a shorthand way of balancing a complex equation involving reduction.

Thanks to the double bond in the carbonyl group, aldehydes (like ketones) undergo *addition reactions*. These are *nucleophilic addition reactions*.

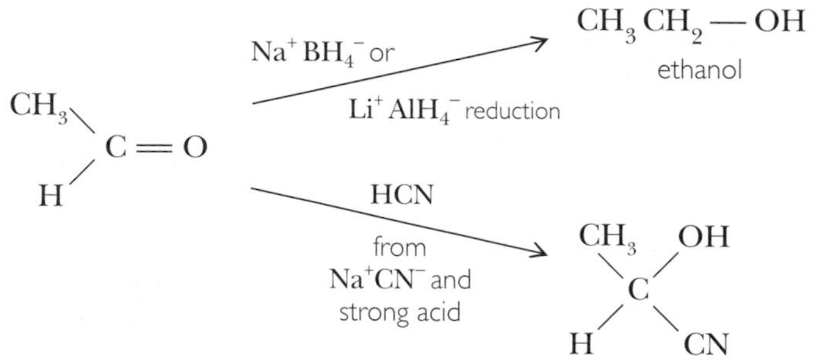

Addition reactions of ethanal

2-hydroxypropanenitrile
(which hydrolyses to
2-hydroxypropanoic acid)

In some reactions addition is immediately followed by elimination of water. These are *addition–elimination reactions*.

alicyclic hydrocarbons are *hydrocarbons* with rings of carbon atoms (but no *benzene* rings). Examples are cycloalkanes and cycloalkenes.

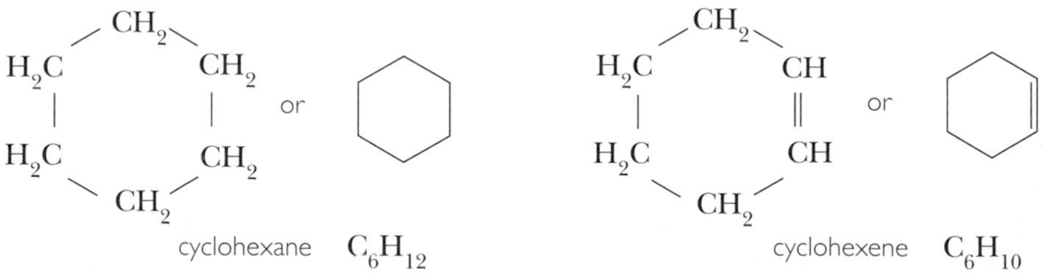

Structures of cyclohexane and cyclohexene

aliphatic hydrocarbons are *hydrocarbons* with no rings of carbon atoms. The chains of carbon atoms may be branched or unbranched. *Alkanes, alkenes* and *alkynes* are all aliphatic compounds.

aliquot: a measured volume taken from a liquid sample for analysis. Typically aliquots are taken by a pipette for *titration*.

alkalis are *bases* which dissolve in water. The common laboratory alkalis are the hydroxides of sodium and potassium, calcium hydroxide (in lime water) and *ammonia*. Alkalis form solutions with a *pH* above 7 so they change the colours of *acid–base indicators*. What alkalis have in common is that they dissolve in water to produce hydroxide (OH^-) ions. Sodium hydroxide (Na^+OH^-) and potassium hydroxide (K^+OH^-) contain hydroxide ions in the solid as well as in solution. Ammonia produces hydroxide ions by reacting with water. Ammonia acting as a base takes *protons* from water molecules.

$$NH_3(g) \ + \ H_2O(g) \rightleftharpoons NH_4^+(aq) \ + \ OH^-(aq)$$

alkali metals: the elements in *group 1* of the periodic table. All the elements react with water to form alkaline solutions of the metal hydroxide. Sodium, for example, reacts to form *sodium hydroxide*. It is important to remember that the hydroxides are alkaline because of the presence of OH^- ions. Alkali metal ions themselves, such as sodium ions, Na^+, do *not* make solutions alkaline.

alkaline cell: an *electrochemical cell* in which the electrolyte is the alkali, potassium hydroxide. Alkaline cells are suitable for toys and cassette recorders with a motor drawing a heavy or continuous current.

alkaline earth metals: the elements in *group 2* of the periodic table. The alkaline earths are the *oxides* of these metals which are much less soluble in water than the corresponding group 1 oxides.

alkaloids are organic nitrogen compounds extracted from plants. Alkaloids can have a powerful effect on the human nervous system and can be very poisonous. Caffeine, the stimulant in tea and coffee, is an alkaloid. Other alkaloids are the *drugs* quinine, morphine and codeine.

Structure of caffeine

alkanes are the *hydrocarbons* which make up most of crude oil and natural gas. Alkanes are *saturated compounds* with the general formula C_nH_{2n+2}. The carbon atoms in alkane molecules may be in straight chains or branched chains but all the bonds are single bonds.

The names of branched alkanes are based on the longest straight chain in the molecule with the positions of the side chain alkyl groups identified by numbering the carbon atoms.

Name	Molecular formula	Structure
methane	CH_4	$$H-\overset{\displaystyle H}{\underset{\displaystyle H}{\overset{\textstyle |}{\underset{\textstyle |}{C}}}}-H$$
ethane	C_2H_6	$$H-\overset{\displaystyle H}{\underset{\displaystyle H}{\overset{\textstyle |}{\underset{\textstyle |}{C}}}}-\overset{\displaystyle H}{\underset{\displaystyle H}{\overset{\textstyle |}{\underset{\textstyle |}{C}}}}-H$$
propane	C_3H_8	$$H-\overset{\displaystyle H}{\underset{\displaystyle H}{\overset{\textstyle |}{\underset{\textstyle |}{C}}}}-\overset{\displaystyle H}{\underset{\displaystyle H}{\overset{\textstyle |}{\underset{\textstyle |}{C}}}}-\overset{\displaystyle H}{\underset{\displaystyle H}{\overset{\textstyle |}{\underset{\textstyle |}{C}}}}-H$$

$$H-\overset{\displaystyle H}{\underset{\displaystyle H}{C}}-\overset{\displaystyle CH_3}{\underset{\displaystyle CH_3}{C}}-\overset{\displaystyle H}{\underset{\displaystyle H}{C}}-\overset{\displaystyle H}{\underset{\displaystyle H}{C}}-H$$

2,2-dimethylbutane

$$H-\overset{\displaystyle H}{\underset{\displaystyle H}{C}}-\overset{\displaystyle CH_3}{\underset{\displaystyle H}{C}}-\overset{\displaystyle CH_3}{\underset{\displaystyle H}{C}}-\overset{\displaystyle H}{\underset{\displaystyle H}{C}}-H$$

2,3-dimethylbutane

Names and structures of some branched alkanes

Alkane molecules are non-*polar* so they do not mix with, or dissolve in, *polar solvents* such as water. The molecules are only held together by weak *intermolecular forces* (*van der Waals forces*). The longer the molecules the greater the attraction between them. The boiling points rise as the number of carbon atoms per molecule increases. Alkanes in the range C_1 to C_4 are gases at room temperature and pressure. Under the same conditions, alkanes in the range C_5 to C_{17} are liquids while those with more than 17 carbon atoms per molecule are solids. Liquid alkanes with longer chain lengths are *viscous liquids* used as lubricants.

The *bond enthalpies* for $C-C$ and $C-H$ bonds are high so the bonds are relatively hard to break. Also the bonds are not polar. This means that alkanes are very unreactive towards reagents in water such as acids and alkalis, as well as oxidising and reducing reagents. Three important reactions of alkanes are: *combustion, halogenation* and cracking (see *thermal cracking* and *catalytic cracking*). Cracking and the halogenation of alkanes are examples of *free-radical chain reactions*.

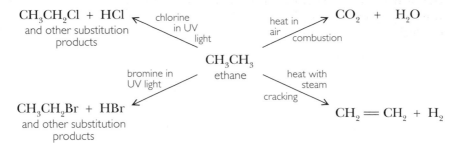

Reactions of alkanes with ethane as the example

alkenes, such as ethene and propene, are products of *cracking* oil fractions. They are important starting points for making other chemicals because of the reactivity of the double bonds in their molecules.

Alkenes are *unsaturated compounds*. They are hydrocarbons with the general formula C_nH_{2n}. The characteristic *functional group* of the alkenes is a carbon–carbon double bond. The presence of the double bond makes alkenes more reactive than *alkanes*.

The name of an alkene is based on the name of the corresponding alkane with the ending change to 'ene'. Where necessary, a number in the name shows the position of the double bond in the structure, as in the two structural *isomers* but-1-ene and but-2-ene. Counting starts from the end of the chain that will give the lowest possible number in the name. The number in the name shows the first of the two atoms connected by the double bond. In but-1-ene, for example, the double bond is between the first and the second atoms in the carbon chain.

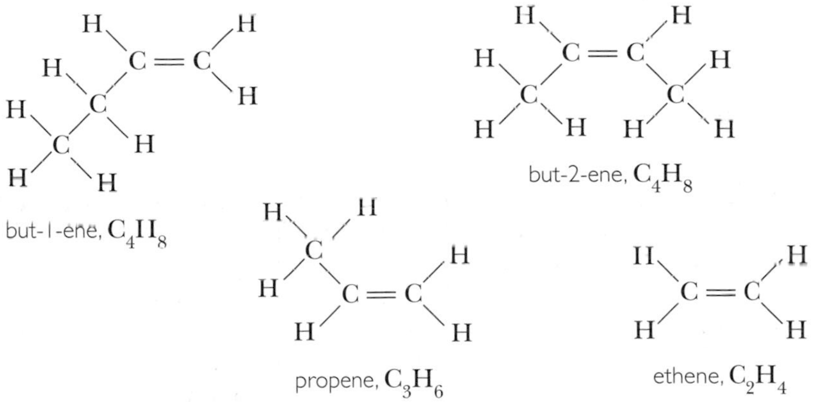

but-1-ene, C_4H_8

but-2-ene, C_4H_8

propene, C_3H_6

ethene, C_2H_4

Names and structures of alkenes

The *pi (π) bonds* in alkenes prevent free rotation about the double bonds. This means that some alkenes can show *geometrical isomerism*.

The boiling points of alkenes increase as the number of carbon atoms in the molecules increase. Ethene, propene and the butenes are gases at room temperature. Alkenes with more than four carbon atoms are liquids or even solids. Alkenes, like other hydrocarbons, do not mix with or dissolve in water.

The characteristic reactions of alkenes are *addition reactions* (see below). The reactions with reagents such as bromine and hydrogen bromide are *electrophilic addition reactions.* The high pressure process for making poly(ethene) involves a *free radical chain reaction.*

Addition reactions of ethene

Markovnikov's rule helps to predict the main product of an addition reaction when an unsymmetrical molecule adds to an unsymmetrical alkene.

Potassium manganate(VII) oxidises alkenes. The products depend on the conditions. A dilute, acidified solution of potassium manganate(VII) converts an alkene to a diol at room temperature.

The reaction of ethene with dilute, acidified manganate(VII) ions producing ethane-1,2-diol

A solution of manganate(VII) ions is purple. The colour disappears as it reacts with an alkene. So the reaction with cold $MnO_4^-(aq)$ ions can be used to distinguish unsaturated and saturated hydrocarbons. Another test for unsaturated hydrocarbons uses a solution of bromine (see *organic analysis*).

A hot, concentrated solution of acidic potassium manganate(VII) breaks apart the double bond in an alkene.

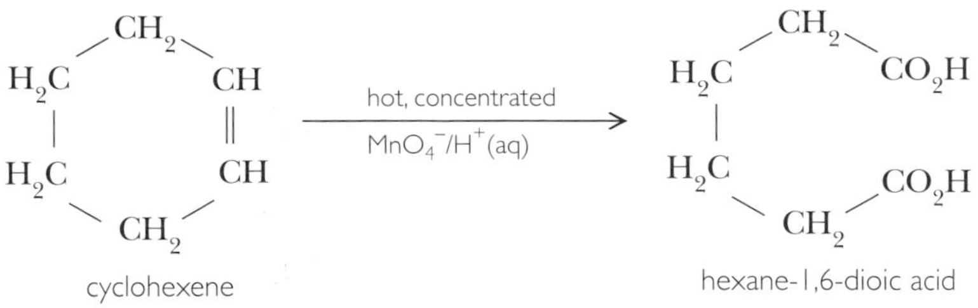

cyclohexene →(hot, concentrated $MnO_4^-/H^+(aq)$)→ hexane-1,6-dioic acid

Using hot, acidic potassium manganate(VII) to break a carbon chain at a double bond. In this example the reaction converts cyclohexene to hexane-1,6-dioic acid.

alkoxides are metal *salts* of alcohols. Ethanol, for example, reacts with *sodium* to produce sodium ethoxide.

$$C_2H_5OH(l) + Na(s) \longrightarrow C_2H_5O^-Na^+(s) + H_2(g)$$

The ethoxide ion is a very *strong base*. It rapidly gains a proton if water is added to turn back to ethanol.

alkyl group: a group of carbon and hydrogen atoms which forms part of the structure of a molecule. The simplest example is the methyl group CH_3— which is methane with one hydrogen atom removed. In general, alkyl groups are *alkane* molecules minus one hydrogen atom.

alkyl group	formula
methyl	CH_3-
ethyl	CH_3CH_2-
propyl	$CH_3CH_2CH_2-$
butyl	$CH_3CH_2CH_2CH_2-$

A useful shorthand for any alkyl group is the capital letter R. Further alkyl groups are then represented by R' or R". So, for example a tertiary *amine* with three different alkyl groups attached to the nitrogen atom can be written as:

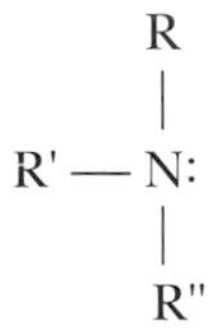

$$\begin{array}{c} R \\ | \\ R' - N: \\ | \\ R" \end{array}$$

General formula for a tertiary amine

alkylation is a reaction which introduces an *alkyl group* into a molecule by *addition* or *substitution*. Industrially the alkylation of alkenes produces branched alkanes needed to raise the octane number of *petrol*. See the diagram at the top of page 22.

The *Friedel–Crafts reaction* adds alkyl groups to benzene rings in *arenes*.

Important aluminium compounds include *aluminium oxide* and hydroxide, *aluminium chloride* and salts containing the hydrated *aluminium(III) ion*. In all its compounds aluminium is in the *oxidation state* +3. The relatively high charge and small size of the Al^{3+} ion state helps to account for the differences between aluminium compounds and the compounds of metals in group 1 and group 2. In some ways aluminium compounds have properties more characteristic of non-metals than metals. Compared with the larger Na^+ ions, Al^{3+} ions have a strong tendency to polarise neighbouring anions, giving rise to *polar covalent* rather than *ionic bonding*. Only the fluoride and oxide are ionic. (See also *Fajan's rules*.)

aluminium extraction: *aluminium* is obtained by *electrolysis* of a solution of aluminium oxide in molten cryolite, Na_3AlF_6. Pure aluminium oxide for the process is obtained by purifying *bauxite*.

The discovery that aluminium oxide dissolves in molten cryolite was essential to the development of the process because the pure oxide melts at 2015°C – much too high for economic industrial processing.

Electrolysis takes place in carbon-lined steel tanks called 'pots'. The carbon lining is the *cathode* of the cell. The *anodes* are blocks of carbon. The currents used are high, of the order of 100 000 A, so the process is generally carried out where electricity is relatively cheap, often close to a source of hydroelectric power.

Reduction at the cathode: $Al^{3+} + 3e^- \longrightarrow Al$

The aluminium is liquid at the temperature of the molten electrolyte (970°C) and it collects at the bottom of the pot. The molten metal is tapped off from time to time.

Oxidation at the anode: $2O^{2-} \longrightarrow O_2 + 4e^-$

Much of the oxygen reacts with the carbon of the anodes, forming carbon dioxide. The anodes burn away and have to be replaced regularly. The energy from the burning anodes helps to heat the electrolyte, keeping it at the working temperature at around 960°C.

Waste gas from the cell contains fluorides and has to be thoroughly cleaned to avoid pollution of the region surrounding the plant.

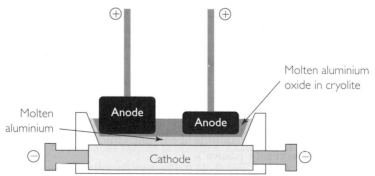

Cross sectional diagram of an electrolysis cell for extracting aluminium

aluminium oxide (Al_2O_3) is a white solid with a very high melting point. Aluminium oxide exists naturally as a group of minerals called corundum. Corundum is an important abrasive; its hardness is 9 on the *Mohs scale*. Emery is a greyish-black variety of corundum made of aluminium oxide mixed with iron oxide minerals. Some gemstones are varieties of corundum including rubies which are red because of the presence of some chromium(III) ions in place of aluminium(III) in the crystal structure. Sapphires are blue because some of the aluminium ions are replaced by a mixture of titanium(IV) and cobalt(II) or iron(II) ions.

Aluminium oxide is widely used as an engineering *ceramic*. It is used to make the insulator in sparking plugs. A translucent, polycrystalline form of the compound encloses high-pressure, high-temperature sodium street lamps which give a much whiter light than the yellow low-pressure lamps.

Aluminium oxide is *amphoteric* but the pure solid is insoluble and this property is more easily demonstrated by adding alkali to a solution of aluminium(III) ions. The amphoteric behaviour is used in the production of the pure oxide from *bauxite*.

aluminium chloride ($AlCl_3$) is an off-white solid which *sublimes* on heating and fumes in moist air because of *hydrolysis* to hydrogen chloride and the hydroxide. Aluminium chloride is manufactured on a large scale by injecting a stream of chlorine gas into molten aluminium at 800°C. At this temperature the product vaporises and is collected as a solid in condensers.

Aluminium atoms in $AlCl_3$ have only six electrons and have a strong tendency to accept two more. As a result, aluminium chloride vapour contains Al_2Cl_6 *dimers*.

Solid $AlCl_3$ has a layer lattice with bonding intermediate between ionic and covalent.

A dot and cross diagram for $AlCl_3$ and the bonding in an Al_2Cl_6 dimer. Lone pairs on chlorine atoms form *dative covalent bonds* with aluminium atoms.

$AlCl_3$ is a strong acceptor of electron pairs, making it a powerful *Lewis acid*.

Aluminium chloride is the catalyst for *Friedel–Crafts* and related reactions. In industry this means that it is an important catalyst for the manufacture of a wide range of useful products including *polymers*, pigments, pharmaceuticals, *dyes* and *detergents*.

aluminium(III) ions are hydrated in aqueous solution and are present as a *complex ion* $[Al(H_2O)_6]^{3+}$. Six water molecules form *co-ordinate bonds* with each metal ion (see the diagram on page 26).

The electrons in the water molecules are pulled towards the highly polarising aluminium ion, making it easier for the water molecules linked to the aluminium ion to give away protons. The hydrated aluminium ion is as strong an *acid* as ethanoic acid. The hydrated ion gives protons to water molecules, forming *oxonium ions*. The solution is acidic enough to release carbon dioxide when added to sodium carbonate.

$$[Al(H_2O)_6]^{3+} + H_2O \rightleftharpoons [Al(H_2O)_5OH]^{2+} + H_3O^+$$
$$[Al(H_2O)_5OH]^{2+} + H_2O \rightleftharpoons [Al(H_2O)_4(OH)_2]^+ + H_3O^+$$

Adding a base such as hydroxide ions to a solution of aluminium ions removes a third proton, producing an uncharged complex. The uncharged complex is much less soluble in water and precipitates as a white jelly-like precipitate – hydrated aluminium hydroxide.

$$[Al(H_2O)_4(OH)_2]^+ + OH^- \rightleftharpoons [Al(H_2O)_3(OH)_3](s) + H_2O(l)$$

Adding excess alkali removes yet another proton, now producing a negatively charged ion which is soluble in water so that the precipitate redissolves. This demonstrates the *amphoteric* properties of aluminium hydroxide.

$$[Al(H_2O)_3(OH)_3](s) + OH^- \rightleftharpoons [Al(H_2O)_2(OH)_4]^-(aq) + H_2O(l)$$

All these changes are reversed by adding a solution of a strong acid such as hydrochloric acid which turns the hydroxide ions in the complex back to water molecules.

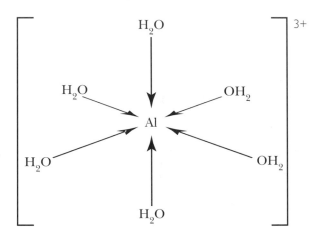

Structure of a hydrated aluminium ion showing the octahedral arrangement of water molecules. This is the hexaaquoaluminate(III) ion.

alums are *double salts* which crystallise easily from solutions in water forming attractive octahedral crystals. Alums have the general formula $M(I)M(III)(SO_4)_2.12H_2O$. $M(I)$ is an ion with a 1+ charge such as Na^+, K^+ or NH_4^+ and $M(III)$ is a metal ion with a 3+ charge such as Al^{3+}, Cr^{3+} or Fe^{3+}.

amalgams are solutions of metals in *mercury*. Dentists make amalgams for filling teeth. They grind up the compound Ag_3Sn with mercury. This produces an amalgam which is workable so that it can be shaped easily to fill a cavity. The amalgam soon hardens as the mercury reacts with the silver and the tin. As it hardens the mixture expands to fit tightly in the tooth cavity.

amides are organic nitrogen compounds derived from *carboxylic acids* by replacing the — OH group with an — NH_2 group.

Amides are white crystalline solids at room temperature, apart from methanamide which is a liquid. The simple amides are soluble in water.

ethanamide   benzamide

Names and structures of amides

Amides form rapidly at room temperature when *acyl chlorides* or *acid anhydrides* react with ammonia. *Esters* react more slowly to form amides.

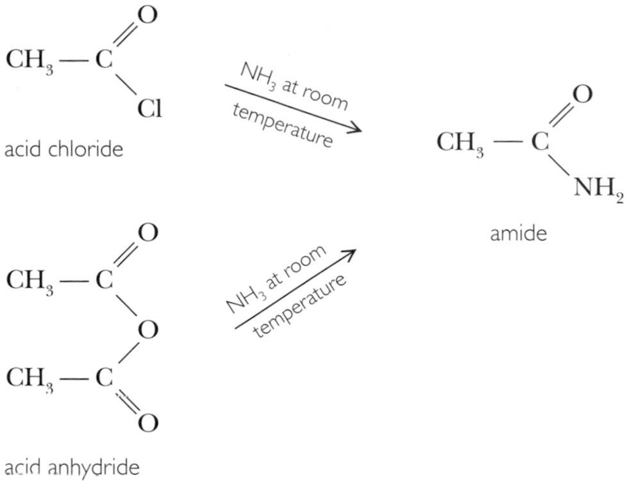

$$CH_3 - C \overset{O}{\underset{Cl}{\big\langle}}$$

acid chloride

NH₃ at room temperature →

$$CH_3 - C \overset{O}{\underset{NH_2}{\big\langle}}$$

amide

$$CH_3 - C \overset{O}{\underset{O}{\big\langle}} \\ CH_3 - C \overset{}{\underset{O}{\big\langle}}$$

acid anhydride

NH₃ at room temperature →

Reactions which produce amides

Another way of making amides is to convert a carboxylic acid to an ammonium salt and then to dehydrate the salt by heating it under reflux for several hours.

$$CH_3CO_2H + NH_3 \longrightarrow CH_3CO_2^-NH_4^+ \longrightarrow CH_3CONH_2 + H_2O$$

carboxylic   ammonia   ammonium salt   amide
acid

Unlike *amines*, amides are very weak bases. They give neutral solutions in water and do not form salts with acids such as hydrochloric acid. The lone pair of electrons on the nitrogen atom is delocalised with the carbonyl group making it unavailable to form a dative bond to a proton.

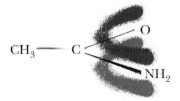

Delocalisation of electrons in the amide group

Reduction converts an amide to an amine. A suitable procedure is to treat the amide with *lithium tetrahydridoaluminate(III)* in dry ethoxyethane and then to add water.

$$CH_3CONH_2 \xrightarrow[\text{2. }H_2O]{\text{1. LiAlH}_4 \text{ in ether}} CH_3CH_2NH_2$$

Heating with phosphorus(V) oxide dehydrates an amide to a *nitrile*.

Heating with dilute acid or dilute alkali hydrolyses amides to the corresponding acid

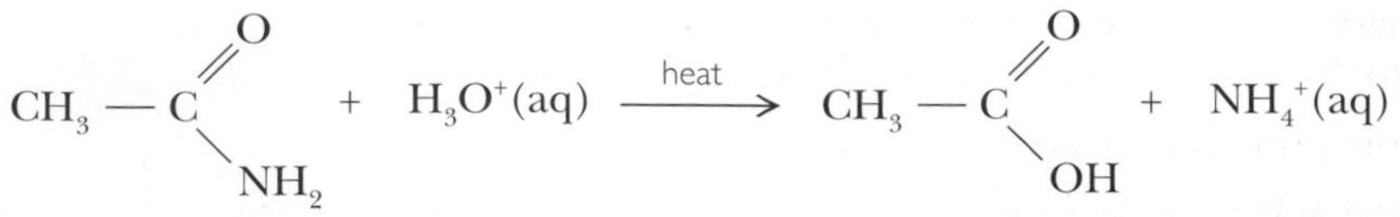

$$CH_3 - \overset{\overset{\displaystyle O}{\|}}{C} \diagdown NH_2 \quad + \quad H_3O^+(aq) \quad \xrightarrow{\text{heat}} \quad CH_3 - \overset{\overset{\displaystyle O}{\|}}{C} \diagdown OH \quad + \quad NH_4^+(aq)$$

Hydrolysis of ethanamide

(See also the *Hofmann degradation.*)

amines are nitrogen compounds in which one or more of the hydrogen atoms in ammonia (NH_3) is replaced by an *alkyl group* or *aryl group*. The number of alkyl groups determines whether the compound is a primary amines, a secondary amines or a tertiary amine.

The names of amines are based on the nature and number of alkyl groups attached to the nitrogen atom.

Amines can be very smelly. Ethylamine has a fishy smell. Animal flesh smells putrescent because it gives off diamines.

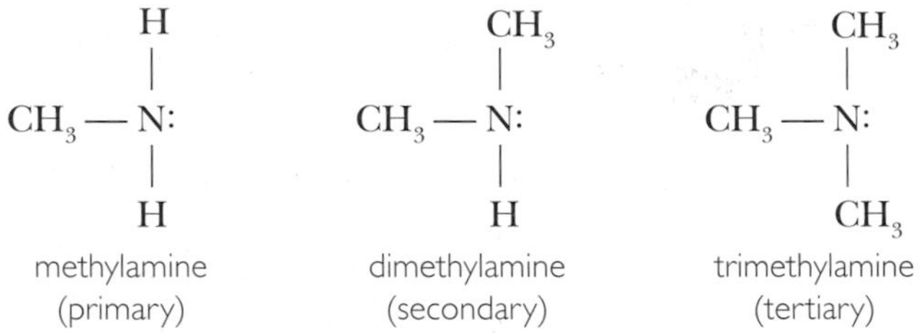

$$\begin{array}{ccc}
H & CH_3 & CH_3 \\
| & | & | \\
CH_3 - N: & CH_3 - N: & CH_3 - N: \\
| & | & | \\
H & H & CH_3 \\
\text{methylamine} & \text{dimethylamine} & \text{trimethylamine} \\
\text{(primary)} & \text{(secondary)} & \text{(tertiary)}
\end{array}$$

Names and structures of primary, secondary and tertiary amines. Note the meaning of *primary, secondary and tertiary* here – it is not the same as for alcohols and halogenoalkanes.

Two methods of making primary amines are:

- the reduction of a nitrile
- the substitution reaction of ammonia with a halogenoalkane.

Primary amines, like ammonia, can acts as *bases*, form *complex ions* with metal ions and react as *nucleophiles.*

Acid anhydrides and acyl chlorides both acylate primary amines.

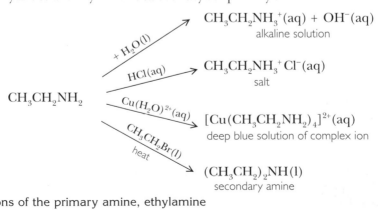

$CH_3CH_2NH_2$

$+ H_2O(l)$ → $CH_3CH_2NH_3^+(aq) + OH^-(aq)$
alkaline solution

$HCl(aq)$ → $CH_3CH_2NH_3^+ Cl^-(aq)$
salt

$Cu(H_2O)_2^{2+}(aq)$ → $[Cu(CH_3CH_2NH_2)_4]^{2+}(aq)$
deep blue solution of complex ion

$CH_3CH_2Br(l)$ heat → $(CH_3CH_2)_2NH(l)$
secondary amine

Reactions of the primary amine, ethylamine

Primary alkyl amines are stronger bases than ammonia (see *base dissociation constants*). This can be explained in terms of the *inductive effect*.

For the properties of aryl (aromatic) amines see *phenylamine*.

amino acids are the compounds which join together in long chains to make *proteins*. They are carbon compounds with two *functional groups*; a basic amino group and a carboxylic acid group. Proteins consist of long chains of amino acids. There are about 20 different amino acids which link together to make proteins. Chemists sometimes call them α-amino acids because the amino group is attached to the alpha carbon atom – the one next to the carboxylic acid group.

Amino acids are crystalline solids. They are very soluble in water and crystallise as *zwitterions*.

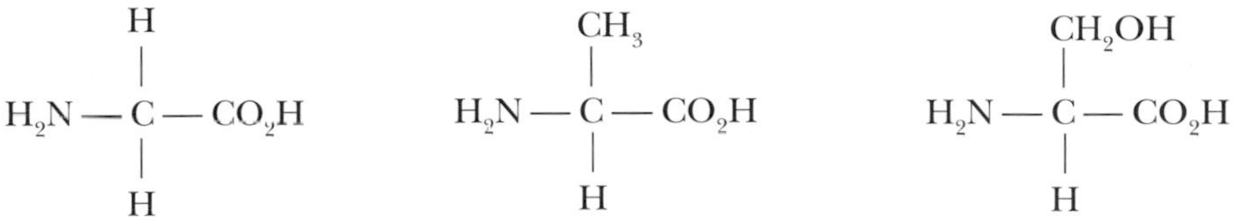

$$H_2N - \underset{\underset{H}{|}}{\overset{\overset{H}{|}}{C}} - CO_2H \qquad H_2N - \underset{\underset{H}{|}}{\overset{\overset{CH_3}{|}}{C}} - CO_2H \qquad H_2N - \underset{\underset{H}{|}}{\overset{\overset{CH_2OH}{|}}{C}} - CO_2H$$

glycine   alanine   serine

Examples of amino acids in proteins. Note that the amino group is attached to the carbon atom next to the carboxylic acid group.

Each of the amino acids in a protein has a central carbon atom attached to four other groups so, except for glycine which has two hydrogen atoms, these are *chiral* molecules which can exist in mirror image forms. The natural amino acids all take the L-form.

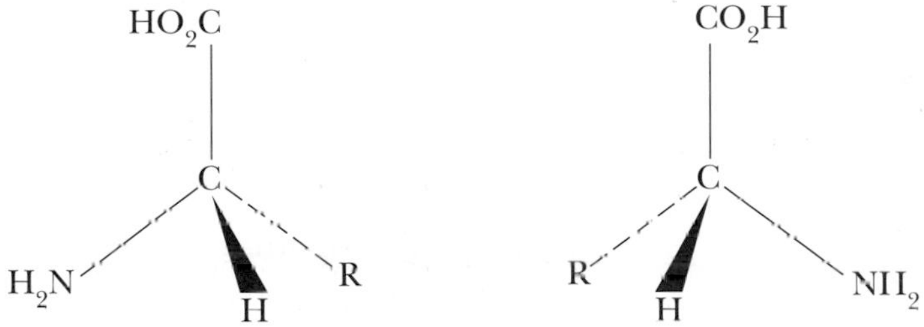

L-amino acid
(occurs naturally)

D-amino acid
(made in laboratory)

Mirror image forms of an amino acid

amino group: the $-NH_2$ functional group found in primary *amines*, and *amino acids*. As in *ammonia*, there is a *lone pair of electrons* on the nitrogen atom. The amino group, like ammonia, can act as a *base*, form *complex ions* with metal ions and react as a *nucleophile*.

ammonia (NH$_3$) is the simplest compound of nitrogen and hydrogen. It is a colourless gas with a pungent smell. It is the only common alkaline gas. The molecule has a pyramidal shape:

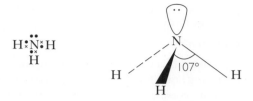

$$H \overset{\cdot\cdot}{\underset{\underset{H}{\times\times}}{\overset{\cdot\cdot}{N}}} H$$

Dot and cross diagram for ammonia and the shape of an ammonia molecule. There are three bonding pairs and one lone pair in the outer shell the nitrogen atom. The four electron pairs point to the vertices of a distorted tetrahedron. The bond angle is less than the regular tetrahedral angle (109.5°) because repulsion between the lone pair and the bonding pairs is greater than the repulsion between bonding pairs (see *shapes of molecules*).

The properties of ammonia are affected by *hydrogen bonding*. Ammonia molecules can form hydrogen bonds with each other and with water molecules. As expected for a compound with small molecules, ammonia is a gas at room temperature but because of hydrogen bonding it is quite easily liquefied by cooling or increasing the pressure. Ammonia is also very soluble in water because of hydrogen bonding.

Many of the important properties of ammonia involve the lone pair of electrons on the nitrogen atom. The lone pair means that ammonia is:

- a *weak base:* $NH_3(g) + HCl(g) \longrightarrow NH_4^+Cl^-(s)$
- a *ligand* in *complex ions:*
  $$[Cu(H_2O)_6]^{2+}(aq) + 4NH_3(aq) \longrightarrow [Cu(NH_3)_4(H_2O)_2]^{2+}(aq) + 4H_2O(aq)$$
- a *nucleophile:* $CH_3CH_2CH_2Br + NH_3(aq) \longrightarrow CH_3CH_2CH_2NH_3^+Br^-(aq)$.

Other reactions of ammonia involve the $N\!-\!H$ bonds. In particular, the reactions in which ammonia acts as a *reducing agent*. Ammonia burns in oxygen to form nitrogen and steam. It reacts with chlorine to form nitrogen and ammonium chloride. It reduces metal oxides, such as copper(II) oxide.

$$3CuO(s) + 2NH_3(g) \longrightarrow 3Cu(s) + N_2(g) + H_2O(l)$$

The catalytic oxidation of ammonia is important in *nitric acid manufacture*.

ammonia manufacture: the Haber process for the synthesis of ammonia combines nitrogen with hydrogen in the presence of an iron catalyst.

$$N_2(g) + 3H_2(g) \rightleftharpoons 2NH_3(g) \quad \Delta H = -92.4 \text{ kJ mol}^{-1}$$

Hydrogen for the process comes from natural gas and steam (see *steam reforming*). The nitrogen comes from the air.

The reaction is very slow at room temperature. Raising the temperature increases the rate of reaction but the *reversible* reaction is *exothermic* so, according to *Le Chatelier's principle*, the higher the temperature the lower the yield of ammonia at equilibrium. A catalyst makes it possible for the reaction to go fast enough without the temperature being so high that the yield is too low. Also, according to Le Chatelier's principle, increasing the pressure raises the percentage of ammonia at equilibrium.

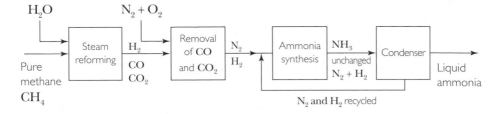

Flow diagram for the synthesis of ammonia

The process typically operates at pressures between 70 atmospheres and 200 atmospheres with temperatures in the range 400°C to 600°C.

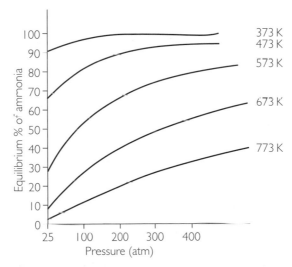

Graph showing how the equilibrium yield of ammonia varies with pressure and temperature

The main uses of ammonia are in making:

- *fertilisers* (80%)
- nylon (7%)
- *nitric acid* (5%).

ammoniacal silver nitrate: see *Tollens' reagent, aldehydes* and *ketones.*

ammonium hydroxide is the traditional name for 'ammonia solution' or 'aqueous ammonia'. The older name was used because a solution of *ammonia* in water contains some ammonium ions and hydroxide ions.

$$NH_3(aq) + H_2O(l) \rightleftharpoons NH_4^+(aq) + OH^-(aq)$$

However ammonia is a *weak base*, so most of the ammonia in the solution is not ionised. Aqueous ammonia is a better name, because almost all the ammonia is present as dissolved molecules linked to water molecules by *hydrogen bonds.*

ammonium salts contain the ammonium ion (NH_4^+), formed when *ammonia* reacts with acids. Examples are ammonium chloride, ammonium sulfate and ammonium nitrate.

$$2NH_3(aq) + H_2SO_4(aq) \longrightarrow (NH_4)_2SO_4(aq)$$
$$NH_3(aq) + HNO_3(aq) \longrightarrow NH_4NO_3(aq)$$

An ammonium ion forms when the *lone pair* on the nitrogen atom in ammonia forms a dative bond with a hydrogen ion (proton) from an acid. The ammonium ion has four bonding electron pairs around the nitrogen atom so it has a tetrahedral shape.

Ammonium salts such as ammonium sulfate and ammonium nitrate are soluble nitrogen compounds manufactured and used on a large scale as *fertilisers*. Ammonium nitrate is also used as an *explosive*.

Formation of an ammonium ion and the shape of the ammonium ion. Once formed, all four bonds are the same.

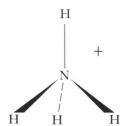

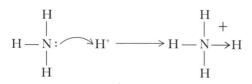

amorphous solids have no regular crystal structure; their particles are in a random jumble. *Glasses* are amorphous solids.

amount of substance: the measurement which allows chemists to find formulae, write equations and make fair comparisons between equal numbers of atoms, molecules or ions.

Amount of substance is a *physical quantity* (symbol n) which is measured in the unit *mole* (symbol mol). In chemistry the word 'amount' has this precise meaning. There is no measuring instrument for determining amounts directly, unlike balances for determining *masses* in kilogrammes or graduated glassware for measuring *volumes* in dm^3 (litres). Instead, chemists first measure masses or volumes and then calculate the amount in moles given the *molar mass, concentration* or *molar volume* of the specified substance or *entity*.

For any pure substance:

$$\text{amount of substance (mol)} = \frac{\text{mass of substance (g)}}{\text{molar mass (g mol}^{-1})}$$

For a solution:

$$\text{amount of substance (mol)} = \text{volume of solution (dm}^3) \times \text{concentration of solution (mol dm}^{-3})$$

For a gas:

$$\text{amount of substance (mol)} =$$

$$\frac{\text{volume of a gas at a specified temperature and pressure (m}^3)}{\text{molar volume of any gas under the same conditions (m}^3\text{ mol}^{-1})}$$

(See also the *Avogadro constant.*)

ampere: the SI unit of current. A current of one ampere is a flow of one *coulomb* of charge per second. 1 A = 1 C/s.

amphetamines are *drugs* which stimulate the central nervous system and make people feel more alert. For a while amphetamines, such as benzedrine, were used medicinally until it was found that they cause strong psychological dependence.

$$\langle\!\bigcirc\!\rangle\!-\!CH_2\!-\!\underset{\displaystyle CH_3}{\overset{\displaystyle H}{\underset{\displaystyle |}{\overset{\displaystyle |}{C}}}}\!-\!\overset{\displaystyle H}{\overset{\displaystyle |}{N}}\!-\!H$$

Molecular structure of benzedrine

amphoteric compounds are substances which can behave both as *acids* and as *bases*. Water is an example. Water can both accept *protons* from acids forming *oxonium ions* and donate protons to stronger *bases* forming hydroxide ions.

Water acting as a base (taking protons from an acid):

$$H_2O(l) + H^+ \rightleftharpoons H_3O^+(aq)$$
from acid

Water acting as an acid (giving protons to a base):

$$H_2O(l) \rightleftharpoons OH^-(aq) + H^+$$
to base

Alternatively, these substances are described as being amphiprotic. Other examples are the hydrogencarbonate ion (HCO_3^-) and *amino acids*.

amphoteric oxides are oxides which react both like *acidic oxides* and *basic oxides*. Examples are: ZnO, Al_2O_3, Cr_2O_3, SnO, SnO_2, PbO, PbO_2. The anhydrous oxides are often relatively *inert* to aqueous reagents so that it is easier to demonstrate their amphoteric behaviour on a test-tube scale with freshly formed samples of the metal hydroxides. Zinc hydroxide, for example, dissolves in acids to form *salts* (so acting as a *base*), it also dissolves in a solution of a strong alkali, forming zincate ions (thus acting as an *acid*).

$$Zn(OH)_2(s) + 2H^+(aq) \rightleftharpoons Zn^{2+}(aq) + 2H_2O(l)$$
$$Zn(OH)_2(s) + 2OH^-(aq) \rightleftharpoons ZnO_2^{2-}(aq) + 2H_2O(l)$$

The reaction of an amphoteric hydroxide with acids and bases can also be described in terms of the addition or removal of protons from a complex of the metal ion, such as $Al(H_2O)_3(OH)_3$. (See *aluminium(III) ions*.)

anaerobic respiration: see *respiration*.

anaesthetics: total anaesthetics are used in medicine to induce pain-free sleep during surgery. Early anaesthetics were trichloromethane (chloroform), ethoxyethane (ether) and dinitrogen oxide (laughing gas). From 1956 the most commonly used anaesthetic was halothane ($CF_3CHBrCl$). Unfortunately the liver metabolises halothane producing toxic products which can cause hepatitis. To avoid these side effects new anaesthetics were developed based on *ethers*. Ethers are highly flammable but can be converted to safe and effective anaesthetics by replacing hydrogen atoms with fluorine atoms. The latest anaesthetics were introduced in the mid 1990s. They include Sevoflurane, $FCH_2 - O - CH(CF_3)_2$.

Local anaesthetics, such as Novocaine, for minor surgery and dentistry, prevent pain just in the region of the body where they are applied or injected.

analgesics: drugs which relieve pain. There are mild analgesics such as *aspirin, paracetamol* or *ibuprofen*. Other analgesics are powerful, like the *narcotic* morphine from the opium poppy.

analytical chemistry is concerned with determining the *qualitative* and *quantitative* composition of substances.

anhydrides: see *acid anhydrides*.

anhydrous salts are the compounds left after removing the water from a *hydrated* salt. Hydrated copper(II) sulfate, for example, is blue. Heating drives off the *water of crystallisation* as steam leaving a white solid, anhydrous copper(II) sulfate.

$$CuSO_4.5H_2O(s) \longrightarrow CuSO_4(s) + 5H_2O(l)$$

aniline is the traditional name for *phenylamine*. Aniline and related compounds are used to make *azo dyes* which are therefore also called aniline dyes.

anions are negative ions attracted to the *anode* during *electrolysis*.

charge	anion	symbol
1–	bromide	Br^-
	chloride	Cl^-
	ethanoate	$CH_3CO_2^-$
	hydrogencarbonate	HCO_3^-
	hydroxide	OH^-
	iodide	I^-
	nitrate [nitrate(V)]	NO_3^-
	nitrite [nitrate(III)]	NO_2^-
	manganate(VII)	MnO_4^-
2–	carbonate	CO_3^{2-}
	oxide	O^{2-}
	sulfate [sulfate(VI)]	SO_4^{2-}
	sulfide	S^{2-}
	sulfite [sulfate(IV)]	SO_3^{2-}
	thiosulfate	$S_2O_3^{2-}$
	dichromate(VI)	$Cr_2O_7^{2-}$
3–	nitride	N^{3-}
	phosphate	PO_4^{3-}

anion tests are used in *qualitative analysis* to identify the negative ions in salts. The tests use reagents which can produce colour changes, gases and precipitates. Knowledge of the chemistry makes it possible to interpret the changes and identify the ions. See Table on page 35.

anisotropic solids have properties which depend on the direction in which they are measured. Examples of highly anisotropic substances are fibrous materials such as *asbestos*, and layer structures such as mica. The carbon atoms in graphite are arranged in layers (see *allotropes*). The properties of large single crystals of graphite show the differences between measurements parallel to the planes of atoms and at right-angles to the planes of atoms.

Anion tests (see page 34)

Test	Observations	Inference
Test for carbonate, sulfite and nitrite Add dilute hydrochloric acid to the solid salt. Warm gently if there is no reaction at first.	Gas which turns limewater milky white.	carbon dioxide from a carbonate
	Gas which is acidic, has a pungent smell and turns acid-dichromate paper from orange to green.	sulfur dioxide from a sulfite
	Colourless gas given off which turns brown where it meets the air.	nitrogen oxide (NO) from a nitrite turning to nitrogen dioxide (NO_2)
Test for halide ions Make a solution of the salt. Acidify with nitric acid, then add silver nitrate solution. Test the solubility of the precipitate in ammonia solution.	White precipitate soluble in dilute ammonia solution.	chloride
	Cream precipitate soluble in concentrated ammonia solution.	bromide
	Yellow precipitate insoluble in excess ammonia.	iodide
Test for sulfate and sulfite ions Make a solution of the salt. Add a solution of barium nitrate or chloride. If a precipitate forms add dilute nitric acid.	White precipitate which redissolves on adding acid.	sulfite
	White precipitate which does not redissolve in acid.	sulfate
Test for nitrates Make a solution of the salt, add sodium hydroxide solution and then a piece of aluminium foil or a little Devarda alloy. Heat.	Alkaline gas evolved which turns red litmus blue. Pungent smell.	ammonia from a nitrate (or nitrite)
Test for chromate ions Make a solution of the salt. Divide into three: • add dilute acid • add a solution of barium nitrate • add a solution of lead nitrate	Yellow solution turns orange. Yellow precipitates forms with barium ions and lead ions.	yellow chromate ions turning to orange dichromate precipitates of insoluble barium and lead chromates chromate

annealing is a process of controlled heating and cooling used to modify the properties of a material, for which purpose it is kept for a time in a furnace at a temperature below its melting point. Annealing is used to remove internal stresses from *glass* objects after blowing or casting. Annealing is an important part of the heat treatment used to modify *crystal structures of metals* and hence control their properties.

anode: the *electrode* at which *oxidation* takes place. During *electrolysis* the external power supply removes electrons from the anode so that it becomes the positive electrode, attracting negative ions; the latter lose electrons (oxidation), thus turning into atoms or molecules.

The term anode is also sometimes used in *electrochemical cells*, where the oxidation process at this electrode in one of the *half-cells* forces electrons onto the electrode, which becomes the negative terminal of the cell.

Note that electrons flow out of the anode to the external circuit both during electrolysis and when a current is drawn from a chemical cell.

anodising is an electrolytic process for thickening the oxide layer on the surface of *aluminium*. It is an example of anodic *oxidation*. In the process, an aluminium sheet or component is the *anode* of an *electrolysis* cell containing sulfuric acid. The newly formed oxide layer can absorb dyes so that the surface of anodised aluminium is easily coloured. The thicker oxide film helps to protect the metal, so anodised aluminium is more resistant to *corrosion*.

antacids are ingredients of indigestion tablets taken to neutralise acid in the stomach. After a meal the cells lining the stomach produce gastric juice which contains, among other things, a mixture of hydrochloric acid and enzymes to digest food. Antacids are *bases* which neutralise acids. Examples are sodium hydrogencarbonate, calcium carbonate, aluminium hydroxide and magnesium hydroxide.

antibiotics are chemicals which prevent *bacteria* growing or kill them. Thus antibiotics can cure bacterial diseases. Most antibiotics are the products of micro-organisms. Some of the antibiotics used in medicine are natural antibiotics which have been modified chemically, such as the semi-synthetic penicillins. Other antibiotics are now made entirely by chemical synthesis. The misuse of antibiotics in medicine and agriculture has led to the emergence of antibiotic resistant bacteria.

antifreeze is a chemical added to the cooling system of engines to prevent the water freezing and damaging the engine during frosty weather. Antifreeze lowers the *freezing point* of water well below 0°C. Ethane-1,2-diol is used as antifreeze because it mixes freely with water but has a higher boiling point (198°C) so that it does not evaporate from the coolant when the engine is hot. Ethane-1,2-diol has a high boiling point and mixes freely with water because it has two —OH groups allowing extensive *hydrogen bonding* both between molecules of the diol and between the diol and water molecules. Another advantage of adding ethane-1,2-diol to the water in cooling systems is that it helps to prevent *corrosion* of iron and steel.

anti-knock additives prevent knocking in engines by raising the octane number of *petrol* and preventing the fuel from igniting too early. From the 1920s the main anti-knock additive was a volatile lead compound tetraethyl lead, $Pb(C_2H_5)_4$. Lead compounds in exhaust gases are toxic and so poison the metal catalyst in *catalytic* *converters.* So petrol companies now produce high-octane fuels by refining the fuel further and adding alcohols and *ethers* such as MTBE.

antiseptics are chemicals which kill micro-organisms, but, unlike *disinfectants*, can safely be used on the skin. *Phenol* was the first antiseptic used in surgery, by Joseph Lister in 1857. Lister found that a spray of phenol controlled the growth of bacteria in open wounds but it also made the wounds difficult to heal so he moved on to other methods. Dettol and TCP are improved antiseptics related to phenol.

aqua regia was well known to alchemists for its ability to dissolve noble metals such as gold. Hence the name, which is Latin, meaning 'royal water'. It is a mixture of three parts concentrated *hydrochloric acid* with one part concentrated *nitric acid.* The nitric acid oxidises gold atoms to Au^{3+} ions which then form complex ions, $[AuCl_4]^-$, with the high concentration of chloride ions.

aqueous solution: a solution of one or more *solutes* dissolved in water. The state symbol (aq) in *chemical equations* shows that a *species* is in aqueous solution.

arenes are *hydrocarbons* such as *benzene* with rings of carbon atoms stabilised by *delocalisation of electrons.* Traditionally, chemists have called the arenes 'aromatic hydrocarbons', ever since *Kekulé* was struck by the fragrant smell of oils such as benzene. In the modern name, the 'ar-' comes from aromatic and the ending '-ene' means that these hydrocarbons are unsaturated compounds like the *alkenes.*

benzene   methylbenzene   1,3-dimethylbenzene

Structures of arenes. The circle in the benzene ring represents six *delocalised electrons.* This representation explains the shape and *stability of benzene.* The Kekulé structure is more helpful when describing the mechanism of the reactions of benzene.

As well as arenes related to benzene, there are others with fused ring systems such as naphthalene and anthracene.

Structure of naphthalene with fused rings, showing both the Kekulé structure and delocalisation. The delocalised electrons are shared across both the rings.

Arenes are non-polar. The forces between the molecules are weak *van der Waals forces.* The boiling points of arenes depend on the sizes of the molecules. The bigger the molecules the higher the boiling points. Benzene and methylbenzene are liquids at room temperature. Naphthalene is a solid.

The important reactions of benzene and other arenes are very different from *alkenes* because of the delocalised electrons. Instead of *electrophilic addition* reactions, the useful changes are *electrophilic substitution* reactions.

Heating a mixture of an arene bearing a hydrocarbon side chain with alkaline potassium manganate(VII) oxidises the side chain to a carboxylic acid group.

CH$_3$ $\xrightarrow[\text{heat}]{\text{KMnO}_4/\text{OH}^-}$ CO$_2$H

methylbenzene   benzoic acid

Arenes Oxidation of the hydrocarbon side chain of methyl benzene

argon (Ar) is the third member of the family of *noble gases* coming below neon in the periodic table with the *electron configuration* [Ne]3s^{2}3p^6.

The element is a colourless, odourless gas consisting of single atoms.

Argon makes up about 0.9% of dry *air* but it was not identified until the late nineteenth century because of its *inertness*.

About 30 000 tonnes of argon are produced in the UK by the *fractional distillation* of liquid air. 90% of the gas is used to provide an inert atmosphere for processes such as welding; the other 10% is used to fill filament light bulbs.

aromatic hydrocarbons: the traditional term for *arenes*.

aromatic nitro compounds: see *benzene and phenylamine*.

aromatic substitution: another name for *electrophilic substitution* in arenes.

Arrhenius equation: an equation which describes how the *rate constant* for a reaction varies with temperature and makes it possible to determine the *activation energy* for the reaction. The *collision theory* of reaction rates helps to account for the form of the equation.

Arrhenius equation: $k = Ae^{-E_a/RT}$

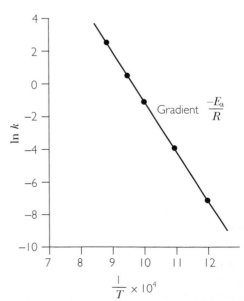

Arrhenius equation Plot of ln k against $1/T$ for a reaction

k is the rate constant for the reaction and E_a is the activation energy, R is the gas constant and T the temperature in kelvin. A is another constant.

Putting the equation into logarithmic form (using *logarithms* to base e) makes it easier to use.

$$\ln k = \ln A - \frac{E_a}{RT} = \text{constant} - \frac{E_a}{RT}$$

This equation gives a straight line if $\ln k$ is plotted against $1/T$. The gradient of the line is $-E_a/R$.

arsenic (As) is a greyish-white *metalloid* which forms very toxic compounds. The element comes below phosphorus in group 5 of the *periodic table* and has the *electron configuration* $[Ar]3d^{10}4s^24p^3$.

aryl group: a group of carbon and hydrogen atoms which forms part of the structure of a molecule. The simplest example is the *phenyl group*, C_6H_5 —, which is *benzene* with one hydrogen atom removed. In general aryl groups are arene molecules minus one hydrogen atom.

asbestos is the name given to various fibrous *silicate* minerals in which SiO_4 groups are linked into long chains. These minerals have very high melting points and they do not burn. The *fibres* are flexible enough to be woven into fabrics which are heat resistant, fire-proof and do not conduct electricity. Breathing asbestos fibres can cause lung disease many years later including asbestosis and mesothelioma (a form of cancer). For this reason the use of the asbestos is being phased out in many parts of the world, especially blue asbestos which is particularly hazardous.

ascorbic acid (vitamin C) is a white crystalline compound found dissolved in the juices of fresh fruit and vegetables. Lack of ascorbic acid in the diet leads to scurvy. Ascorbic acid is relatively soluble in water and may be washed out of foods on cooking. It is thermally unstable and is progressively destroyed during cooking.

Structure of ascorbic acid

The ascorbic acid molecule does not contain a free *carboxylic acid* group because its carboxyl group reacts with its hydroxyl group, eliminating water to form a ring compound. Ascorbic acid is a good *reducing agent*. It is added to some processed foods as preservative; as an antioxidant, it prevents oxidation of other food components.

aspirin is the common name for the most widely used medical drug. Chemically it is the ethanoyl (acetyl) *ester* of 2-hydroxybenzoic acid (salicylic acid). (See *acylation*.)

Aspirin is a mild painkiller (*analgesic*); it also helps to lower fevers and reduce inflammation of joints. Large-scale studies have shown that people who have had a heart attack are less likely to have a second attack if they take half an aspirin tablet a day.

Children under 12 years old should not take aspirin because of the slight risk of a rare illness called Reye's syndrome, which damages the liver and brain and can be fatal.

assaying is a process of *quantitative analysis* which determines how much of a given sample is the substance indicated by its name. So a piece of gold is assayed to see what proportion of the metal sample is gold. Gold and silver articles are assayed before being hallmarked. A range of assaying techniques has been developed to determine the purity or biological activity of drugs, enzymes and other substances which affect living things.

astatine (At) is the rarest element in the *halogen* family. It is the element below iodine in group 7 of the periodic table. It has the *electron configuration* $[Xe]4f^{14}5d^{10}6s^26p^5$. There are twenty known *isotopes* of astatine all of which are highly *radioactive*. The longest lived is astatine-210 with a *half-life* of 8.3 hours. The element was discovered in 1940 when it was prepared artificially by bombarding bismuth with *alpha particles* in a cyclotron.

asymmetric molecules are molecules with no centres, axes or planes of symmetry. Asymmetric molecules are *chiral* and can exist in distinct mirror image forms giving rise to *optical isomerism*. Any carbon atom with four different groups or atoms attached to it is asymmetric.

The mirror image forms of the asymmetric molecule, lactic acid

atactic polymer: a form of *addition polymer*, such as poly(propene), in which the side groups along the *polymer* chain are randomly orientated. Atactic poly(propene) is an *amorphous*, rubbery polymer of little value unlike *isotactic* poly(propene).

Part of a chain of atactic poly(propene)

atmosphere: the zone around the Earth which contains all the gases in the *air*. By convention the upper limit of the atmosphere is taken to be 1000 km above the Earth's surface. However, because of gravity, most of the air is concentrated in the lower regions. About 50% by mass lies within 5.6 km of the surface and 99% within 40 km. Most of the climate and weather processes take place in the troposphere which extends up to about 17 km.

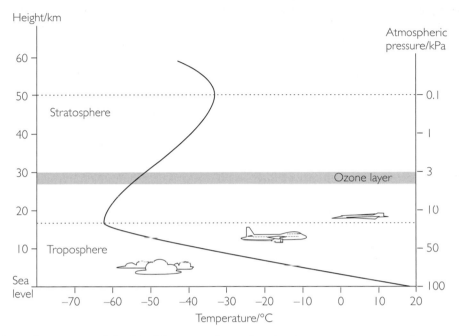

Vertical structure of the Earth's atmosphere showing how temperature and pressure vary with altitude

atmospheric pressure usually means the *pressure* of the atmosphere at ground level. Standard atmospheric pressure (1 atm) = 101 325 Nm^{-2} = 101.325 kPa. The choice of this value dates back to the days when atmospheric pressure was normally measured by a column of mercury in a barometer. A pressure of one atmosphere was defined as the pressure needed to support a column of mercury 760 mm high.

atom: the smallest particles of an element. An atom consists of a tiny nucleus surrounded by a cloud of electrons. The nucleus consists of protons and neutrons. All the atoms of the same element have the same *atomic (proton) number*. The number of neutrons may vary and this accounts for the existence of *isotopes*. (See also *fundamental particles.*)

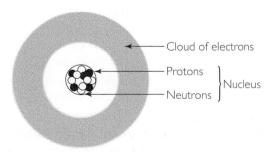

Atomic structure

Chemical reactions involve the electrons in the outer *shells* of atoms. As a result the *electron configuration* of an element helps to account for its chemical behaviour.

Radioactivity and the changes in nuclear reactors involve the nuclei of atoms. When describing nuclear changes it helps to use symbols which show the *mass number* and atomic number.

Mass number — $_{Z}^{A}X$ — Element symbol
Atomic number

Symbols for atoms. The atomic number gives the number of protons in the nucleus. In a neutral atom the number of electrons is the same. The difference between mass number and atomic number gives the number of neutrons in the nucleus.

atomic absorption spectroscopy is a technique of *quantitative analysis* for measuring the amount of an element in a sample. The procedure depends on the fact that the atoms absorb radiation at particular frequencies.

The sample is vaporised and split into gaseous atoms by a very hot flame at around 2000°C. Light from a special lamp produces radiation with a wavelength absorbed by the element to be measured.

By measuring the absorbance at a particular wavelength it is possible to determine the amount of certain elements in the sample. The procedure is *calibrated* using samples of known concentration.

Atomic absorption spectroscopy provides a sensitive and reliable method for determining more than 60 elements. Examples include the determination of traces of mercury in the environment, lead in blood, or heavy metals in the effluent from a factory. The technique is widely used in commercial laboratories, for example in the water industry.

atomic emission spectroscopy is a very sensitive method of analysis used to identify and measure the elements in a sample. Atomic emission spectra of pure elements also provide the evidence needed to determine the *energy levels* and *electron configurations* in atoms.

Carrying out a *flame test* and examining the coloured light from the flame with a hand-held spectroscope is a simple demonstration of this type of spectroscopy.

The first step in a more sophisticated analysis is to vaporise the sample using a very hot flame or an electric spark to atomise the sample and produce a gas consisting of free atoms. Energy from the flame or spark also excites some of the electrons in the atoms so that they jump to higher energy levels. As the electrons drop back to lower levels they emit radiation. The radiation passes through a prism or diffraction grating to produce the *atomic spectrum* of the elements in the sample.

Important applications of atomic emission spectroscopy include the measurement of sodium, potassium, lithium and calcium ions in biological fluids such as blood and urine.

atomic number is the number of protons in the nucleus of an *atom*. The alternative term is proton number.

atomic orbitals are the sub-divisions of the electron shells in atoms. The main shells divide into sub-shells labelled s, p, d and f. The sub-shells are further divided into atomic orbitals.

Each orbital is defined by its:

- energy
- shape
- direction in space.

There is one orbital in the first shell, four in the second shell, nine in the third shell and 16 in the fourth shell. Each orbital can contain up to two electrons.

The study of *atomic spectra* makes it possible to determine the energies of atomic orbitals. The terms 'energy level' and 'orbital' are often used interchangeably.

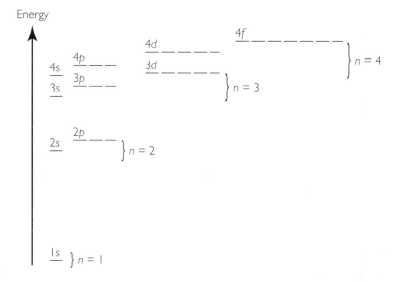

The energies of atomic orbitals in atoms. The labels s, p, d and f are left over from early studies of atomic spectra which used the words sharp, principal, diffuse and fundamental to describe different series of lines. These terms have no significance in an advanced level course. The pattern of energy levels accounts for the electron configurations of atoms and the arrangement of elements in the periodic table.

The shapes of orbitals are derived from theory. The shapes are determined by solving a mathematical equation (the Schrödinger wave equation) which makes it possible to calculate the probability of finding an electron at any point in an atom.

The probable location of the electrons in the first shell is spherical. It is an example of an *s-orbital* (1s).

The four orbitals in the second shell are made up of one *s*-orbital (2s) and three dumbell shaped *p-orbitals*. The three *p*-orbitals ($2p_x$, $2p_y$, $2p_z$) are arranged at right angles to each other.

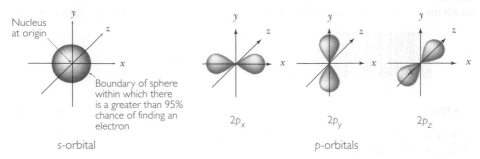

s-orbital   p-orbitals

The shapes of s and p atomic orbitals

atomic radius: this is a measure of the size of an atom in a crystal or molecule. Chemists use *X-ray crystallography* and other techniques to measure the distance between the nuclei of atoms. The *atomic radius* of an atom depends on the type of bonding and on the number of bonds. Usually atomic radii for metals are calculated from the distances between atoms in metal crystals (metallic radii). Atomic radii for non-metals are calculated from the lengths of covalent bonds in crystals or molecules (*covalent radii*).

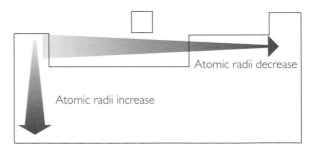

Atomic radii in a molecule and in a crystal

Atomic radii increase down any group in the periodic table as the number of electron shells increases. In *group 1* the metallic radii rise from 0.157 nm for lithium to 0.272 nm for caesium.

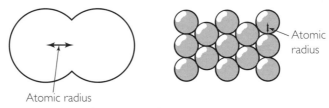

Trends in atomic radii in the periodic table

Atomic radii decrease across from left to right across a period. Across the period Na to Ar, atomic radii fall from 0.191 nm for sodium to 0.099 nm for chlorine. From one element to the next across a period the charge on the nucleus increases by one as the number of electrons in the same outer shell increases by one. *Shielding* by electrons in the same shell is limited so the 'effective nuclear charge' increases and the electrons are drawn more tightly to the nucleus.

atomic spectrum: the pattern of lines seen with a spectroscope when the atoms of an element emit radiation. Atoms emit radiation when excited by heat or electricity.

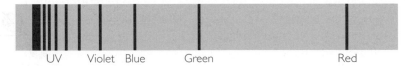

UV   Violet   Blue   Green   Red

The line emission spectrum for hydrogen in the visible and ultraviolet (UV) regions

Some elements emit radiation in the visible region of the spectrum and these are the elements which give coloured flames in *flame tests.*

Quantum theory explains why atomic emission spectra consist of a series of sharp lines. Each line in an emission spectrum corresponds to an energy jump of a definite size as electrons drop back from a higher to a lower energy level.

Electron jumps between energy levels in the hydrogen atom giving rise to the visible lines in the emission spectrum

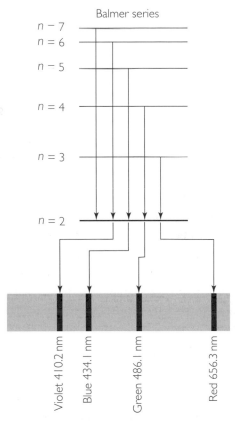

Atoms also absorb radiation. If radiation passes through the vapour of an element and is viewed through a spectroscope, the absorption spectrum appears as dark lines where particular frequencies are absorbed from the continuous spectrum.

atomic theory: the theory that all matter is made up of very tiny particles called atoms. Modern atomic theory was foreshadowed in ancient Greek philosophy but chemists now trace the development of the theory back to the work of John Dalton.

It is the ideas developed between 1800 and 1930 that are significant for a chemistry course at Advanced level. Studies of the structure of atomic nuclei continue with the help of huge particle accelerators such as those of CERN in Switzerland.

1804   John Dalton's theory of atoms as solid spheres explains the differences between elements and compounds. In this theory all the atoms of the same element weigh the same.

1896   Henri Becquerel discovers *radioactivity* leading to the work of Marie and Pierre Curie who separate and identify more radioactive elements.

1897   J J Thompson discovers the electron showing that atoms are themselves made up of even smaller particles.

1911   Hans Geiger and E Marsden's *alpha particle* scattering experiment leads to the Rutherford model with a nucleus surrounded by orbiting electrons.

1913   Niels Bohr model explains the *atomic spectrum* of hydrogen by proposing that the electrons orbiting the nucleus could only have definite energies. According to *quantum theory* atoms absorb and emit radiation at particular frequencies depending on the size of the energy jumps between *energy levels*.

1919   Francis Aston uses his mass spectrograph to demonstrate the existence of *isotopes* which had earlier been proposed by Soddy to describe atoms of the same element which have different masses.

1923   Louis-Victor de Broglie puts forward the theory that electrons have wave-like properties as well as particle properties. This theory is confirmed by experiment in 1927 when scientists in the USA and UK obtained diffraction effects with beams of electrons.

1926   Erwin Schrödinger publishes his mathematical theory of wave mechanics. The Schrödinger wave equation explains the pattern of energy levels in the hydrogen atoms and gives rise to the picture of electron in *atomic orbitals* with particular energies and shapes.

1932   James Chadwick discovers the neutron, making it possible to explain the existence of isotopes and radioactivity.

atomisation is a process in which an element or compound is converted into gaseous atoms. (See *enthalpy change of atomisation*.)

ATP (adenosine triphosphate) plays a crucial role in the energy transfer processes of living organisms. *Hydrolysis* of ATP, under the control of *enzymes*, splits off one of the three phosphate groups, releasing 31 kJ mol^{-1} to drive biochemical processes and for muscle movement.

Structure of ATP

attacking group is a term used to describe the mechanism of chemical reactions. *Nucleophiles* and *electrophiles* are examples of attacking groups.

aufbau principle: the principle that the *electron configurations* of atoms build up according to a set of rules.

The three rules are that:

- electrons go into the *orbital* at the lowest available *energy level*
- each orbital can only contain at most two electrons (with opposite *spins*)
- where there are two or more orbitals at the same energy, they fill singly before the electrons pair up.

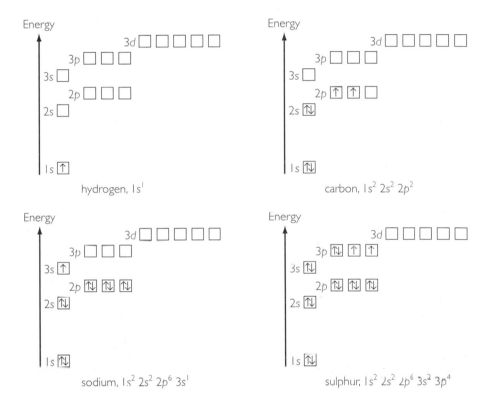

Aufbau principle Electrons in energy levels for four atoms to show the application of the aufbau principle

autocatalysis is catalysis of a reaction by one of its products. An autocatalytic reaction starts slowly but then speeds up as the catalytic product starts to form. The oxidation of ethanedioate ions by manganate(VII) ions in acid solution is catalysed by manganate(II) ions, thus:

$$2MnO_4^-(aq) + 5C_2O_4^{2-}(aq) + 16H^+(aq) \longrightarrow 10CO_2(g) + 2Mn^{2+}(aq) + 8H_2O(l)$$

At first bubbles of carbon dioxide appear very slowly but once the reaction produces some manganese(II) ions it speeds up and gas is evolved more rapidly. This is an example of *homogeneous catalysis*.

auxochrome: see *chromophore*.

Avogadro constant: the number of atoms, molecules, ions or other chemical *entities* in one mole of a substance. The Avogadro constant $(L) = 6.02 \times 10^{23}$ mol^{-1}.

amount of substance (mol) $\times$ Avogadro constant (mol^{-1}) = no. of specified entities

So 0.25 mol carbon atoms contains 1.50×10^{23} atoms while 0.5 mol oxygen molecules contains 3.10×10^{23} molecules.

Avogadro's law states that equal *volumes* of gases under the same conditions of temperature and *pressure* contain equal numbers of molecules. This means that one *mole* of any gas occupies the same volume under the same conditions. The volume occupied by a gas at 273 K (0°C) and one atmosphere pressure (101.3 kPa) is about 22.4 dm^3. These are the conditions of standard temperature and pressure (*stp*). In a warmish laboratory at 298 K (25°C) the volume of one mole of gas is about 24 dm^3.

Amadeo Avogadro (1776–1856) was the Italian scientists who proposed this law to account for *Gay-Lussac's law* of combining volumes.

azeotropic mixture: a mixture of liquids which on *boiling* produces a *vapour* with the same composition as the liquid. This means both that the liquids cannot be separated by fractional distillation and that azeotropes are constant boiling mixtures. A mixture containing 95.6% ethanol with 4.4% water is an azeotrope which boils at 78.2°C. As a result it is not possible to produce pure ethanol (anhydrous or absolute ethanol) by distillation alone.

azo compound: a compound with the group $-N = N-$ in its structure.

azo dyes are made by coupling a *diazonium salt* with one of a variety of coupling agents. Benzene diazonium chloride, for example, couples with naphthalene-2-ol to make a yellow azo compound.

benzene
diazonium
chloride

naphthalene-2-ol

an azo dye

Equation for a coupling reaction to make an azo dye

Very many azo *dyes* are manufactured by linking combinations of 50 diazonium salts with 52 coupling agents. Most of the dyes are red, orange or yellow. The *acid–base indicator* methyl orange is an azo dye.

back titration is an analytical technique used when the reaction between the standard solution and the substance to be analysed is slow. The procedure is to add a measured excess of the standard solution, allow time for the reaction to finish and then to use a titration with a second standard solution to measure how much of the first standard solution remains unused.

This technique is used to measure amounts of ammonium salts. An excess of a standard solution of sodium hydroxide is added to a sample and the mixture heated to drive off ammonia. In the process some of the sodium hydroxide is used up.

$$NH_4^+(aq) + OH^-(aq) \longrightarrow NH_3(g) + H_2O(l)$$

After all the ammonia has been driven off by boiling, a titration with standard hydrochloric acid measures the amount of unused sodium hydroxide remaining. This shows how much of the alkali was used to react with the ammonium salt in the sample and hence the amount of ammonium salt in the sample.

bacteria are a group of micro-organisms. Some bacteria are harmful and responsible for diseases such as cholera, tuberculosis and food poisoning. Most bacteria play a vital part in ecological processes. Soil bacteria, for example, help to break down the remains of living things releasing the nutrient elements such as nitrogen, in a form which plants can use for growth.

Bacteria are the basis of much traditional and modern biotechnology. In a traditional process, for example, bacteria are used to make yoghurt. In new methods they are altered by genetic engineering to produce substances such as human insulin for diabetics.

bactericide: a chemical which kills bacteria. See *chlorine water treatment* and *water treatment*.

Bakelite is the phenol–methanal plastic discovered by the Flemish-born chemist Leo Baekland in 1905 who was then living in the USA. This was the start of the modern plastics industry. Bakelite is a cross-linked, *thermosetting polymer* formed when phenol and methanal (formaldehyde) are mixed and heated with an acid *catalyst*.

balanced equations show the amounts (in moles) of reactants and products involved in chemical reactions. There is no change in the total number of atoms of each element as reactants turn into products during a chemical reaction. Four steps lead to the balanced equation for a reaction. Take for example the reaction of sodium with water.

> **Step 1** Identify the reactants and products by name:
>
> sodium reacts with water to form sodium hydroxide and hydrogen.

> **Step 2** Write down the correct formulae for reactants and products:
>
> $Na + H_2O \longrightarrow NaOH + H_2$

> **Step 3** Balance the numbers of atoms of each element by inspection and by writing numbers in front of the formulae as necessary. In this

Base	K_b/mol dm^{-3}
ethylamine, $C_2H_5NH_2$	5.4×10^{-4}
methylamine, CH_3NH_2	4.3×10^{-4}
ammonia, NH_3	1.8×10^{-5}
phenylamine, $C_6H_5NH_2$	3.8×10^{-10}

basic oxide: an oxide of a *metal* which reacts with acids to form salts and water. Copper(II) oxide, for example reacts with acids to produce copper(II) salts.

$$CuO(s) + H_2SO_4(aq) \longrightarrow CuSO_4(aq) + H_2O(l)$$

Note that it is the oxide ion in the basic oxide which acts as a base by taking a hydrogen ion from the acid.

$$O^{2-} + 2H^+ \longrightarrow H_2O$$

Basic oxides which dissolve in water are *alkalis*. As the compound dissolves, the oxide ion, acting as a base, takes a hydrogen ion from water and forms a hydroxide ion.

$$CaO(s) + H_2O(l) \longrightarrow Ca(OH)_2(aq)$$

batch process: a process which produces a specified amount of a product in a single operation. In industry a batch processes is used to manufacture chemicals needed in relatively small amounts so that a *continuous process* is not worthwhile. Batch processing is typically used to make fine chemicals on a scale of up to 100 tonnes per year. An industrial batch process is essentially a large scale version of a synthesis carried out in laboratory glassware. After producing a batch of a chemical the apparatus is cleaned and used again.

Some metals are extracted by batch processes (see *titanium extraction*).

battery: two or more *electrochemical cells* connected in series make up a battery of cells. This is the strictly correct meaning of the term battery. It is a term which scientists borrowed from the army where a line of guns is still called a battery. A car battery consists of six rechargeable *lead–acid cells* connected in series to give a 12 volt supply. In everyday conversation, however, it is common to use the word 'battery' when referring to a single cell.

bauxite: an ore of aluminium consisting of impure aluminium oxide. Impurities are removed by heating powdered bauxite with sodium hydroxide solution. Aluminium oxide, which is *amphoteric*, dissolves but other oxides such as iron(III) oxide and titanium(IV) oxide do not. After filtering, *seed crystals* are added and hydrated aluminium oxide crystallises as the solution cools. Heating the hydrate crystals at 1000°C produces anhydrous aluminium oxide ready for *aluminium extraction*.

becquerel (Bq): the *SI unit* for measuring radioactivity. If one atomic nucleus decays per second the activity is one becquerel.

bench reagents are solutions traditionally kept on the laboratory bench for general chemical use such as *qualitative analysis*. The concentrations of bench reagents are only approximate. The concentration of bench acids and alkalis is often around 2 mol dm^{-3} for hydrochloric acid, nitric acid, sodium hydroxide and ammonia solutions but 1 mol dm^{-3} for sulfuric acid (which produces two moles of hydrogen ions per mole of acid).

Benedict's solution is a reagent for detecting *reducing sugars*. The deep-blue reagent contains copper(II) ions complexed with citrate ions in alkaline solution. A reducing sugar reduces the reagent to copper(I) oxide. The blue colour goes and a reddish-brown precipitate forms. The reagent is similar to Fehling's solution, but safer to use because it is less *corrosive*. Benedict's solution is also more stable and does not have to be stored as two reagents.

benzene (C_6H_6) is the simplest *arene* hydrocarbon. It is a non-polar liquid at room temperature and was once widely used as a solvent until it was known to be *carcinogenic*.

Three ways of representing the structure of benzene. Note that all the atoms in a benzene molecule are in the same plane. The molecule is planar.

At first sight benzene is an unsaturated compound with three double which should be highly reactive like the *alkenes*. Benzene is far more stable and less reactive than the *Kekulé structure* suggests (see *stability of benzene*).

The 'three double bonds' in the benzene ring are not isolated. The electron clouds in these double bonds overlap and merge to form a double doughnut-shaped electron cloud of *delocalised electrons* that sandwiches the six-sided carbon skeleton.

Like other arenes, benzene burns with a very smoky, yellow flame. The characteristic reactions of benzene are *electrophilic substitution* reactions.

Summary of the substitution reactions of benzene (see also *sulfonation of benzene*)

benzene derivates consist of a *benzene* ring with one or more side chains or functional groups substituted for the hydrogen atoms of the ring. Chemists continue to use a mixture of systematic and traditional names for these derivatives.

The properties of functional groups such as — OH in *phenol* or — NH_2 in phenylamine are modified when bonded directly to a benzene ring. These functional groups show their normal behaviour, however, when substituted into a side chain.

Biopol was the first totally biodegradable plastic to be manufactured for use as a plastic. Biopol is a trade name for a *polyester* made by bacteria. The bacteria make the natural polyester as an energy store. Manufacturers have developed techniques for growing the bacteria on a large scale, extracting the polymer and converting it to a useful plastic. The plastic is much more expensive than poly(ethene) but microorganisms in the soil can break it down by hydrolysing the ester links.

biofuels are fuels produced from vegetable matter and organic wastes. Biofuels include *wood* and *biogas*.

Vegetable oils have always been used for heating and lighting on a domestic scale though in industrialised countries they have been largely replaced by natural gas and other fossil fuels.

There is increasing commercial interest in biofuels. Countries such as Italy and Australia are making significant use of diesel fuel manufactured from the oils from crops such as rape. Brazil is the leading country using ethanol as a fuel for motor vehicles. The ethanol is manufactured by *fermentation* of sugars from sugar cane.

Concern about the enhanced *greenhouse effect* is encouraging people to consider biofuels. Crops take in carbon dioxide from the air by *photosynthesis* as they grow, making sugars or oil. The carbon dioxide returns to the air as the biofuels burn. It appears that the use of biofuels would have no overall effect on the level of carbon dioxide in the atmosphere. This, however, ignores the energy from burning *fossil fuels* during the planting, harvesting and processing of the crop. *Fertilisers* may also be needed for growth and these need fossil fuels for their manufacture.

biogas is methane gas produced by the action of bacteria on animal and plant wastes under anaerobic conditions. Many sewage works in the UK produce biogas. Biogas also forms in *landfill* sites as the waste rots down. The gas can be hazard unless collected and burnt. Small scale biogas digesters are used in many parts of the world to supply nearby homes with gas for cooking.

biological oxygen demand (BOD) is a measure of water *pollution* by organic matter from sources such as sewage. BOD is the quantity of dissolved oxygen (measured in mg dm^{-3}) removed from a sample of water by micro-organisms when incubated for five days at 20°C. A sample of water containing much organic matter will support large numbers of bacteria which use dissolved oxygen as they break down the pollutants.

biopolymers are naturally occurring *polymers* in living things. They include *polysaccharides*, *proteins* and *nucleic acids*.

biotechnology makes use of micro-organisms or *enzymes* to produce useful products. Traditional biotechnology uses yeast to make beer, bread and wine as well as bacteria to make yoghurt. Modern biotechnology, based on genetic engineering, makes it possible to use genetically modified bacteria to produce medically important substances such as insulin, growth hormone and the blood clotting agent, factor VIII.

Biuret test: a test to detect *proteins*. The procedure is to add sodium hydroxide to a test sample followed by a few drops of copper(II) sulfate solution. The solution turns mauve if protein is present. The test actually detects *peptide* bonds.

blank determinations are used in *quantitative analysis* to eliminate errors. The analyst works through the whole procedure with a blank solution containing the solvent and all the reagents but none of the sample. In a *titration* a blank determination makes it possible to allow for the volume of reagent needed to change the colour of the indicator at the *end-point*. Blank determinations in colorimetry and spectroscopy help to detect, and allow for, errors caused by contamination of reagents or the effects of the apparatus used.

blast furnace: a furnace for extracting metals, which depends on a blast of pre-heated air entering near the bottom of the furnace. *Iron extraction* is a large scale, *continuous process* in blast furnaces.

bleaching is a process for destroying unwanted colours by *oxidation* or *reduction*. Domestic liquid bleach is a solution of sodium chlorate(I) formed by the reaction of *chlorine* with cold sodium hydroxide solution. Bleaching powder is calcium chlorate(I), $Ca(OCl)_2$ made by absorbing chlorine gas in calcium hydroxide. These are oxidising bleaches. Another oxidising bleach is *hydrogen peroxide.*

Oxidising bleaches not only remove colour but also kill micro-organisms.

Sulfur dioxide and sodium sulfite are reducing bleaches. Paper bleached white by a reducing bleach may turn yellow with age as oxygen in the air reverses the process.

Colour in organic compounds is caused by delocalised electrons in sequences of alternating double and single bonds (*conjugated systems*). Oxidising bleaches break some of the double bonds. Reducing bleaches convert double bonds to single bonds.

body-centered cubic structure is a crystal structure with an atom at each corner of a cube surrounding one atom at the centre of the cube. This structure is more open than the two *close-packed structures.*

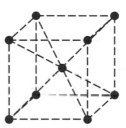

Body-centred cubic structure. Each metal atom is surrounded by eight nearest neighbours.

Metals with the body-centered cubic structure are the *group 1* metals lithium, sodium and potassium, also the *d-block* metals chromium, vanadium and tungsten.

boiling: a liquid boils when it is hot enough for bubbles of vapour to form within the body of the liquid. This happens when the *vapour pressure* of the liquid equals the external *pressure.*

The boiling point of a liquid varies with pressure. Raising the external pressure raises the boiling point. Boiling points are usually measured at atmospheric pressure. The normal boiling point is the temperature at which the vapour pressure of the liquid equals one atmosphere (101.3 kPa).

The boiling point of a mixture of two liquids varies with the composition of the mixture. This has to be taken into account when separating mixtures by *fractional distillation*.

Boltzmann distribution: see *Maxwell–Boltzmann distribution*.

bomb calorimeter: an apparatus for measuring energy changes when compounds burn.

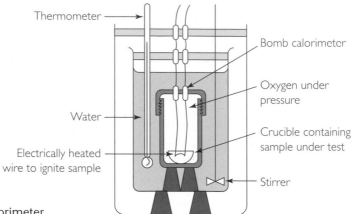

Thermometer

Bomb calorimeter

Oxygen under pressure

Water

Crucible containing sample under test

Electrically heated wire to ignite sample

Stirrer

A bomb calorimeter

A measured amount of a sample burns in oxygen under pressure and the temperature rise of the whole apparatus is measured. The calorimeter is calibrated using benzoic acid for which the standard *enthalpy change of combustion* is known accurately.

A bomb calorimeter operates at constant volume so a correction is needed to convert the results to *enthalpy changes* at constant pressure.

bond angle is the angle between two *covalent bonds* in a molecule or *giant structure*. X-ray and electron diffraction studies make it possible to measure bond angles accurately. The results show that *covalent bonds* have a definite direction as well as length.

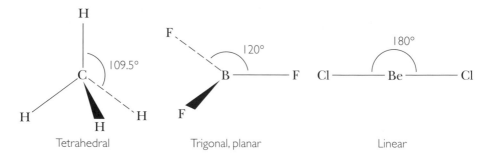

Tetrahedral   Trigonal, planar   Linear

Molecules showing the bond angles

Electron pair repulsion theory makes it possible to predict the *shapes of molecules* and bond angles with surprising accuracy.

bond breaking: see *heterolytic fission* and *homolytic fission*. Note that it takes energy to break bonds. Bond breaking is an *endothermic change*.

bond dissociation enthalpy is the *enthalpy change* on breaking one mole of a covalent bond in a gaseous molecule. Energy is needed to break covalent bonds. Bond breaking is *endothermic* so the enthalpy change is positive.

Chemists use spectroscopy to measure bond dissociation enthalpies. The bond dissociation enthalpies for the hydrogen halides are as follows:

$$HX(g) \longrightarrow H(g) + X(g) \qquad \Delta H = +562 \text{ kJ mol}^{-1} \text{ for HF}$$
$$\Delta H = +432 \text{ kJ mol}^{-1} \text{ for HCl}$$
$$\Delta H = +366 \text{ kJ mol}^{-1} \text{ for HBr}$$
$$\Delta H = +299 \text{ kJ mol}^{-1} \text{ for HI}$$

For these compounds the bonds get easier to break as the halogen atoms get larger and the bonds get longer down group 7.

In molecules with two or more bonds between similar atoms, the successive bond dissociation enthalpies are not the same. In water, for example the energy needed to break the first O — H bond in H — O — H(g) is 498 kJ mol^{-1} but the energy needed to break the second O — H bond in O — H(g) is only 428 kJ mol^{-1}.

bond enthalpies (or bond energies) are the average values of *bond dissociation enthalpies* used in approximate calculations to estimate enthalpy changes for reactions. The average values of bond enthalpies take into account the facts that:

- the successive bond dissociation enthalpies are not the same in compounds such as water or methane
- the bond dissociation enthalpy for a specific covalent bond varies slightly from one molecule to another.

Worked example:

Use mean bond enthalpies to estimate the enthalpy of formation of *hydrazine*.

Note on the method

Write out the equation showing all the atoms and bonds in the molecules to make it easier to count the numbers of bonds broken and formed.

Look up the mean bond energies in a book of data. The symbol $E(N - H)$ stands for the bond energy of a covalent bond between a nitrogen atom and a hydrogen atom.

Answer

The equation for the reaction:

$$N \equiv N \ + \ 2\,H - H \ \longrightarrow \ \begin{matrix} H & & H \\ \diagdown & & \diagup \\ & N - N & \\ \diagup & & \diagdown \\ H & & H \end{matrix}$$

The energy needed to break the bonds in the reactants

$= E(N \equiv N) \text{ kJ mol}^{-1} + 2E(H - H) \text{ kJ mol}^{-1}$

$= 945 \text{ kJ mol}^{-1} + 2 \times 436 \text{ kJ mol}^{-1}$

$= 1817 \text{ kJ mol}^{-1}$

The energy given out when the product forms

$= E(N—N)$ kJ mol^{-1} + $4E(N—H)$ kJ mol^{-1}

$= 158$ kJ mol^{-1} + 4×391 kJ mol^{-1}

$= 1722$ kJ mol^{-1}

More energy is needed to break bonds than is given out when bonds are formed so the reaction is endothermic and the enthalpy change is positive.

$\Delta H = +1817$ kJ mol^{-1} − 1722 kJ mol^{-1} = $+95$ kJ mol^{-1}

bond length: the distance between the nuclei of two atoms linked by one or more *covalent bonds.* For the same two atoms, *triple bonds* are shorter than *double bonds* which in turn are shorter than single bonds.

Bond	Bond length/nm
C — C	0.154
C = C	0.134
C ≡ C	0.120
C ---- C in benzene	0.140

Bond lengths which are intermediate between single and double bonds, as in *benzene,* are a sign of electron *delocalisation.*

bond polarity is the extent to which the shared pair of electrons in a *covalent bond* is attracted towards one of the atoms joined by the bond. Bonding electrons are pulled towards the more *electronegative* atom. The more electronegative atom carries a partial negative charge ($\delta-$) the other atom carries a partial positive charge ($\delta+$). (See also *polar covalent bonds.*)

$$\overset{\delta+ \quad \delta-}{H—Cl}$$

$$\overset{H}{\underset{H}{\diagdown}}C\overset{\delta+ \quad \delta-}{=}O$$

$$\overset{2\delta-}{O} \quad \underset{\delta+ \qquad \delta+}{H \quad H}$$

Examples of molecules with polar covalent bonds

bond rotation is possible about single covalent bonds but is prevented under normal conditions by *double bonds* or *triple bonds.* The lack of rotation about double bonds gives rise to *geometric isomerism.*

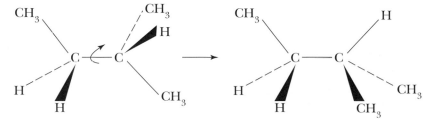

The groups at the end of a single C — C bond can rotate freely relative to each other giving rise to different *conformations of molecules*

bonding: see *metallic bonds, covalent bonds, ionic bonding, hydrogen bonding* and *inter-molecular forces.*

bonding molecular orbital: a region in space in a molecule where there is a high probability of finding bonding electrons. Molecular orbitals, like *atomic orbitals,* are solutions to the Schrödinger wave equation. At a simplified, descriptive level, a molecular orbital forms by overlap of atomic orbitals of the atoms linked by a covalent bond.

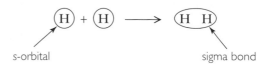

Formation of a sigma orbital   s-orbital   sigma bond

Examples of molecular orbitals are *sigma (σ) bonds* and *pi (π) bonds.*

Sigma and pi bonds in ethene   p-orbital

borax (Na$_2$B$_4$O$_7$.10H$_2$O) is a naturally occurring mineral which has been used as a flux in metallurgy for hundreds of years. Molten borax cleans the surface of hot metals by dissolving metal oxides. This makes for good contact between metal surfaces when metals are welded or soldered together. Most borax is now used to make *borosilicate glass.*

Adding acid converts borax to boric acid (H$_3$BO$_3$) which is separated by cooling the solution in ice and filtering off the white crystals. Boric acid is a weak acid and mild antiseptic used in eye lotions and other medicines.

Born–Haber cycle: a thermochemical cycle which can be used to calculate the *lattice energy (enthalpy)* for a compound of a metal with a non-metal.

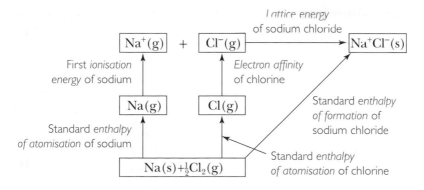

Born–Haber cycle for sodium chloride (see the separate entries in this book for the definitions of the terms)

All but one of the terms in the cycle can be measured experimentally. Applying *Hess's law* makes it possible to calculate the one unknown term, which is the lattice energy (enthalpy).

A Born–Haber cycle can help reveal the factors that determine the *stability of compounds*. For a stable compound, such as magnesium chloride, the lattice energy must be bigger than the total energy needed to produce gaseous ions from the elements. Removing two electrons from a magnesium atom is highly endothermic. What accounts for the stability of magnesium chloride ($MgCl_2$) is the large, negative lattice energy when the gaseous ions assemble themselves into a crystal lattice. $MgCl_3$ is not stable because of the very large third ionisation energy of magnesium. MgCl might be stable but not as stable as $MgCl_2$. $MgCl_2$ has a much larger lattice energy because an Mg^{2+} ion not only has a larger charge but is also smaller than a Mg^+ ion because it has lost both electrons in the outer shell.

A Born–Haber can also help to determine whether the bonding in a compound is truly ionic. The experimental lattice energy calculated from a Born–Haber cycle can be compared with the theoretical value calculated using the laws of electrostatics and assuming that the only bonding in the crystal is ionic.

Compound	Experimental lattice energy from a Born–Haber cycle/kJ mol^{-1}	Theoretical lattice energy calculated assuming that the only bonding is ionic/kJ mol^{-1}
NaCl	−780	−770
KCl	−711	−702
AgCl	−905	−833

The values in the table show that ionic bonding can account for the lattice energies of sodium and potassium chlorides. The lattice energy of silver chloride is greater than can be explained by ionic bonding. There must be a contribution from covalent bonding.

(Note that chemists also use a second type of Born–Haber cycle to analyse the energy changes when an ionic crystal dissolves in water – see *enthalpy change of solution*.)

boron (B) is a hard, black solid with a high melting point and the only non-metal in *group 3* of the periodic table. Its *electron configuration* is $[He]2s^22p^1$. Boron is in oxidation state +3 in all its compounds. The bonding between boron and other elements is covalent.

Boron has become important as an element since the development of the nuclear industry. The *isotope* boron-10 absorbs neutrons strongly so the element was scattered over the Chernobyl nuclear reactor after the explosion to quench chain reactions.

Boron trichloride (BCl_3) is a colourless liquid and like *aluminium chloride*, it is a strong electron pair acceptor (*Lewis acid*).

Sodium tetrahydridoborate(III) ($NaBH_4$) is a useful reducing agent.

borosilicate glass is used to make oven glassware and laboratory equipment because it has a higher melting point than cheaper, soda lime glass and is much less likely to crack as it heats up or cools down. Borosilicate glass has a high refractive index making is useful for the production of lenses.

Boyle's law states that the volume of a fixed amount of gas is inversely proportional to its pressure at constant temperature.

$$p \propto \frac{1}{V} \text{ at constant temperature for a fixed amount of a gas}$$

Hence $p_1 V_1 = p_2 V_2$ = constant

Brady's reagent is a solution of 2,4-dintrophenylhydrazine in acid, used to detect and identify *carbonyl compounds*. Mixing an aldehyde or ketone with the reagent produces an yellow–orange precipitate. The *addition–elimination reaction* produces a 2,4-dinitrophenylhydrazone. The solid *derivative* can be filtered off, *recrystallised* and identified by measuring its melting point. This makes it possible to identify the original carbonyl compound.

Formation of the 2,4-dinitrophenylhydrazone derivative of propanone

brasses are *alloys* of copper (60–80%) and zinc (20–40%). Brass is easily worked, has an attractive gold colour and does not corrode.

breathalyser: an instrument for estimating the concentration of alcohol in blood. A breathalyser measures the ethanol in a sample of air from the lungs. There is an equilibrium between ethanol dissolved in blood and ethanol vapour in the air in the lungs and the constant ratio at body temperature makes it possible to infer the blood alcohol concentration.

The first successful breathalyser was based on the chemical reduction of orange dichromate(VI) ions to green chromate(III) ions as they oxidise ethanol to ethanal. The driver breathed through a tube containing the orange crystals and the extent to which they turned yellow was *calibrated* to measure the blood alcohol concentration.

Many of the roadside breathalysers are now *fuel cells*. At one electrode ethanol is oxidised to ethanoic acid while at the other electrode oxygen is reduced to water. The higher the concentration of ethanol in the driver's breath, the greater the voltage of the cell. This method is not considered accurate enough for prosecution in court. Suspects are taken to a police station where the alcohol concentration in their breath is measured using an infra-red absorption spectrometer. If spectroscopy shows that drivers are above the limit, they have the right to ask for an analysis of ethanol in a blood sample, determined by *gas-liquid chromatography*.

Thomas Lowry (1874–1936) of the University of Cambridge, but the two worked independently.

The Brønsted–Lowry theory is more general than the Arrhenius theory, because it also covers reactions which happen without water. According to the theory, hydrogen chloride molecules give hydrogen ions to water molecules when they dissolve in water producing hydrated hydrogen ions called *oxonium ions*. The water acts as a base:

$$HCl(g) \ + \ H_2O(l) \ \longrightarrow \ H_3O^+(aq) \ + \ Cl^-(aq)$$

An example of an acid–base reaction in the absence of water is the formation of a white smoke of ammonium chloride when *ammonia* gas mixes with hydrogen chloride gas.

$$NH_3(g) \ + \ HCl(g) \ \longrightarrow \ NH_4^+ \, Cl^-(s)$$

bronze is an *alloy* of copper with up to 12% tin. Bronze is a strong, hardwearing alloy with good resistance to corrosion. It is used to make gear wheels, bearings, propellers for ships, statues and coins.

Brownian motion is the rapid random motion of minute particles suspended in a liquid or gas. The phenomenon was first seen by the British botanist, Robert Brown looking at pollen grains in water through a microscope. Brownian motion is evidence for the *kinetic theory*. The erratic random motion of pollen, smoke particles or other *colloid*-sized specks of matter is due to continuous bombardment by fast moving molecules which are too small to be seen through a light microscope.

Buchner flask and funnel: the apparatus used for filtering solutions especially when isolating and purifying solid products of reactions by *recrystallisation*. A pump (usually a water pump) lowers the pressure inside the flask so that a pressure difference across the filter paper speeds up filtration.

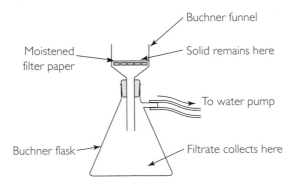

Apparatus for Buchner filtration

buckminsterfullerene: a form of carbon consisting of C_{60} particles (see *allotropes*) and named after the architect Richard Buckminster Fuller who designed the geodesic dome for Expo '67 in Montreal. The popular name for C_{60} molecules is 'buckyballs'.

Buckminsterfullerene was discovered in 1985 by Harry Kroto and his team at the University of Sussex. It was the first of a family of *fullerenes*.

C_{60} molecules are formed as a carbon vapour mixed with helium slowly cools and condenses. C_{60} is special because it is the only structure with pentagons and hexagons which is smooth because every atom has the same curvature as every other atom. Carbon atoms spontaneously assemble themselves into C_{60} and other *fullerenes*, including *nanotubes*.

bucky tubes: see *nanotubes*.

buffer solutions are mixtures of molecules and ions in solution which help to keep *pH* more or less constant. A buffer solution cannot prevent pH changes but it evens out the large changes in pH which can happen without a buffer.

Many household products include buffer mixtures, for instance washing powders, shampoos, skin creams and a variety of medicines including eye drops. Advertisers sometimes claim that their products are '*pH* balanced' if the formulation includes a buffer solution.

Buffers are important in living organisms. The pH of blood, for example, is closely controlled by buffers within the narrow range 7.38 to 7.42. Chemists use buffers when they want to investigate chemical reactions at a fixed pH.

Buffers are equilibrium systems which illustrate the practical importance of *Le Chatelier's principle*. A typical buffer mixture consists of a solution of a *weak acid* and one of its salts. For example a mixture of ethanoic acid and sodium ethanoate. There must be plenty of the acid and its salt.

By choosing the right weak acid, it is possible to prepare buffers at any pH value throughout the pH scale. If the concentrations of the weak acid and its salt are the same, then the pH of the buffer is equal to pK_a for the acid.

$$CH_3CO_2H \quad + \quad H_2O \quad \rightleftharpoons \quad H_3O^+ \quad + \quad CH_3CO_2^-$$

acid molecules – a reservoir of H^+ ions	stays roughly constant – so the pH hardly changes	base ions – with the capacity to accept H^+ ions
Plenty of the weak acid to supply more H^+ ions if alkali is added		Plenty of the ions from the salt to combine with extra H^+ ions if acid is added

The action of a buffer solution

More generally the pH of a buffer mixture can be calculated with the logarithmic form of the equilibrium law (see *Henderson–Hasselbalch equation*).

$$pH = pK_a + \lg \frac{[salt]}{[acid]}$$

Diluting a buffer solution with water does not change the ratio of the concentrations of the salt and acid so the pH does not change (unless the dilution is so great that the assumptions made when deriving the equation no longer apply).

Worked example:

What is the pH of a buffer solution containing 0.40 mol dm^{-3} methanoic acid and 1.00 mol dm^{-3} sodium methanoate?

Notes on the answer

Look up the value of pK_a in a book of data. The pK_a of methanoic acid is 3.8.

Answer

$$pH = 3.8 + \lg \frac{1.00}{0.40}$$

$$pH = 4.20$$

---

bumping is violent boiling which shakes the apparatus and can throw liquid from the container in which it is being heated. Adding a few fragments of porous pottery or some jagged anti-bumping granules cuts the risk of bumping by helping the bubbles of vapour to form smoothly.

burettes are graduated tubes, with taps or valves, used to measure the volumes of liquids or solutions during quantitative investigations such as *titrations*. The *accuracy* of a burette, when clean and properly used, depends on the *precision* with which it is manufactured and calibrated. (See *uncertainty of measurement*.)

by-products are unwanted products of chemical synthesis or manufacturing. By-products are formed by side reactions which happen at the same time as the main reaction reducing the yield of the product required.

In the laboratory preparation of 1-bromobutane from butan-1-ol, for example, the reagent is a mixture of sodium bromide and concentrated sulfuric acid. There are two side reactions which cut the yield. In the presence of acid, some of the butan-1-ol dehydrates to an alkene and some reacts to form an ether.

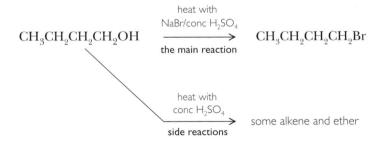

The main reaction and the side reactions

caesium chloride structure is the cubic crystal structure of the ionic compound caesium chloride (CsCl). In general it is a structure for a compound M⁺X⁻ in which each positive ion is surrounded by eight nearest neighbours at the corners of a cube and each negative ion is similarly surrounded by eight positive ions. So the *co-ordination numbers* for both elements is 8.

○ Cl

● Cs

Structure of caesium chloride showing 8:8 co-ordination. The structure consists of a simple cubic array of positive ions interpenetrating a cubic array of negative ions.

Other compounds with this structure include CsBr, CsI and NH_4Cl.

This structure is only possible in compounds with a positive ion which is relatively large so it can be in contact with eight neighbouring negative ions. Caesium, in period 6 of the periodic table, forms larger ions than, for example, sodium. (See also *unit cell*.)

caffeine is a *drug* found in tea, coffee, chocolate and cola drinks. Caffeine gives people a 'lift' and helps them to be mentally alert. The drug is used medically to stimulate the nervous, respiratory and cardiovascular systems. Caffeine is also added to medicines to counteract sleepiness caused by other ingredients.

Structure of caffeine
(1,3,5-trimethylxanthine)

Caffeine can be isolated from tea or coffee as a white crystalline solid by *solvent extraction*.

calcium (Ca) is a reactive, metallic element in *group 2* of the periodic table. Its electron configuration is $[Ar]4s^2$. Samples of the silvery metal usually look grey because they are covered with a layer of calcium oxide.

Calcium burns brightly in air with a brilliant red flame forming the white solid calcium oxide (CaO). The metal also reacts with cold water producing hydrogen and a white precipitate of calcium hydroxide.

$$Ca(s) + 2H_2O(l) \longrightarrow Ca(OH)_2(s) + H_2(g)$$

Calcium forms ionic compounds with non metals in which the metal is in the +2 oxidation state as Ca^{2+}.

Calcium oxide is a white solid made by heating *calcium carbonate*. It is a *basic oxide*. Its reaction with cold water to make calcium hydroxide is highly *exothermic*. Calcium hydroxide is only sparingly soluble in water forming an alkaline solution often called limewater.

Anhydrous calcium chloride ($CaCl_2$) is a cheap drying agent. The chloride crystallises from solution as a hydrate, $CaCl_2.6H_2O$.

calcium carbonate occurs naturally as limestone, chalk and marble. Limestone is an important mineral. Some of the rock is quarried for road building and construction. Pure limestone is also used in the chemical industry. Heating limestone in a furnace at 1200 K converts it to calcium oxide (quicklime). The reaction of quicklime with water produces calcium oxide (slaked lime).

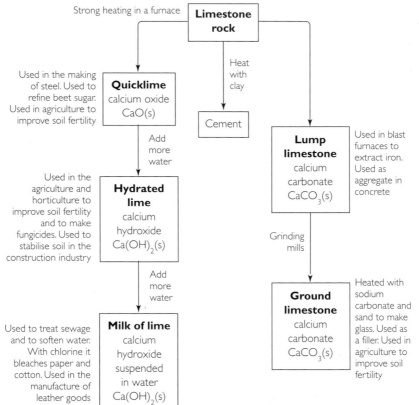

Products from limestone and their uses

calibration: a procedure used in quantitative analysis. Calibration makes it possible to convert the measurement of temperature, volume or light absorption into an accurate *concentration* in moles per litre. When using colorimeters or spectrometers, for example, it is common to plot a calibration curve. Instrument readings are taken for a series of *standard solutions* of known concentration.

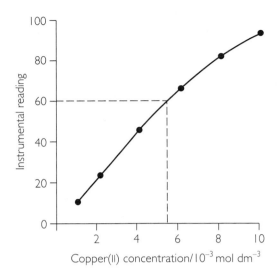

Instrumental reading

Copper(II) concentration/10^{-3} mol dm^{-3}

Calibration curve used to determine the concentration of copper(II) ions in aqueous solution obtained by taking instrument readings with a series of standard solutions. The instrument reading for an unknown sample is 60.0. The concentration of the sample can be read from the graph. The concentration of copper(II) ions in the sample is 5.5×10^{-3} mol dm^{-3}.

calorimeter: an apparatus for measuring the energy change during a chemical reaction. Typically a calorimeter is insulated from its surroundings and contains water. The energy from the reaction heats up the water and rest of the apparatus. An accurate thermometer measures the temperature rise. The apparatus is *calibrated* to determine its overall *specific heat capacity* by measuring the temperature rise for a reaction with a known *enthalpy change*.

Enthalpy changes of combustion are measured using a *bomb calorimeter*.

Enthalpy changes for reactions in solution can be compared quickly using an expanded polystyrene cup with a lid, as the calorimeter. Expanded polystyrene is an excellent insulator and has a negligible specific heat capacity. If solutions are dilute it is sufficiently accurate to assume that their density and specific heat capacity are the same as those of water.

Worked example:

When 50 cm^3 of 2.0 mol dm^{-3} hydrochloric acid mixes with 50 cm^3 2.0 mol dm^{-3} sodium hydroxide in a polystyrene cup the temperature rise is 13.5°C. What is the enthalpy change for the neutralisation reaction?

Notes on the method

Assume that the density of the solutions is the same as water (1 g cm^{-3}) and that for both, the specific heating capacity, like water, is 4.18 J g^{-1} K^{-1}.

Note that the total volume of solution is 100 cm^3 so the mass of solution heating up can be taken as 100 g.

Carbon dioxide is an acidic oxide. It dissolves in water forming carbonic acid which is a weak acid.

$$CO_2(g) + H_2O(l) \rightleftharpoons H_2CO_3(aq)$$

The gas is strongly absorbed by alkalis such as potassium hydroxide

The simplest chloride of carbon, tetrachloromethane (CCl_4), is a colourless liquid with tetrahedral molecules. Unlike many other non-metal chlorides it is not hydrolysed by water or alkalis (see *hydrolysis of non-metal chlorides*). Carbon does not react directly with chlorine, so tetrachloromethane is made by the reaction of carbon disulfide (CS_2) with chlorine.

carbon cycle: the cycling of the element carbon in the natural environment as carbon compounds move between the main reservoirs of carbon in the atmosphere, the oceans and on land (see *environmental chemistry*).

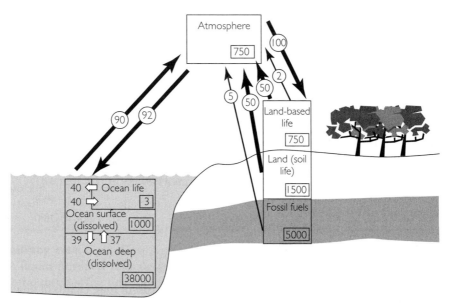

A highly simplified carbon cycle with estimates of the quantities in the main reservoirs where the figures are in gigatonnes of carbon (1 Gt = 1000 million tonnes). Also shown are estimates of the flows of carbon in gigatonnes per year (Gt yr^{-1}).

carbon monoxide is a colourless gas with no smell, formed by the incomplete combustion of fuels containing carbon. The gas is highly poisonous because it is held much more strongly by blood haemoglobin than oxygen. Death results when most of the haemoglobin is blocked by carbon monoxide. Lower levels of exposure to the gas reduce the ability of blood to carry oxygen and this puts a strain on the heart.

Carbon monoxide is also an important industrial chemical, both as a reducing agent and as a starting point for the synthesis of organic chemicals. Carbon monoxide is used as a reducing agent in metallurgy, for example during *iron extraction* in a blast furnace.

Steam reforming is the source of synthesis gas which is a mixture of carbon monoxide and hydrogen.

Carbon monoxide is used on a large scale to manufacture *methanol*. A mixture of carbon monoxide and hydrogen gases combines to make methanol at about 250°C over a catalyst made from copper and zinc oxide. This is an example of *heterogeneous catalysis*.

$$CO(g) + 3H_2(g) \rightleftharpoons CH_3OH(l)$$

carbonium ions are ions in which a carbon atom carries a positive charge (see *carbocations*).

carbonyl compounds contain the carbonyl group ($C = O$). The two main classes of carbonyl compounds are the *aldehydes* and the *ketones*. The $C = O$ double bond is polar with the electrons drawn towards the more electronegative oxygen atom. The characteristic reactions of carbonyl compounds are *nucleophilic addition reactions* and *addition–elimination reactions*.

The carbonyl group in aldehydes and ketones

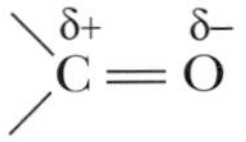

$$\begin{array}{c} \diagdown \quad \delta+ \quad \delta- \\ C = O \\ \diagup \end{array}$$

carborundum is the common name for silicon carbide (SiC), a shiny, black solid made by heating silicon dioxide with carbon at 2000°C. It has the same crystal structure as diamond but with every other carbon atom replaced by a silicon atom. Carborundum is a useful abrasive being harder than corundum (*aluminium oxide*) but not as hard as diamond. Like diamond, carborundum has a very high melting point making it a useful *refractory*.

carboxylic acids are compounds with the formula $R - CO_2H$ where R represents an *alkyl group, aryl group* or a hydrogen atom. The carboxylic acid group ($- CO_2H$) is the *functional group* which gives the acids their characteristic properties.

Carboxylic acids are named by changing the ending of the corresponding *alkane* to '-oic' acid. So ethane becomes ethanoic acid.

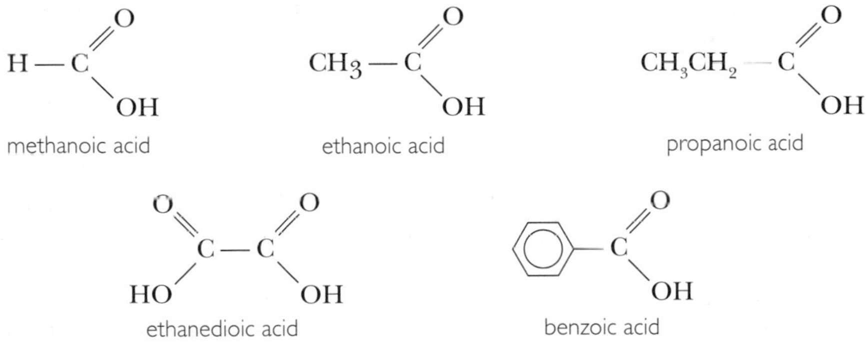

methanoic acid   ethanoic acid   propanoic acid

ethanedioic acid   benzoic acid

Names and structures of carboxylic acids

Even the simplest acids, such as methanoic acid and ethanoic acid, are liquids at room temperature because of *hydrogen bonding* between the carboxylic acid groups. Also, because of hydrogen bonding, these acids mix freely with water. Hydrogen bonding leads to *dimerisation* of carboxylic acids in non-polar solvents.

Benzoic acid is a solid at room temperature. It is only very slightly soluble in cold water but more soluble in hot water.

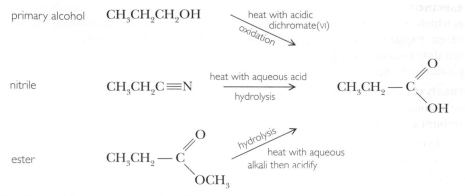

primary alcohol $CH_3CH_2CH_2OH$ — heat with acidic dichromate(VI), oxidation →

nitrile $CH_3CH_2C{\equiv}N$ — heat with aqueous acid, hydrolysis → $CH_3CH_2-C{\bigg\langle}^{O}_{OH}$

ester $CH_3CH_2-C{\bigg\langle}^{O}_{OCH_3}$ — hydrolysis, heat with aqueous alkali then acidify →

Reactions which produce carboxylic acids

Carboxylic acids are *weak acids*. The —OH group in the carboxylic acid group is more acidic than the —OH group in *alcohols* because the carboxylate ion formed is stabilised by *delocalisation of electrons*. Carboxylic acids are sufficiently acidic to produce carbon dioxide when added to a solution of sodium carbonate (or sodium hydrogencarbonate). This reaction distinguishes carboxylic acids from weaker acids such as *phenols*.

$$CH_3-C{\bigg\langle}^{O}_{OH} + H_2O \rightleftharpoons CH_3-C{\bigg\langle}^{O}_{O}{}^{-} + H_3O^{+}$$

Ionisation of ethanoic acid. Delocalisation of electrons stabilises the carboxylate ion and favours ionisation of the acid.

Carboxylic acids form a number of related compounds (derivatives): *acyl chlorides, acid anhydrides, esters* and *amides*. The reactions of ethanoic acid are typical of carboxylic acids.

$CH_3-C{\bigg\langle}^{O}_{OH}$

NaOH(aq) → $CH_3-C{\bigg\langle}^{O}_{O^-\ Na^+}$ sodium ethanoate (salt)

C_2H_5OH with acid catalyst → $CH_3-C{\bigg\langle}^{O}_{OC_2H_5}$ ethyl ethanoate (ester)

PCl_5 at room temperature → $CH_3-C{\bigg\langle}^{O}_{Cl}$ ethanoyl chloride (acyl chloride)

$LiAlH_4$ in dry ether → CH_3-CH_2OH ethanol (alcohol)

Reactions of ethanoic acid

carcinogens are compounds which can cause cancer. In a healthy body, the rate at which cells divide to produce new cells is strictly controlled. Cancer is a disorder which happens when some cells somehow escape from the normal control and multiply to produce a growth, or tumour. Chemicals which cause cancer probably produce changes (mutations) in the genes that control cell division.

catalysts speed up the rates of chemical reactions without themselves changing permanently. Catalysts can be recovered at the end of the reaction. Often a small amount of catalyst is effective.

Catalysts work by providing for the reaction an alternative pathway with a lower *activation energy*. A lower activation energy means that there is an increased proportion of molecules with enough energy to react at a particular temperature.

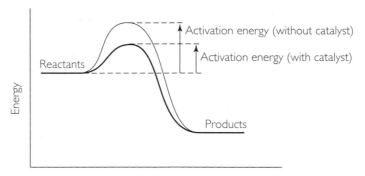

Catalysts Reaction profile showing the effect of a catalyst on the activation energy of a reaction

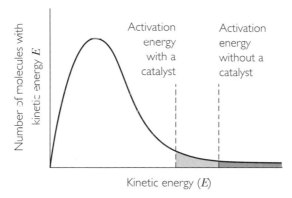

Catalysts Distribution of molecular energies showing how the proportion of molecules able to react increases when a catalyst lowers the activation energy

Often a catalyst changes the mechanism of the reaction and makes it more productive by increasing the yield of the desired product and reducing waste.

Many catalysts are specific to a particular reaction. This is especially true of *enzymes*. If the catalyst is in the same phase as the reactants it is a *homogeneous catalyst*. If the catalyst is in a different phase it is a *heterogeneous catalyst*.

Catalysts speed up reactions but they do not change the position of equilibrium for a reversible reaction.

Here are some highlights in the development of industrial catalysts:

- 1908 Fritz Haber discovered how to make *ammonia* from nitrogen and hydrogen with a modified iron catalyst.
- 1912 Paul Sabatier first used a nickel catalyst to *hydrogenate* unsaturated *vegetable oils* and turn them to solid *fats* for margarine.
- 1930 Eugene Houdry developed *catalytic cracking* of oil fractions to make *petrol*.
- 1942 Vladimir Ipatieff and Herman Pines found a catalytic method of alkylating *hydrocarbons* to produce branched hydrocarbons with high octane numbers to prevent *knocking* in petrol engines.
- 1976 General Motors and the Ford Motor Corporation developed *catalytic converters* to cut pollution from motor vehicles.

catalyst poisons inactivate *catalysts*, especially heterogeneous catalysts. Lead compounds in *petrol*, for example, poison the catalyst in catalytic converters and stop it working. Carbon monoxide poisons the iron catalyst used for *ammonia manufacture*.

catalytic converter: a device in the exhaust system of a car which contains a *catalyst* to covert pollutants in the exhaust gases to less harmful substances. Car exhausts pollute the air because the engine does not burn all the fuel and because the temperature and pressure in the cylinders are high enough for nitrogen from the air to react with oxygen.

Pollutant	Origin of the pollutant
carbon dioxide, CO_2	complete combustion of hydrocarbons in petrol
carbon monoxide, CO	incomplete combustion of fuel
hydrocarbons, C_xH_y	unburnt fuel
nitrogen oxides, NO_x	reaction of nitrogen and oxygen from the air in the hot engine
lead compounds	from anti-knock additives in leaded petrol

Unleaded petrol must be used in cars fitted with a catalytic converter because lead would poison the catalyst and stop it working.

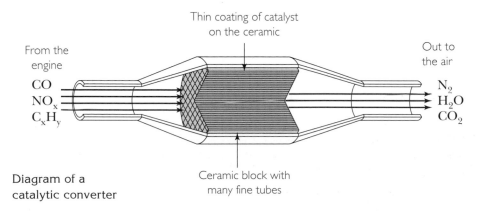

From the engine
CO
NO_x
C_xH_y

Thin coating of catalyst on the ceramic

Out to the air
N_2
H_2O
CO_2

Ceramic block with many fine tubes

Diagram of a catalytic converter

The catalyst is a finely divided alloy of platinum and rhodium supported on an inert ceramic, pierced with many fine tubes to give a very large surface area. Once the catalyst is hot enough it converts the pollutants to steam, carbon dioxide and nitrogen.

catalytic cracking: a process in an oil refinery for converting fractions from the *fractional distillation of oil* into more useful products by breaking up larger molecules into smaller ones. The problem in refining is to produce the oil products in the amounts needed by customers. Generally crude oil contains too much of the high-boiling fractions (bigger molecules) and not enough of the low-boiling fractions (smaller molecules) needed for fuels such as *petrol*.

The *heterogeneous catalyst* is a *zeolite*. Cracking takes place on the surface of the catalyst. The mechanism involves *heterolytic bond* breaking and the intermediates are *carbocations*.

Catalytic cracking is a *continuous process*. The finely powdered catalyst made of silicon and aluminium oxides gradually gets coated with carbon so it circulates through a regenerator where the carbon burns away in a stream of air.

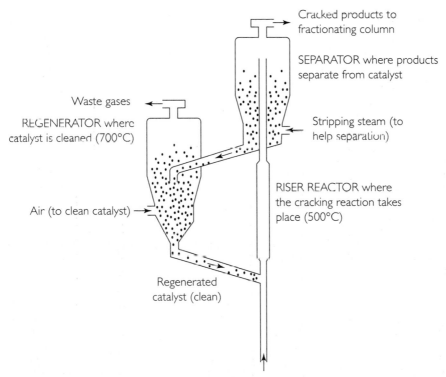

Catalytic cracking The catalyst powder flows to the vertical reactor where cracking takes place. The cracked vapours pass to a fractionating column while the catalyst flows to the regenerator.

catenation: the ability of the atoms of an element to join together in long chains or rings. *Carbon* atoms have an exceptional ability to catenate – hence the wide range of organic compounds. *Silicon*, the second element in group 4, does not have the same ability to catenate. Silicon forms *hydrides*, Si_nH_{2n+2}, analogous to the *alkanes* but the largest known molecule has n equal to eight, and only silane (SiH_4) is stable for any length of time.

Sulfur atoms can catenate. Both the rhombic and monoclinic *allotropes* consist of S_8 rings. Plastic sulfur is a tangled mass of long chains of sulfur atoms but it rapidly reverts to the rhombic form at room temperature.

cathode the *electrode* at which *reduction* takes place. During *electrolysis* the external power supply adds *electrons* to the cathode so that it becomes the negative electrode attracting positive ions which gain electrons (reduction) turning into atoms or molecules.

The term cathode is also sometimes used in *electrochemical cell* where the reduction process at this electrode in one of the *half-cells* takes electrons from the electrode which becomes the positive terminal of the cell.

Note that electrons flow into the cathode from the external circuit both during electrolysis and when a current is drawn from a chemical cell.

cathodic protection: an electrochemical method of preventing *corrosion* used with pipelines, oil rigs and the hulls of ships. Steel corrodes where it is oxidised. This happens wherever the metal is an *anode*. In anodic regions the iron atoms give up electrons turning into ions. Attaching a more reactive metal, such as zinc or magnesium, to iron creates an electrochemical cell in which the iron is the *cathode* so the more reactive metal corrodes.

cations: are *positive ions* attracted to the *cathode* during *electrolysis*. (See *cation tests* for examples of common cations.)

cation tests are used in qualitative analysis to identify the positive ions in salts. See the table at the top of page 81. Adding sodium hydroxide produces a precipitate if the metal hydroxide is insoluble. The precipitate dissolves in excess of the alkali if the hydroxide is *amphoteric.*

Adding *ammonia* solution also precipitates insoluble hydroxides. These redissolve in excess if the metal ion forms stable *complex ions* with ammonia molecules.

These tests can be supplemented with *flame tests* to identify metal ions such as sodium and potassium which do not give precipitates.

cell: see *electrochemical cell.*

Celsius temperature scale: a scale in which the melting point of ice is set at $0°C$ and the boiling point of water at $100°C$. The scale is based on the absolute or *Kelvin* temperature scale.

Temperature in $°C$ = absolute temperature (K) – 273 K

cement is manufactured by heating clay with powdered limestone in a rotating kiln and then grinding the lumps of product to a fine powder. Cement is a complex mixture of calcium silicates and calcium aluminosilicates which becomes hydrated and sets when mixed with water. Stirring cement with sand, gravel and water makes concrete.

Cation tests (see page 80)

Positive ion in solution	Observations on adding sodium hydroxide solution drop by drop and then in excess	Observations on adding ammonia solution drop by drop and then in excess
calcium, Ca^{2+}	white precipitate but only if the calcium ion concentration is high	no precipitate
magnesium, Mg^{2+}	white precipitate insoluble in excess reagent	white precipitate insoluble in excess reagent
barium, Ba^{2+}	no precipitate	no precipitate
aluminium, Al^{3+}	white precipitate which dissolves in excess reagent	white precipitate insoluble in excess reagent
chromium(III), Cr^{3+}	green precipitate which dissolves in excess to form a dark green solution	green precipitate insoluble in excess reagent
manganese(II), Mn^{2+}	off-white precipitate insoluble in excess reagent	off-white precipitate insoluble in excess reagent
iron(II), Fe^{2+}	green precipitate insoluble in excess reagent	green precipitate insoluble in excess reagent
iron(III), Fe^{3+}	browny-red precipitate insoluble in excess reagent	browny-red precipitate insoluble in excess reagent
copper(II), Cu^{2+}	pale blue precipitate insoluble in excess	pale blue precipitate dissolving in excess to form a dark blue solution
zinc, Zn^{2+}	white precipitate which dissolves in excess reagent	white precipitate which dissolves in excess reagent
lead, Pb^{2+}	white precipitate which dissolves in excess reagent	white precipitate insoluble in excess reagent
ammonium, NH_4^+	alkaline gas (ammonia) given off on heating	no visible change

ceramics are materials such as pottery, glasses, cement, concrete and graphite. Ceramics also include a wide range of crystalline materials such as *carborundum*, silicon nitride, *aluminium oxide* and magnetic *ferrites*. What all these materials have in common is that they are non-metallic, inorganic materials which are heated to a high temperature in a furnace at some stage during their manufacture.

Typical advantages of ceramics are that they are:

- very hard
- strong in compression
- chemically *inert*
- *refractories*
- electrical insulators.

Typical disadvantages of many ceramics are that they are:

- weak in tension
- brittle
- liable to crack if there is a sudden temperature change.

CFCs (chlorofluorocarbons) such as CCl_3F, CCl_2F_2 and CCl_2FCClF_2 are compounds containing carbon, chlorine and fluorine. They have advantages: they are unreactive, do not burn and are not *toxic*. Also it is possible to make CFCs with different boiling points to suit different applications. These properties make CFCs ideal as the working fluid in refrigerators and air conditioning units. They are also used to make the bubbles in expanded plastics and insulating foams. CFCs make good solvents for dry cleaning and removing grease from electronic equipment.

The problem with CFCs is that they escape into the *atmosphere* where they are so stable that they last for many years, long enough for them to diffuse up to the stratosphere. In the stratosphere the intense ultraviolet light from the Sun splits CFCs into *free radicals* including chlorine atoms. Chlorine atoms react with *ozone*:

$$Cl\bullet + O_3 \longrightarrow ClO\bullet + O_2$$
$$ClO\bullet + O\bullet \longrightarrow Cl\bullet + O_2$$

The first reaction is much faster than other reactions in the stratosphere. The second reaction involves oxygen atoms which are common in the stratosphere and this recreates the chlorine atom. This means in effect that one chlorine atom can rapidly destroy many ozone molecules.

Now that the damaging effects of CFCs are known they are being phased out where possible. The hunt is on for alternative compounds with the desirable properties of CFCs but without the environmental problems.

changes of state are changes from one state of matter to another. The following are all examples of changes of state:

- a solid melting to a liquid
- a liquid freezing to a solid
- a liquid evaporating and becoming a gas
- a gas condensing to a liquid.

Energy is taken in from the surroundings during melting and evaporating (they are *endothermic* processes). The energy is needed to break the bonds between atoms, molecules or ions.

Energy is given out to the surroundings during freezing or condensing (they are *exothermic* processes). Energy is given out as the bonds between particles reform.
A few solids turn directly to gas when heated at atmospheric pressure. On cooling the vapour turns directly back to a solid. This is *sublimation*.

charcoal is a form of *carbon* made by heating wood or bones in the absence of air. Charcoal consists of minute graphite crystals. Heating charcoal in steam at about 1000°C produces activated charcoal by driving out volatile compounds from the pores of the solid. Activated charcoal is an excellent absorbent used to filter out impurities from gases and solutions.

Charles' law states that the volume of a fixed amount of gas at constant pressure is proportional to its absolute temperature.

$$V \propto T \text{ or } \frac{V}{T} = \text{constant}$$

The law follows from the *ideal gas* equation $pV = nRT$ if p and n are constant. Real gases deviate from ideal gas behaviour. Note that the law only applies if there is no chemical change altering the number of gas molecules as the gas gets hotter or colder.

chelates are *complex ions* in which each *ligand* molecule or ion forms more than one *dative covalent bond* with the central metal ion. Chelates are formed by *bidentate* and *poly-dentate* ligands such as *edta*. The term chelate comes from the Greek word for a crab's claw, reflecting the claw-like way in which chelating ligands grab hold of metal ions. Powerful chelating agents trap metal ions and effectively isolate them in solution.

Chelate complexes are generally more stable than complexes formed by monoden-tate ligands (see *stability constants*). The chelate effect can be explained by considering the *entropy* change for the ligand exchange reactions involved. When a bidentate ligand replaces monodentate ligands in a complex there is an increase in the number of molecules and ions. The value of ΔS_{system} is positive, increasing the tendency for the reaction to happen.

$$[Ni(NH_3)_6]^{2+}(aq) + 3H_2NCH_2CH_2NH_2(aq) \longrightarrow [Ni(H_2NCH_2CH_2NH_2)_3]^{2+}(aq) + 6NH_3(aq)$$

chemical equations describe what happens during reactions by identifying the reactants and products. State symbols show the *states of matter* of the chemicals involved. Equations may also show the reaction conditions by including information about temperature, pressure and catalysts above or below the arrow leading from reactants to products. There are various types of chemical equation.

- Word equations simply name the reactants and products:
    hydrogen + oxygen $\longrightarrow$ water

- *Full, balanced equations* with formulae are used to calculate the amounts of reactants and products:
    $2H_2(g) + O_2(g) \longrightarrow 2H_2O(l)$

- *Ionic equations* are balanced equations leaving out any *spectator ions* which do not change:
    $H^+(aq) + OH^-(aq) \longrightarrow H_2O(l)$

- *Half equations* are balanced equations which show part of a *redox* reaction:
    $2H^+(aq) + 2e^- \longrightarrow H_2(g)$

- Unbalanced symbol equations show just the main starting chemical and the main product with the conditions for reaction written by the arrow:

    $$C_2H_5OH \xrightarrow[\text{heat}]{Cr_2O_7^{2-}/H^+} CH_3CO_2H$$

- Mechanistic equations show intermediate steps in the *mechanism* for a reaction:
    $H_2 + Br\bullet \longrightarrow HBr + H\bullet$

chemical industry: the industry which converts raw materials such as crude oil, natural gas and minerals into useful products such as pharmaceuticals, *fertilisers*, *detergents*, paints and *dyes*.

A chemical plant consists not only of the reaction vessels and equipment for separating and purifying products, but also the storage vessels, pumps and pipes, sources of energy and heat exchangers together with the control room.

Chemical industry

Main raw materials	Large scale processes and products	Uses of the products
Crude oil and *natural* gas for the petrochemical industry	*Fractional distillation, thermal cracking, steam reforming* and *isomerisation* and many other processes to produce the building blocks for the industry such as *ethene*, propene and *benzene*	Manufacture of *polymers*, solvents, pharmaceuticals, *agrochemicals, dyes*, pigments and *detergents*
Salt (sodium chloride) and limestone for the chlor-alkali industry	*Electrolysis of brine* for making chlorine, sodium hydroxide and hydrogen. Solvay process to make sodium carbonate	Manufacture of *bleaches, disinfectants*, solvents, some polymers and in the *glass* and paper industries
Sulfur from underground deposits of the element or from the purification of oil and gas plus oxygen from the air	Contact process for *sulfuric acid manufacture*	Manufacture of *paints*, pigments, *fertilisers*, detergents, *plastics* and many uses in the chemical, metallurgical and petrochemical industries
Nitrogen from the air, natural gas and oil fractions	Haber process for *ammonia manufacture* and catalytic oxidation of ammonia for *nitric acid manufacture*	Manufacture of fertilisers, dyes, pigments, detergents, *explosives*, plastics and *fibres*
Calcium phosphate rock	Treatment of phosphate rock with concentrated sulfuric acid to make *phosphoric(V) acid* and phosphates	Fertiliser industry, manufacture of washing powders, toothpaste, food industry, enamels and glazes
Fluorite (calcium fluoride)	Action of concentrated sulfuric acid on fluorite to make hydrogen fluoride; electrolysis of fluorides in hydrogen fluoride to make fluorine	Etching and polishing glass and integrated circuits; manufacture of fluorocarbons and hydrofluorocarbons (to replace *CFCs*); pharmaceuticals and the polymer ptfe.

Bulk chemicals are manufactured on a scale of thousands or even millions of tonnes per year. Examples are ethene, sulfuric acid, ammonia and chlorine. They are mainly used as the starting point for making other substances. Fine chemicals such as pesticides and pharmaceuticals are made on a much smaller scale – a few tonnes or hundreds of tonnes per year.

Speciality chemicals are manufactured for their particular properties as thickeners, stabilisers, flame retardants and so on.

chemotherapy is the use of chemicals to treat disease. Chemotherapy began with the work of Paul Ehrlich (1854–1915) who had the idea that it might be possible to find chemicals which kill the micro-organisms which cause disease without harming healthy living cells. Paul Ehrlich worked as an assistant to Robert Koch (1843–1910) who pioneered the use of *dyes* to stain and identify *bacteria*. Ehrlich was particularly interested in selective dyes. Some dyes, for example, take well on cotton but not on wool. Ehrlich found that methylene blue would dye nerve cells well but not other parts of the body. This inspired him to search for chemical 'magic bullets' to target micro-organisms and diseased cells. After a long series of experiments Ehrlich and his Japanese colleague, Sahachiro Hata finally discovered an arsenic compound which cured syphilis. For the first time a synthetic chemical was used to cure a bacterial disease.

The high hopes raised by Ehrlich's work led to many disappointments until 1932 when Gerhard Domagk was involved in testing the medical effects of new dyes produced by a German chemical company. He found that the dye Prontosil red was remarkably effective against streptococcal infections in mice. Domagk saved his daughter's life with the new drug when she accidentally picked up a serious infection by pricking her finger in his laboratory. This led to the development of sulfonamide drugs which were used to treat bacterial diseases until the discovery of *antibiotics*.

Today chemotherapy is widely used to treat cancer but few drugs are 'magic bullets'. If they are effective in destroying cancerous cells they are generally toxic and damage other parts of the body too (see *cisplatin*).

chiral compounds are *asymmetric* so that they have mirror-image forms which are not identical. The commonest chiral compounds have a carbon atoms attached to four different atoms or groups. The two mirror image forms are known as enantiomers.

Mirror-image forms of 2-hydroxypropanoic acid (lactic acid)

Enantiomers behave identically in ordinary chemical reactions and their main physical properties are the same. They differ in their effect on polarised light – they are optically active. One mirror-image form rotates the plane of polarised light in one direction. The other form has the opposite effect. They are *optical isomers*.

Chirality is very important in living organisms. Living cells are full of messenger and carrier molecules which interact selectively with the active sites and receptors in other molecules such as *enzymes*. The messenger and carrier molecules are all chiral and the body works with only one of the mirror-image forms. In most living things all the *amino acids*, for example, are L-amino acids.

The importance of chirality in living things was brought home to people forcibly by the impact of the thalidomide tragedy. Thalidomide was a sedative used in the early 1960s. Doctors believed that it was very safe and prescribed it widely. Sadly it was soon discovered that thalidomide could harm babies if taken by mothers during the early months of pregnancy.

Thalidomide is chiral. One isomer is the effective sedative. The other isomer causes malformations in babies (it is *teratogenic*). Ever since the thalidomide affair, the pharmaceutical and agrochemical industries have to test the mirror image forms of chiral chemicals separately before they can be used as drugs or agrochemicals.

Some complex ions are chiral.

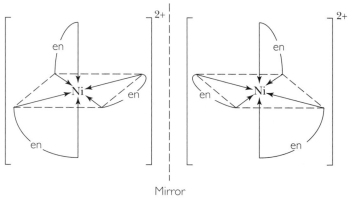

Mirror-image forms of a complex between nickel(II) ions and 1,2-diaminoethane, en, (see *bidentate ligand*). Three dimensional models make it easier to see that they are non-superimposable.

chlorates(I) and (V): see *chlorine oxoanions*.

chlorination is a reaction which replaces a hydrogen atom in an organic molecule with a chlorine atom. One example is the chlorination of *alkanes* – a *free-radical chain reaction* initiated by ultraviolet radiation.

Another example is the chlorination of *benzene* by chlorine in the presence of an iron chloride catalyst. This is an *electrophilic substitution* reaction.

chlorine (Cl) occurs as a greenish, toxic gas consisting of Cl_2 molecules. It is the second element in the family of non metals called the halogens (*group* 7). Its electron configuration is $[Ne]3s^23p^5$.

Chlorine is manufactured on a large scale by the *electrolysis of brine*.

Chlorine is a powerful *oxidising agent* which reacts directly with most elements. In its compounds chlorine is usually present in the −1 oxidation states but chlorine can be oxidised to positive oxidation states by oxygen and fluorine.

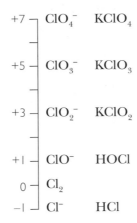

+7	ClO_4^-	$KClO_4$
+5	ClO_3^-	$KClO_3$
+3	ClO_2^-	$KClO_2$
+1	ClO^-	$HOCl$
0	Cl_2	
-1	Cl^-	HCl

Oxidation states of chlorine

Chlorine forms ionic chlorides with metals.

Formation of ions in magnesium chloride when chlorine reacts with magnesium

Chlorine forms covalent, molecular chlorides with most non-metals but it does not react directly with carbon, oxygen or nitrogen. Hydrogen burns in chlorine to produce the colourless, acidic gas hydrogen chloride, HCl. Hydrogen chloride is very soluble in water forming *hydrochloric acid.*

Reaction of hydrogen with chlorine to produce hydrogen chloride molecules

Chlorine dissolves in water. It reacts reversibly with water forming a mixture of weak chloric(I) acid and strong hydrochloric acid. This is an example of a *disproportionation reaction.*

$$Cl_2(aq) + H_2O(l) \rightleftharpoons HOCl(aq) + Cl^-(aq) + H^+(aq)$$

Chlorine oxidises a range of ions or molecules in solution:

- iron(II) ions to iron(III) ions
- bromide ions to bromine
- iodide ions to iodine
- sulfite ions to sulfate ions
- thiosulfate ions to sulfate ions
- hydrogen sulfide to sulfur.

chlorine oxoanions form when chlorine reacts with water and alkalis. In these compounds chlorine is oxidised to positive oxidation states through a series of *disproportionation reactions.*

When chlorine dissolves in potassium (or sodium) hydroxide solution at room temperature it produces chlorate(I) and chloride ions.

$$Cl_2(aq) + 2OH^-(aq) \longrightarrow ClO^-(aq) + Cl^-(aq) + H_2O(l)$$

On heating the chlorate(I) ions disproportionates to chlorate(V) and chloride ions:

$$3ClO^-(aq) \longrightarrow ClO_3^-(aq) + 2Cl^-(aq)$$

Potassium chlorate(V) can be crystallised from the solution. Careful heating just above the melting point converts potassium chlorate(V) to potassium chlorate(VII) and potassium chloride.

$$KClO_3(s) \longrightarrow 3\,KClO_4(s) + KCl(s)$$

chlorine water treatment: a life saving application of *chlorine* chemistry to kill micro-organisms in drinking water and swimming pools. Since the nineteenth century, the treatment of drinking water with chlorine has helped to control the spread of diseases such as typhoid and cholera. In Europe today it is chlorine which makes safe almost all drinking water.

Chlorine disinfects tap water by forming chloric(I) acid, HOCl.

$$Cl_2(aq) + H_2O(l) \rightleftharpoons HOCl(aq) + H^+(aq) + Cl^-(aq)$$

Choric(I) acid is a powerful oxidising agent and a weak acid. It is an effective disinfectant because the molecule can pass through the cell walls of bacteria, unlike the negatively charged ClO$^-$ ions. Once inside the cell the HOCl molecules break it open and kill the organism by oxidising and chlorinating within the cell.

Chlorine gas is very hazardous, so for household cleaning it is dissolved in sodium hydroxide to make domestic bleach, sodium chlorate(I). Sodium chlorate ionises fully in water.

$$NaOCl(aq) \longrightarrow Na^+(aq) + OCl^-(aq)$$

Choric(I) acid is a *weak acid* so in bleach solution some of the chloric(I) ions take hydrogen ions from water molecules and turn into the un-ionised acid.

$$OCl^-(aq) + H^+(aq) \rightleftharpoons HOCl(aq)$$

cholesterol is a *steroid* which plays an important part in metabolism. It is a part of cell membranes and the precursor of *steroid* hormones such as testosterone and progesterone.

Structure of cholesterol

High levels of cholesterol in the blood may lead to deposits building up in arteries resulting in heart disease. Cholesterol levels are monitored in some older people and in those thought likely to suffer from heart disease.

chromatography is a method for separating and identifying the chemicals in a mixture. Chromatography can be used to:

- separate and identify the chemicals in a mixture
- check the purity of a chemical product

- identify impurities in a product
- purify a chemical product (on a laboratory or industrial scale).

All types of chromatography have a stationary phase (a solid or a liquid held by a solid) and a mobile phase (a liquid or gas). The components of a mixture separate as the mobile phase moves through the stationary phase. Components which tend to mix with the mobile phase move faster. Components which tend to be held by the stationary phase move slower.

The basic principle underlying the separation is:

- *adsorption* when the stationary phase is a solid
- *partition* when the stationary phase is a liquid held as a thin layer on the surface of a solid.

There are several types of chromatography including: *liquid chromatography, high-performance liquid chromatography* (hplc), *paper chromatography, thin-layer chromatography* (tlc) and *gas–liquid chromatography* (glc).

chromium (Cr) is a hard, silvery *d-block* metal with the electron configuration $[Ar]3d^5 4s^1$. This electron configuration is an exception to the normal $[Ar]3d^x 4s^2$ pattern for the first series of *d*-block elements. Energetically it is more favourable to have one electron in each *d-orbital* and thus to half-fill the *d*-sub-shell.

In solution chromium forms ion in the +2, +3 and +6 *oxidation states.*

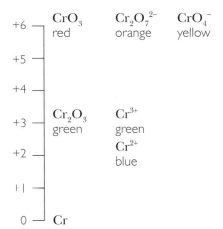

Oxidation states of chromium

There is an equilibrium between yellow chromate(VI) ions and orange dichromate(VI) ions in aqueous solution. The position of equilibrium depends on the pH. In acid the hydrogen ion concentration is high so the solution is orange because dichromate ions predominate. Adding alkali removes hydrogen ions and turns the solution yellow as chromate ions form. These shifts in the position of equilibrium are as predicted by *Le Chatelier's principle.* Note that this is not a redox reaction. Chromium is in the +6 state on both sides of the equation.

$$2CrO_4^-(aq) + 2H^+(aq) \rightleftharpoons Cr_2O_7^{2-}(aq) + H_2O(l)$$

Dichromate ions in acid solution are used to oxidise *alcohols* and *aldehydes.* Paper moistened with dichromate(VI) solution is used in the *gas test* for sulfur dioxide. The paper turns from orange to green as dichromate(VI) ions are reduced to chromium(III) ions. Potassium dichromate(VI) is used as a *primary standard* in *redox titrations*.

$$Cr_2O_7^{2-}(aq) + 14H^+(aq) + 6e^- \longrightarrow 2Cr^{3+}(aq) + 7H_2O(l)$$

Chromium(III) ions in solution exist as aquo complexes: $[Cr(H_2O)_6]^{3+}$. The hydrated ions are acidic and behave in a very similar way to aluminium(III) ions.

The removal of successive protons from the hydrated chromium(III) ion, $[Cr(H_2O)_6]^{3+}$, produces a neutral complex, $[Cr(H_2O)_3(OH)_3]$, which is sparingly soluble and precipitates. The removal of further protons produces a negative ion, $[Cr(OH)_6]^{3-}$, and the ionic compound redissolves. This is possible because of the high polarising power of the small, highly charged chromium(III) ions.

Under alkaline conditions *hydrogen peroxide* oxidises chromium(III) to chromium(VI).

Zinc reduces a green solution of chromium(III) to a blue solution of chromium(II) ions. Chromium(II) is a powerful *reducing agent* and is quickly converted to chromium(II) by oxygen in the air.

chromophore: the part of a molecule which gives rise to its colour. Chromophores in carbon compounds have an extended system of *delocalised electrons*. Families of dyes with the same chromophore are made by attaching different *functional groups* to modify the properties of the dyes.

Functional groups linked to a chromophore may act as:

- auxochromes which change the colour of the dye
- solubilising groups which make the dye more soluble in water and help it stick to the fibres of a textile.

chrysoidine

alizarin yellow

Two dyes based on the same chromophore

cisplatin is an anticancer drug consisting of a complex ion of platinum, $Pt(NH_3)_2Cl_2$. The ion is planar and can exists as *cis* and *trans* isomers (see *geometrical isomerism.*). The *cis* isomer inhibits cell division but does not prevent cell growth. This makes it a useful treatment for cancer. Unfortunately cisplatin is toxic and has unpleasant side effects.

Structure of the cis and trans isomers of $Pt(NH_3)_2Cl_2$

cis isomer

trans isomer

cis–trans isomerism: see *geometrical isomerism.*

clay minerals are hydrated aluminosilicates formed by the *weathering* of feldspars in igneous rocks. Kaolinite is the main component of china clay which is mined on a large scale in Devon and Cornwall. Much china clay is used as a coating agent and filler. Kaolinite is also the main constituent of ball clay used to make *ceramic* tableware, porcelain and wall tiles.

Kaolinite has a layer structure. Each layer consists of a sheet of *silicate* (SiO_4) tetrahedra interlocked with an aluminate sheet.

close packed structures are found in metal crystals. In a layer of close packed spheres each atom has six other spheres touching it. In three dimensions, layers of close packed atoms stack up in two possible ways. In hexagonal close packing the third layer is directly over the first layer (aba). In cubic close packing it is the fourth layer which corresponds with the first layer (abca). In the two structures each atom touches 12 nearest neighbours so for both the *co-ordination number* is 12.

Hexagonal and cubic close packing of metal atoms

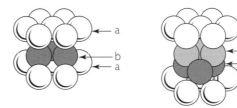

The *unit cell* of the cubic close packed structure is a face-centered cube.

coal is a fossil fuel formed by the action of heat and pressure on the remains of plant buried under sediments. Heating coal in the absence of air at 1000°C drives off coal tar, ammonia and coal gas leaving a residue of coke. Coke is required for a range of industrial processes including *iron extraction* in blast furnaces. Coal tar was a major source of organic chemicals until crude oil took over in the 1940s. Coal tar is a rich source of arenes and was the main source of *benzene* when William Perkin was pioneering the production of synthetic dyes.

cobalt (Co) is a hard, silvery *d-block* metal which is less reactive than iron. It has the electron configuration $[Ar]3d^74s^2$.

Cobalt is an ingredient of alloys *steels* such as the ferromagnetic alloy, Alnico which makes excellent permanent magnets.

In solution cobalt forms ions in the +2 and +3 oxidation states. Cobalt(II) is the more stable state. Anhydrous cobalt(II) chloride is blue but it turns pink on adding water as the cobalt ions are hydrated to the $[Co(H_2O)_6]^{2+}$ ion. This is used as a test to detect water. The granules of the drying agent 'self-indicating *silica gel*' are blue because they contain anhydrous cobalt ions. When the gel's drying action is exhausted the granules turn pink.

A dilute solution of cobalt chloride is pink because the cobalt(II) ions are hydrated. A concentrated solution is blue. A dilute solution also turns blue on adding concentrated hydrochloric acid. The colour change is due to a *ligand* exchange reaction as chloride ions replace the water molecules. The reaction is reversible:

$$[Co(H_2O)_6]^{2+} + 4Cl^-(aq) \rightleftharpoons [CoCl_4]^{2-} + 6H_2O(aq)$$

It is normally very difficult to oxidise aqueous cobalt(II) to cobalt(III) but the reaction goes readily if the cobalt(II) ions are complexed with ammonia molecules. The Co(III) complex with ammonia is more stable than the Co(II) complex. The value for the *standard electrode potential* shows that aqueous Co(III) is a stronger oxidising agent than potassium manganate(VII) in acid solution:

$$[Co(H_2O)_6]^{3+} + e^- \rightleftharpoons [Co(H_2O)_6]^{2+} \qquad E^{\ominus} = +1.82\ V$$

When the two states are complexed with ammonia the standard electrode potential shifts to a value that shows that the Co(III) state is much more stable. Cobalt(II) is now a reducing state and can be oxidised to Co(III) by oxygen or hydrogen peroxide.

$$[Co(NH_3)_6]^{3+} + e^- \rightleftharpoons [Co(NH_3)_6]^{2+} \qquad E^{\ominus} = +0.10\ V$$

colligative properties are the properties of solutions which depend on the concentration of solute particles but not on the nature of the particles. Dissolving salt, sugar or any other solute in water lowers its *vapour pressure*, raises its boiling point, lowers its *freezing point* and increases its *osmotic pressure*. The extent of these changes depends only on the *mole fraction* of solute particles and not on the nature of the dissolved molecules or ions.

collision theory accounts for the effects of concentration, temperature and *catalysts* on *reaction rates*. The idea is that a chemical reaction happens when the molecules or ions of reactant collide, making some bonds break and allowing new bonds to form.

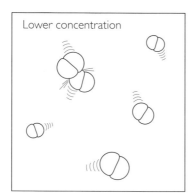

Lower concentration

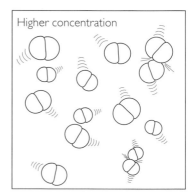

Higher concentration

Raising the concentration means that the reacting particles are closer together. There are more collisions and reactions are faster.

It is not enough for the molecules to collide. In soft collisions the molecules simply bounce off each other. Molecules are in rapid random motion and if every collision led to reaction all reactions would be explosively fast. Only pairs of molecules which collide with enough energy to stretch and break chemical bonds can lead to new products. Reactant molecules have to overcome the *activation energy*.

Collision theory refers to the *Maxwell–Boltzmann distribution* of energies of molecules to explain the effects of temperature and catalysis on reaction rates.

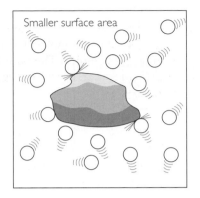 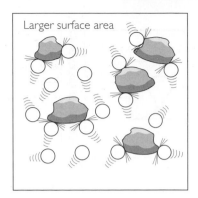

In a *heterogeneous* reaction of a solid with either a liquid or gas the reaction is faster if the solid is broken up into smaller pieces. Crushing the solid increases its surface area, collisions can be more frequent and the rate of reaction is bigger.

colloids consist of fine particles (the disperse *phase*) of one substance finely dispersed in another (the continuous phase). The dispersed particles in liquids and gases are large enough to scatter light but they do not settle out and they show *Brownian motion*. The diameter of colloid particles is typically around 10 to 1000 nm (much larger than atomic diameters which are around 0.2 nm).

Continuous phase	Disperse phase	Type	Example
gas	liquid	liquid aerosol	mist
gas	solid	solid aerosol	smoke
liquid	gas	foam	whipped cream
liquid	liquid	emulsion	hand cream, mayonnaise
liquid	solid	sol	paint, muddy river water, sewage
solid	gas	solid foam	pumice
solid	liquid	gel or solid emulsion	jelly, butter
solid	solid	solid sol	pearl, pigmented plastics

colorimetry is a method for measuring the *concentration* of compounds in solution which can be used with chemicals which are themselves coloured or which give a colour when mixed with a suitable reagent.

The filter lets through light which is strongly absorbed by the solution. The extent of absorption depends on the path length of the light through the solution (which is a constant for the sample chamber used) and on the concentration of the solution.

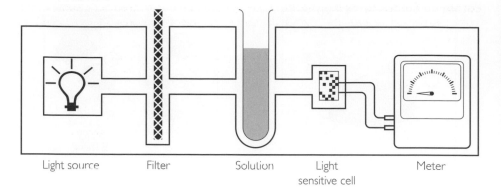

| Light source | Filter | Solution | Light sensitive cell | Meter |

Diagram of a colorimeter

The instrument is *calibrated* with a series of *standard solutions*. It can then be used to measure the concentration of unknown samples.

A colorimeter can be used to determine the formula of a complex ion. The graph shows the results of measuring the absorbance of a series of mixtures of 0.01 mol dm^{-3} $Cu^{2+}(aq)$ ions and 0.01 mol dm^{-3} edta(aq).

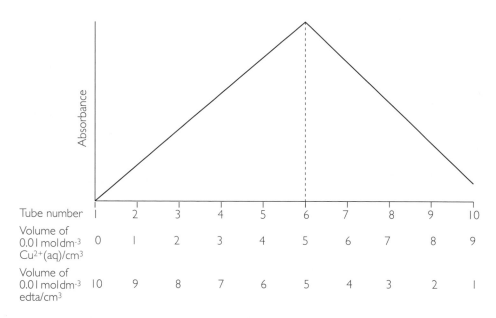

Tube number	1	2	3	4	5	6	7	8	9	10
Volume of 0.01 mol dm^{-3} $Cu^{2+}(aq)/cm^3$	0	1	2	3	4	5	6	7	8	9
Volume of 0.01 mol dm^{-3} edta/cm^3	10	9	8	7	6	5	4	3	2	1

Plot of the results of measuring the absorbance of ten mixtures of 0.01 mol dm^{-3} $Cu^{2+}(aq)$ ions and 0.01 mol dm^{-3} edta(aq)

The peak of absorption corresponds to mixing equal volumes of the two solutions which both have the same molar concentration. This shows that 1 mol $Cu^{2+}(aq)$ forms a complex with 1 mol edta(aq). (For a diagram see *edta*.)

coloured compounds in many instances get their colour by absorbing some of the radiation in the visible region of the *electromagnetic spectrum* with wavelengths between 400 nm and 700 nm.

Colour of compound	Wavelength absorbed/nm	Colour of light absorbed
greenish yellow	400–430	violet
yellow to orange	430–490	blue
red	490–510	blue-green
purple	510–530	green
violet	530–560	yellow-green
blue	560–590	yellow
greenish blue	590–610	orange
blue-green to green	610–700	red

It is the electrons in coloured compounds that absorb radiation as they jump from their normal state to a higher excited state. According to *quantum theory* there is a fixed relationship between the size of the energy jump and the wavelength of the radiation absorbed. In many compounds the jumps are so big that they absorb in the ultraviolet part of the spectrum. Such compounds are colourless.

Colour of transition metal complex ions – the colour of a transition metal complex ion depends on:

- the metal
- the *oxidation state* of the metal
- the *ligand*
- the *co-ordination number*.

Colour in transition metal ions arises from electronic transitions between *d-orbitals*. In a free atom all of the five *d* orbitals have the same energy. When a transition metal ion forms complex ions the *d*-orbitals split into two groups with different energies. The size of the split depends on the number and nature of the ligands in the complex. This helps to account for colour changes.

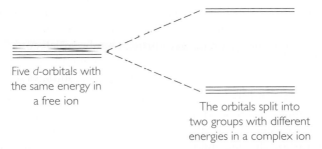

Five *d*-orbitals with the same energy in a free ion

The orbitals split into two groups with different energies in a complex ion

Energy difference between *d*-orbitals in a transition metal complex ion

If all the *d*-orbitals are full there is no possibility of electronics transitions between them. This explains why Zn^{2+} and Cu^+ ions are colourless. Sc^{3+} ions are also colourless because the ion has no *d*-electrons.

Colour in organic compounds – colour in organic compounds is often associated with *delocalised electrons*. The energy jumps between molecular orbitals absorb visible radiation in molecules with extended systems of alternating double and single bonds (see *conjugated systems*).

Skeletal formula of β-carotene, the main orange colour in carrots, used as a colourant in foods, drugs and cosmetics

Other sources of colour – some colours are caused by physical effects rather than electronic transitions. Examples of these effects are:

- light scattering – moonstones, blue skies
- interference – soap bubbles and oil films on water
- diffraction – opal, liquid crystals
- refraction of some wavelengths more than others – 'fire' in gemstones.

combustion is the reaction of a fuel with oxygen to release energy. If there is plenty of air or oxygen, the carbon in the fuel turns to carbon dioxide and hydrogen turns to steam (water).

$$CH_4(g) + 2O_2(g) \longrightarrow CO_2(g) + 2H_2O(l)$$
methane in natural gas

Incomplete combustion in a limited supply of oxygen is hazardous because the products are then soot and the toxic gas *carbon monoxide.*

During combustion, any sulfur in a fuel is oxidised to sulfur dioxide. Nitrogen in a fuel may end up as the element after combustion or, if the temperature is high enough, combined with oxygen to form a mixture of *nitrogen oxides*. The formation of these acidic oxides contributes to *acid rain.*

combustion analysis is a method for determining the *empirical formulae* of organic compounds. A weighed sample of the compound is burnt in excess oxygen mixed with helium. This converts the carbon to carbon dioxide and the hydrogen to water. A catalyst ensures that combustion is complete. The inert helium carries the products of combustion and the excess oxygen through a tube which contains chemicals to remove any volatile halogen, sulfur or phosphorus compounds. Oxides of nitrogen are converted to nitrogen gas and the excess oxygen combines with copper.

The water vapour is absorbed in magnesium chlorate(VII). Carbon dioxide is absorbed in soda lime. In a modern instrument, measurements of the thermal conductivity of the helium before and after absorption make it possible to determine the masses of water, carbon dioxide and nitrogen formed by burning the sample.

From the results it is possible to calculate the *percentage composition* of the compound. Any mass of the sample not accounted for is assumed to be due to oxygen.

Worked example:

Complete combustion of 0.15 g of a liquid compound produced 0.22 g of carbon dioxide and 0.09 g water. What is the empirical formula of the compound?

Notes on the method

The *molar mass* of carbon dioxide, CO_2 = 44 g mol^{-1} of which carbon is 12 g mol^{-1}

The molar mass of water, H_2O = 18 g mol^{-1} of which hydrogen is 2 g mol^{-1}

Answer

The mass of carbon in the sample = $\dfrac{12}{44}$ × 0.22 g = 0.06 g

The mass of hydrogen in the sample = $\dfrac{2}{18}$ × 0.09 g = 0.01 g

The total mass of carbon and hydrogen = 0.07 g in a sample with mass 0.15 g

So the difference gives the mass of oxygen in the sample which is 0.08 g

These are the *amounts* of the elements in the sample:

carbon: 0.06 g ÷ 12 g mol^{-1} = 0.005 mol

hydrogen:0.01 g ÷ 1 g mol^{-1} = 0.01 mol

oxygen: 0.08 g ÷ 16 g mol^{-1} = 0.005 mol

The ratio C:H:O is 1:2:1

The empirical formula of the compound is CH_2O.

---

complex ions consist of a central metal ion linked to a number of molecules or ions with lone pairs of electrons. The surrounding molecules or ions are *ligands* which use lone pairs of electrons to form a *co-ordinate bond* with the metal ion. The number of ligands in a complex ion is typically two, four or six.

Transition metal ions form aquo complexes when dissolved in water. Examples are $[Fe(H_2O)_6]^{2+}$ and $[Cr(H_2O)_6]^{3+}(aq)$.

The overall charge of a complex ion is the sum of the charges on the metal ion and its ligands. (See also *names of complex ions, shapes of complex ions, coloured compounds, co-ordination compounds* and *co-ordination number.*)

complexes with ammonia molecules are useful in *qualitative analysis* in both anion tests and cation tests. Example are the diamminesilver(I) ion and the tetra-ammine copper(II) ion. The *lone pair* on the nitrogen atom of each ammonia molecule forms a *co-ordinate bond* with the central metal ion.

$$[Ag(NH_3)_2]^+ \qquad\qquad\qquad [Co(NH_3)_6]^{2+}$$

diamminesilver(I) ion   hexaamminecobalt(II) ion

Complexes between metal ions and ammonia

Adding ammonia solution to a precipitate of a *silver halide* can distinguish the chloride from the bromide and the iodide. Only silver chloride, the least insoluble, will freely dissolve in dilute ammonia solution to form the diammine complex.

Formation of a complex may produce a colour change or redissolve a precipitate.

Adding ammonia solution to a solution with copper(II) ions first produces a pale blue precipitate of the hydroxide, but with excess ammonia the precipitate dissolves to give a deep blue solution as the tetraammine complex forms.

complex-forming titration is a practical technique used to determine the *concentration* of metal ions. A *titration* measures the volume of a *standard solution* of a complex-forming reagent needed to react exactly with a measured volume of the unknown solution of a metal ions, such as zinc ions.

Complex-forming titrations often use a standard solution of *edta* which forms very stable *complex ions* with metal ions. The procedure is the same as for any other titration.

To find the end-point the analyst adds an indicator and a *buffer solution* which forms a coloured but unstable complex with the metal ion in the flask. A suitable indicator is Eriochrome black T which is blue in solution at pH 10. At the start of the titration it produces a wine-red complex. Edta from a burette forms a more stable complex with zinc ions and so takes the metal ions from the indicator. At the end point all the zinc ions have been complexed by titration with edta. The last drop of edta leaves no Zn^{2+} ions to form the red complex with the indicator so the indicator turns blue again.

Worked example:

An alkaline buffer and a few drops of Eriochrome black T indicator were added to 25.0 cm^3 of a solution of zinc sulfate. In the titration 23.2 cm^3 of 0.010 mol dm^{-3} edta were run in from a burette until the indicator changed from red to blue. What was the concentration of the zinc ions?

Notes on the method

Always start by writing the equation for the reaction. See *titration* for a general method for the calculations.

Remember to convert volumes in cm^3 to volumes in dm^3 by dividing by 1000.

In any titration there is one unknown – in this case the concentration of the zinc ions, c_A.

Answer

The equation for the reaction is:

$Zn^{2+}(aq) + edta^{4-}(aq) \longrightarrow [Zn(edta)]^{2-}(aq)$

The volume of zinc sulfate solution in the flask, $V_A = \dfrac{25.0}{1000}$ dm^3

Let the concentration of zinc ions be c_A.

The volume of hydrochloric acid added from the burette, $V_B = \dfrac{23.2}{1000}$ dm^3

The concentration of hydrochloric acid, c_B = 0.010 mol dm^{-3}

$$\frac{V_A \times c_A}{V_B \times c_B} = \frac{n_A}{n_B}$$

$$\frac{^{25}\!/_{1000} \times c_A}{^{23.2}\!/_{1000} \times 0.010} = \frac{1}{1}$$

Therefore $c_A = \dfrac{23.2 \times 0.010}{25.0} = 0.0928$ mol dm^{-3}

The concentration of the zinc ions was 0.0928 mol dm^{-3}.

composites are made by combining two or more materials to create a new material. A composite combines the desirable properties of its constituents and compensates for their disadvantages. Steel reinforced concrete is a composite material as is galvanised steel. Kitchen worktops are composites consisting of chipboard covered with paper impregnated with a polymer such as a *thermosetting* melamine–methanal resin.

Many important composites consist of fibres of one material, such as *glass*, aramids, or graphite, embedded in a *polymer*, *metal* or *ceramic* matrix. The combinations are designed to give new materials with better properties especially high stiffness per unit weight which is important in road and rail transport, aircraft, sporting goods and many other applications.

Glass fibres in a *polyester* matrix (so-called fibreglass) are used to make boat hulls and car bodies. Parts of aircraft and some sports gear are made from carbon fibres in an epoxy matrix.

concentrations of solutions are usually measured in moles per litre of solution (mol dm^{-3}). There are small volume changes when chemicals dissolve in water so it is important to note that concentrations normally refer to litres of solution not to litres of the solvent.

$$\text{concentration/mol dm}^{-3} = \frac{\text{amount of solute/mol}}{\text{volume of solution/dm}^3}$$

Writing the formula of a chemical in square brackets is the usual shorthand for 'concentration in mol dm^{-3}'. (For example: [CaCl$_2$] – 0.1 mol dm^{3}.)

When ionic crystals dissolve the ions separate and become independent.

$$CaCl_2(s) + aq \longrightarrow Ca^{2+}(aq) + 2Cl^-(aq)$$

So if [CaCl$_2$] = 0.1 mol dm^{-3}, then [Ca^{2+}] = 0.1 mol dm^{-3} but [Cl$^-$] = 0.2 mol dm^{-3}.

Other ways of measuring the concentrations are to use *mole fractions* or *parts per million, ppm*.

Worked example:

What is the concentration of a solution of potassium manganate(VII) made by dissolving 3.95 g of the solid in water and making the solution up to 500 cm^3?

Answer

The molar mass of potassium manganate(VII), (KMnO$_4$) = 158.0 g mol^{-1}

Amount of KMnO$_4$ in solution $= \dfrac{3.95 \text{ g}}{158.0 \text{ g mol}^{-1}} = 0.025$ mol

Volume of the solution $= \dfrac{500}{1000}$ dm^3 = 0.5 dm^3

Concentration of the solution $= \dfrac{0.025 \text{ mol}}{0.5 \text{ dm}^3} = 0.05$ mol dm^{-3}

---

condensation polymers are produced by a series of condensation reactions splitting off water between the functional groups of the *monomers*. Examples of condensation polymers are *polyamides* and *polyesters*. Where each monomer has two function groups this type of polymerisation produces chains. Condensation polymers make good fibres because they form long, straight-chain molecules with few side chains and relatively strong intermolecular forces between polar bonds in neighbouring chains. *Cross linking* is possible if one of the monomers has three functional groups.

condensation reaction: a reaction in which molecules join together by splitting off a small molecule such as water. The *addition–elimination reactions* of carbonyl compounds are examples of condensation reactions. The formation of an ester from an acid and an alcohol is also a condensation reaction.

$$CH_3 - C\!\!\begin{array}{c} O \\ \diagdown \\ OH \end{array} \quad + \quad HO - C_2H_5 \quad \longrightarrow \quad CH_3 - C\!\!\begin{array}{c} O \\ \diagdown \\ O - C_2H_5 \end{array} \quad + \quad H_2O$$

| acid | alcohol | ester | water |

Condensation reaction to form the ester ethyl ethanoate

conditions for organic reactions specify concisely but accurately the reagents, catalysts, temperatures and pressures needed for good yields. Factors to consider include:

- whether a solvent is needed – if so, whether the solvent is water or a non-aqueous solvent such as ethanol
- the concentration of the reagents – whether acids, for example, are dilute or concentrated
- the temperature – whether the mixture has to be cooled, heated or kept at room temperature
- the pressure – many industrial process take place at several times the pressure of the atmosphere
- whether a catalyst is required.

conductiometric titrations: use a conductivity cell and meter to measure the conductivity of the reaction mixture during a titration and hence to determine the end-point (see the diagram on page 101).

The conductivity can change during a titration because there is a change in:

- the ability of the ions present to move through the solution and conduct electricity
- the proportion of electrolytes which are only slightly ionised relative to electrolytes which are fully ionised.

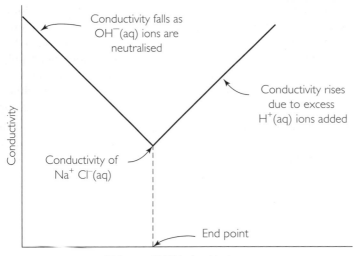

Conductivity falls as OH⁻(aq) ions are neutralised

Conductivity rises due to excess H⁺(aq) ions added

Conductivity of Na⁺ Cl⁻(aq)

End point

Conductivity

Volume of HCl(aq) added

Conductiometric titrations (see page 100) Changes in conductivity during a titration of NaOH(aq) with HCl(aq). In water H^+(aq) and OH^-(aq) ions are very mobile so they conduct electricity very well.

confidence limits are the limits around an experimental mean value within which there is a high probability that the true mean lies.

Statistical analysis shows that there is a 95% probability that the true mean lies within $\pm 1.96\sigma$ of the experimental mean.

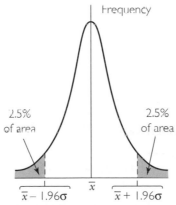

Frequency

2.5% of area

2.5% of area

$\bar{x} - 1.96\sigma$ $\bar{x}$ $\bar{x} + 1.96\sigma$

Normal distribution of a large number of experimental results spread randomly about a mean value $\bar{x}$ where σ is the standard deviation

conformation of molecules: the possible shapes of molecules which can arise because of *rotation about covalent bonds*. Cyclohexane molecules, for example, shift to and fro between 'chair' and 'boat' forms. In both conformations the four single bonds around each carbon atom are arranged tetrahedrally as normal. The chair form is preferred because it is a little more stable. The energy difference between conformations is not large enough for them to be separated as distinct compounds.

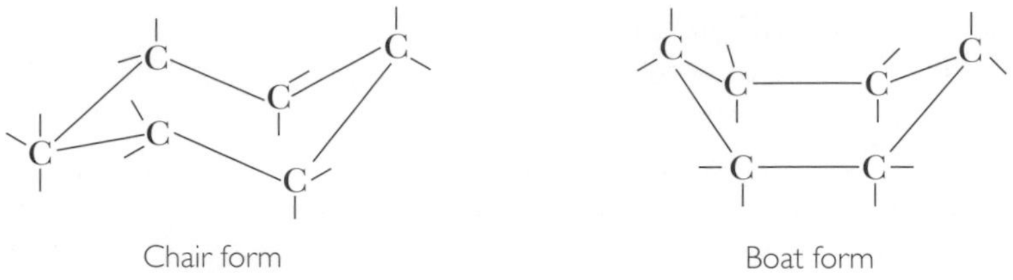

Chair form   Boat form

Chair and boat forms of cyclohexane

conjugate acid–base pairs: see *acid–base equilibria.*

conjugated system: a system of alternate double and single bonds in a molecule. Organic molecules with extended conjugated systems absorb radiation in the visible part of the spectrum and are therefore coloured. Examples are the orange β-carotene molecule in carrots and the indicator phenolphthalein in alkali (see *coloured compounds*).

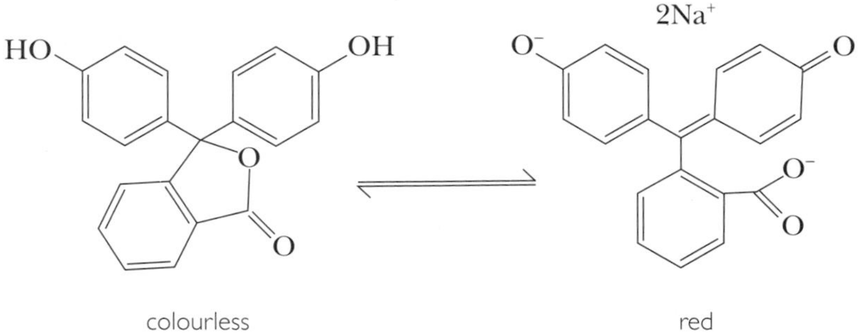

colourless   red

Structures of phenolphthalein in acid and alkali

conservation of mass (law of): the law which chemists now take for granted every time they write a balanced chemical equation. The law states that matter is neither created or destroyed during a chemical reaction.

constant-boiling mixture: see *azeotropic mixture.*

contact process: see *sulfuric acid manufacture.*

continuous processes manufacture chemicals on a large scale in industrial plants which operate for 24 hours in the day. Raw materials are constantly fed into the plant and products continuously removed. Examples of continuous processes are the *fractional distillation of oil,* the Haber process for *ammonia manufacture, iron extraction* in a blast furnace and the contact process for *sulfuric acid manufacture.*

conversion of units is often needed to make sure that the units are consistent before carrying out calculations. Typically the units of measurement (such as cm^3) have to be converted to different units for calculation (in this case dm^3).

Worked example:

In a titration, the volume of acid added from the burette was 23.5 cm^3. What is the volume in litres (dm^3)?

Notes on the method

Multiply by an appropriate factor to convert from the unit used for measurement to the unit needed for the calculation. Find the factor by:

- writing down the relationship between the two units, then by
- writing the relationship as a ratio, so that the units cancel when it multiplies the measurement, such that the original unit is replaced by the unit required.

Answer

$1 \text{ dm}^3 = 1000 \text{ cm}^3$

Hence $23.5 \text{ cm}^3 \times \dfrac{1 \text{ dm}^3}{1000 \text{ cm}^3} = 0.0235 \text{ dm}^3$

---

co-ordinate bond is an alternative name for a *dative covalent bond*. The term co-ordinate bond is often used when describing the dative bonding between ligands and a metal ion in *complex compounds*.

co-ordination compounds contain complexes which may be cations, anions or neutral molecules. In a co-ordination compound, ligand molecules or negative ions form *co-ordinate bonds* with a metal ion.

Examples of co-ordination compounds:

- $K_3[Fe(CN)_6]$ containing the negatively charged complex ion $[Fe(CN)_6]^{3-}$
- $[Ni(NH_3)_6]Cl_2$ containing the positively charged complex ion $[Ni(NH_3)_6]^{2+}$
- $Ni(CO)_4$ a neutral complex between nickel atoms and carbon monoxide molecules.

co-ordination number: the number of nearest neighbours to an atom or ion in a *crystal structure* or the number of *co-ordinate bonds* formed between the ligands and the metal ion in a *complex ion*.

copolymerisation is used to modify the properties of *polymers* by producing polymer chains from a mixture of monomers. ABS, for example, is a rigid, tough plastic widely used for the casing of domestic equipment and for parts of cars. It is a co-polymer of **Acrylonitrile** $(CH_2 = CH - CN)$, **Butadiene** $(CH_2 = CH - CH = CH_2)$ and **Styrene** $(C_6H_5 - CH = CH_2)$

copper (Cu) is a ductile metal with a familiar reddish colour. It has the electron configuration $[Ar]3d^{10}4s^1$. This electron configuration is an exception to the normal $[Ar]3d^44s^2$ pattern for the first series of *d-block elements*. Energetically it is more favourable to have fill the *d*-sub-shell and leave only one electron in the 4s.

Copper is relatively unreactive. It corrodes very slowly in moist air and is not attacked by dilute non-oxidising acids. Copper is a good conductor of electricity; it is widely used in electricity cables and for domestic water pipes.

Copper's mechanical properties are enhanced by making alloys such as *brass* and *bronze*.

Copper forms compounds in the +1 and +2 states. Under normal conditions copper(II) is the stable state in aqueous solution. Copper(I) *disproportionates* in aqueous solution. So when Cu_2O dissolves in dilute sulfuric acid the products are copper(II) sulfate and copper.

$$2Cu^+(aq) \rightleftharpoons Cu^{2+}(aq) + Cu(s)$$

The equilibrium lies well to the right. Copper(I) in the presence of water can exist as very insoluble compounds such as Cu_2O, CuI or $CuCl$. Iodide ions, for example, reduce copper(II) to copper(I) ions which immediately precipitate with more iodide ions as white copper(I) iodide.

$$2Cu^{2+}(aq) + 4I^-(aq) \longrightarrow 2CuI(s) + I_2(s)$$

Copper(I) can exist in aqueous solution as stable complexes such as $[Cu(NH_3)_2]^+$ or $[Cu(CN)_4]^{3-}$.

Fehling's and *Benedict's* reagents contain deep blue copper(II) complexes in alkali. The reagents are used to detect *reducing sugars* and to distinguish *aldehydes* from *ketones*. A reducing sugar or aldehyde reduces the reagent on heating to copper(I) oxide. The blue colour goes and a reddish-brown precipitate forms.

(See *names of complex ions* and *ligand substitution reactions* for more examples of copper complexes. See *carbon monoxide* for an example of copper acting as a catalyst. See *cation tests* and *flame tests* for methods used to detect the presence of copper(II) ions.)

copper refining uses electrolysis to turn copper which is 99.5% pure into 99.99% pure metal. High purity is important especially when copper is to be used as an electrical conductor.

Impure copper is cast into anodes while the cathodes are thin sheets of pure copper. The electrolyte is a mixture of copper(II) sulfate and sulfuric acid.

At the anodes: $Cu(s) \longrightarrow Cu^{2+}(aq) + 2e^-$

At the cathodes: $Cu^{2+}(aq) + 2e^- \longrightarrow Cu(s)$

Valuable impurities such as gold and silver are recovered from the bottom of the electrolysis cell because they do not dissolve in the electrolyte.

corrosion of a metal is a redox process in which oxygen, water and acids attack metals. Most metals are extracted from oxides. Corrosion turns them back into oxides.

The most familiar and economically serious example of corrosion is the rusting of iron. Rusting is an electrochemical reaction.

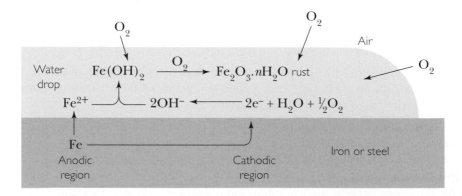

An electrochemical cell on the surface of iron in contact with water and the air.

When iron is in contact with air and water, the regions rich in oxygen are cathodic. In these regions oxygen is reduced to hydroxide ions. Other parts of the metal surface are anodic. In these regions iron is oxidised to iron(II) ions. Iron(II) ions and hydroxide ions diffuse together and form a precipitate of iron(II) hydroxide which is then oxidised to rust, hydrated iron(III) oxide.

COSHH regulations: an acronym of 'Control Of Substances Hazardous to Health', these are designed to protect people who work with materials which can be harmful. The 1988 Regulations apply not just to single chemical compounds but also to mixtures of compounds and micro-organisms. The regulations apply, for example, to:

- toxic, harmful or irritant substances
- substances which cause mutations, cancers or birth defects
- micro-organisms that can transmit disease.

Employers have a duty to prevent or control exposure of their employees to substances hazardous to health by means such as:

- replacing a substance with another which is less hazardous
- changing the method of working
- modifying the process to cut down on hazardous wastes or by-products.

The priority is to control the hazard in ways which avoid the need for personal protective clothing, for example by enclosing the process, improving ventilation or by thorough and regular cleaning.

coulomb (symbol C) is the *SI unit* of electric charge.

electric charge (C) = current (C s^{-1}) × time (s)

A coulomb is the amount of electric charge flowing each second past a point in a circuit when the current is one ampere (1 A = 1 C s^{-1}).

covalent bonds form when atoms share electrons. The atoms are held together by the attraction between the positive charges on their nuclei and the negative charge on the shared electrons.

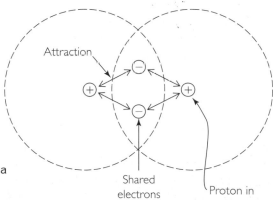

Forces of attraction giving rise to a covalent bond between hydrogen atoms in H$_2$

Attraction

Shared electrons

Proton in nucleus

Covalent bonding holds together the atoms of non-metals in molecules and *giant structures*. Molecules have a definite shape because covalent bonds have a definite length and direction.

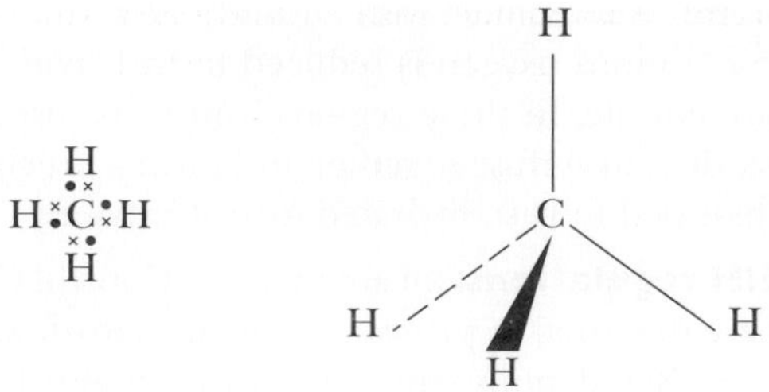

$$H \overset{\bullet}{\underset{\times}{:}} C \overset{\bullet}{\underset{\times}{:}} H$$

Electron sharing in methane and the shape of a methane molecule

One shared pair of electrons gives rise to a single bond. *Double bonds* and *triple bonds* are also possible with two or three shared pairs.

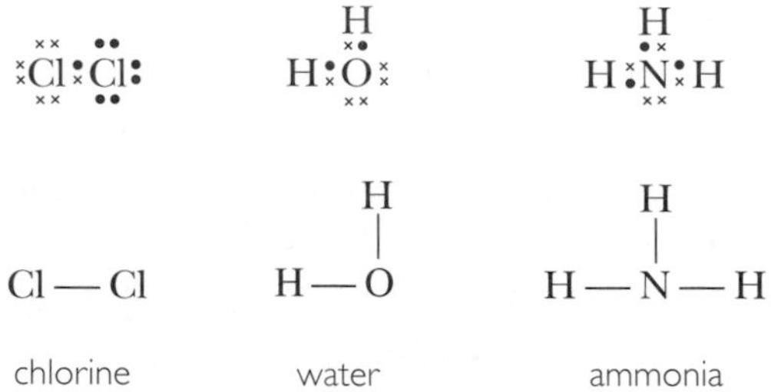

Dot and cross diagrams to show single covalent bonding in molecules. Also shown is a simpler way of showing the bonding in molecules. A line between two symbols represents a covalent bond.

$$\times \overset{\times \times}{\underset{\times \times}{Cl}} \overset{\bullet \bullet}{\underset{\bullet \bullet}{:}} \overset{\bullet \bullet}{\underset{\bullet \bullet}{Cl}} :$$

$$H \overset{\times}{\underset{\times \times}{:}} O \overset{\bullet}{\underset{}{:}}$$

$$H \overset{\bullet}{\underset{\times \times}{:}} N \overset{\bullet}{\underset{}{:}} H$$

Cl—Cl

H—O

H—N—H

chlorine

water

ammonia

The non-metals common in organic chemistry generally form a fixed number of covalent bonds. This helps us to work out the structures of molecules.

Element	Number of covalent bonds	Examples
carbon, C	4	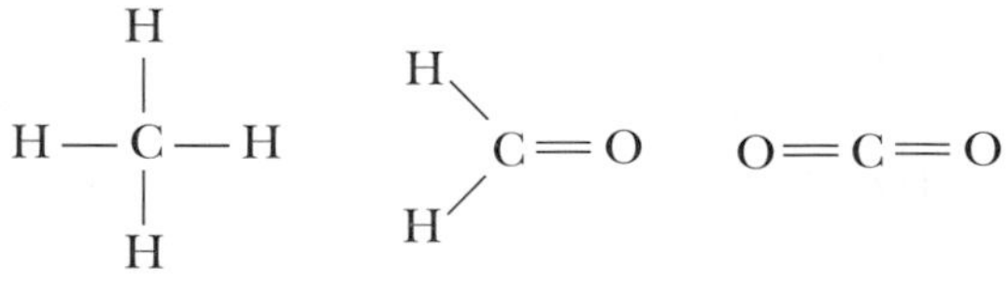 H—C—H (with H above and below); C=O (with H, H); O=C=O
hydrogen, H	1	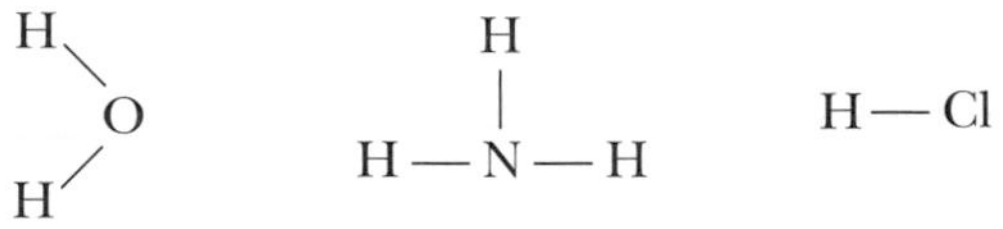 H,H O; H—N—H (with H above); H—Cl
oxygen, O	2	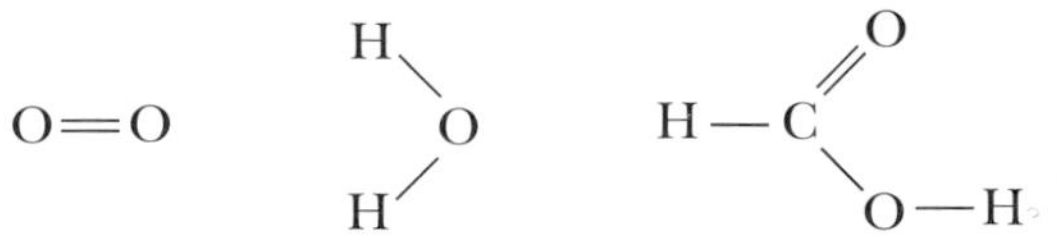 O=O; H,H O; H—C with O (double) and O—H
nitrogen, N	3	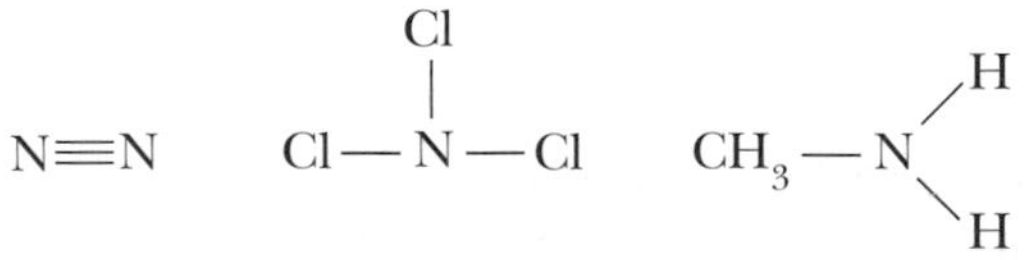 N≡N; Cl—N—Cl (with Cl above); CH₃—N with H, H
halogens, F, Cl, Br, I	1	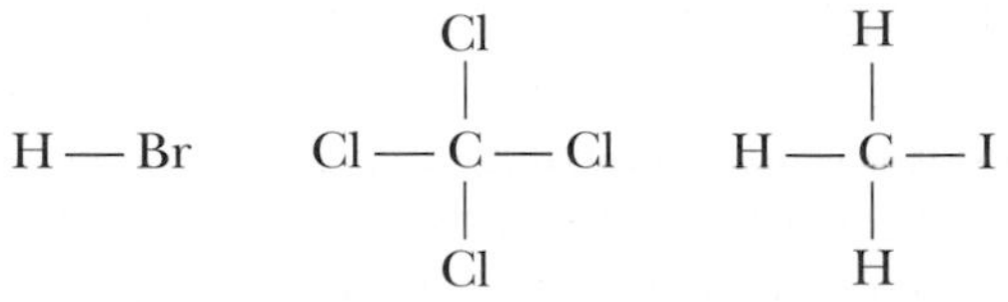 H—Br; Cl—C—Cl (with Cl above and below); H—C—I (with H above and below)

covalent radius: the covalent radius of an element is half the *bond length* when two atoms of the element are linked by a single covalent bond.

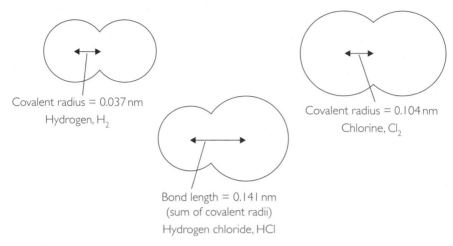

Covalent radius = 0.037 nm
Hydrogen, H_2

Covalent radius = 0.104 nm
Chlorine, Cl_2

Bond length = 0.141 nm
(sum of covalent radii)
Hydrogen chloride, HCl

The length of a single covalent bond between atoms of different elements can be estimated fairly accurately by adding the covalent radii for the two atoms

cracking: see *catalytic cracking* and *thermal cracking*.

critical temperature is the temperature above which it is impossible to liquefy a gas however high the pressure (see *vapours*).

cross-linking is the formation of chemical bonds between polymer chains to modify the properties of polymers such as *thermosetting* plastics as well as natural *rubber* during vulcanising.

Cross-linking determines the three-dimensional shape of *proteins*. Cross-links in protein molecules take various forms including:

- *hydrogen bonding* between amino acid side-chains
- disulfide bridges formed by *covalent bonding* between cystine side-chains.

crude oil is a complex mixture of *hydrocarbon* molecules formed over millions of years when the remains of microscopic sea creatures trapped in sediments were converted by heat and pressure to petroleum. Crude oil is now the main source of *fuels* and organic compounds. The composition of crude oil varies from one oilfield to another. Some crudes contain significant quantities of sulfur and nitrogen compounds as well as traces of various metals. *Fractional distillation of oil* is the first step in refining to produce fuels and *lubricants* as well as feedstocks for the *petrochemical industry*.

crystallinity of polymers arises when the long chain molecules are regular enough to lie more or less parallel to each other in some regions of the solid. The rest of the solid where the chains are tangled together is *amorphous*.

The crystallinity of *polymers* with regular, unbranched chains is generally higher than in polymers with branched or irregular molecules. Relatively strong *intermolecular forces* between chains also favour crystallinity. Highly crystalline polymers are stronger and less flexible than more amorphous polymers.

crystal structures of ionic compounds include the *sodium chloride, caesium chloride* and *fluorite* structures. In an ionic crystal the ions behave like charged spheres in contact. The structures are only stable if each ion is in contact with its nearest neighbours. Caesium ion is large enough to have eight chloride ions around it as in the caesium chloride structure. Sodium ions are smaller and only big enough to touch six neighbouring chloride ions as in the sodium chloride structure.

Compounds with ionic *giant structures* are hard, melt at high temperatures and only conduct electricity when molten or dissolved in water.

The *Born–Haber cycle* makes it possible to decide whether or not the bonding in a crystal is purely ionic due to *electrostatic forces* between ions.

crystal structures of metals: the important metal structures are the two *close packed structures* and the *body-centered cubic* structure.

crystal structures of non-metals: most solid *non-metals,* such as *iodine, sulfur* and white *phosphorus* are molecular so the forces between the particles in the crystals are weak *intermolecular forces.* The crystals easily melt or turn to vapour on gentle heating.

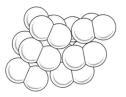

The crystal structure of iodine. Molecules of I_2 held in a regular lattice by intermolecular forces.

An iodine molecule

Some non-metals consist of giant structures of atoms held together by *covalent bonds.* Examples are *carbon* (as graphite or diamond) and *silicon.* Covalent bonds are strong and point in a definite direction so these structures have very high melting points.

Diamond is very hard. Graphite, the other *allotrope* of carbon with a giant structure, has a layer lattice. It too has a high melting point but it can be used as a *lubricant* because the forces between the layers are weak and they can slide over each other.

curly arrows describe the movement of electrons as bonds break and form in the steps of the mechanism of a reaction. A curly arrow with both halves of the arrow head shows the movement of a pair of electrons. The tail of the arrow starts where the electron pair begins. The head of the arrow points to where the electron pair ends up.

Movement of an electron pair during *heterolytic bond breaking*

$$CH_3 - \overset{\overset{\displaystyle CH_3}{|}}{\underset{\underset{\displaystyle CH_3}{|}}{C}} - Br \longrightarrow CH_3 - \overset{\overset{\displaystyle CH_3}{|}}{\underset{\underset{\displaystyle CH_3}{|}}{C}} + \quad + \quad Br^-$$

A curly arrow with only half an arrow head indicates the movement of a single electron.

Movement of single electrons during *homolytic bond breaking*

$$Cl\!:\!Cl \longrightarrow Cl^{\bullet} + Cl^{\bullet}$$

current is a flow of electric charge and is measured in *amperes.* In a metal the flow of charge is carried by electrons. In an electrolyte the charge carriers are negative ions moving towards the anode and positive ions moving towards the cathode.

Dalton's law of partial pressures states that the total pressure of a mixture of gases is the sum of the *partial pressures*. This means that the gases in a mixture do not interfere with each other. Each gas makes its own independent contribution to the total pressure.

dative covalent bond: a covalent bond formed when one atom contributes both of the shared pair of electrons. Chemists call this a 'dative covalent bond' because the word 'dative' means giving and one atom gives both the electrons to make the bond. Once formed there is no difference between a dative bond and any other covalent bond.

Ammonia forms a dative covalent bond with a hydrogen ion when it is acting as a base.

$$H-\underset{\underset{\displaystyle H}{|}}{\overset{\overset{\displaystyle H}{|}}{N}}\text{:}\quad\rightsquigarrow H^{\text{I}}\quad\longrightarrow\quad H-\underset{\underset{\displaystyle H}{|}}{\overset{\overset{\displaystyle H}{|}}{N}}\overset{+}{\rightarrow}II$$

Formation of an ammonium ion

Dative covalent bonding also accounts for *the structures of carbon monoxide*, the Al_2Cl_6 molecules in *aluminium chloride* vapour and the solid chloride of *beryllium*.

$$\underset{\text{oxonium ion}}{H-\overset{\overset{\displaystyle H}{\overset{\displaystyle \uparrow}{|}}\;+}{O}-H}\qquad\qquad\underset{\text{nitric acid}}{\overset{\displaystyle H}{\underset{\displaystyle}{O}}-N\underset{\displaystyle O}{\overset{\displaystyle O}{}}}\qquad\qquad\underset{\text{carbon monoxide}}{C\equiv O}$$

Examples of dative covalent bonds

The alternative name for dative bonding is 'co-ordinate bonding'. Chemists often use this alternative name when describing the bonding in *complex ions* and other *co-ordination compounds*.

***d*-block elements** are the elements in the three horizontal rows of elements in *periods* 4, 5 and 6 for which the last electron added to the atomic structure goes into a *d-orbital*. In period 4, the *d*-block elements run from scandium ($1s^2 2s^2 2p^6 3s^2 3p^6 3d^1 4s^2$) to zinc ($1s^2 2s^2 2p^6 3s^2 3p^6 3d^{10} 4s^2$)

The changes in the properties across a series of *d*-block elements are much less marked than the big changes across a *p-block* series. This is because from one element to the next, as the *atomic number* of the nucleus increases by one, the extra electron goes into the inner *d*-sub-shell. In period 4, the outer shell is always the $4s$ *orbital* which is filled before the $3d$ starts to fill.

		3d	4s
Sc	[Ar]	↑	↑↓
Ti	[Ar]	↑ ↑	↑↓
V	[Ar]	↑ ↑ ↑	↑↓
Cr	[Ar]	↑ ↑ ↑ ↑ ↑	↑
Mn	[Ar]	↑ ↑ ↑ ↑ ↑	↑↓

		3d	4s
Fe	[Ar]	↑↓ ↑ ↑ ↑ ↑	↑↓
Co	[Ar]	↑↓ ↑↓ ↑ ↑ ↑	↑↓
Ni	[Ar]	↑↓ ↑↓ ↑↓ ↑ ↑	↑↓
Cu	[Ar]	↑↓ ↑↓ ↑↓ ↑↓ ↑↓	↑
Zn	[Ar]	↑↓ ↑↓ ↑↓ ↑↓ ↑↓	↑↓

Electron configurations of d-block elements as free atoms. Note that orbitals fill singly before the electrons start to pair up. Note that the configurations *chromium* and *copper* do not fit the general pattern.

The chemistry of an atom is to a large extent determined by its outer electrons because they are the first to get involved in reactions. So the elements Sc to Zn in period 4 are similar in many ways.

All the *d*-block elements are *metals* with useful properties for engineering and construction. Most have high melting points. A plot of *physical properties* against proton number often has two peaks corresponding to the half-filling and then filling of the *d*-shell.

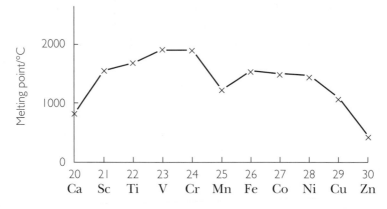

Plot of melting point against atomic number for the elements Ca to Zn

Some, but not all, of the *d*-block elements are classified as *transition elements*.

decantation involves gently pouring off most of a liquid or solution from a solid after centrifuging or simply allowing it to settle to the bottom of a container.

Filtering is often quicker if the solid is first allowed to settle and then most of the liquid decanted through the filter paper before pouring in the bulk of the solid. As a result most of the liquid passes through the filter paper before its pores are clogged by solid particles.

decomposition is a reaction in which compounds break down into simpler substances which may be compounds or elements. Heating is often necessary for decomposition (see *thermal decomposition*).

dehydration removes the elements of water from a compound to form a new compound. Concentrated sulfuric acid is a powerful dehydrating agent. It dehydrates blue copper sulfate crystals, sucrose and ethanol.

$$CuSO_4 \cdot 5H_2O(s) \xrightarrow{-5H_2O} CuSO_4(s)$$
blue   white

$$C_{12}H_{22}O_{11}(s) \xrightarrow{-11H_2O} 12C(s)$$
sucrose, white   black

$$C_2H_5OH(l) \xrightarrow{-H_2O} CH_2 = CH_2(g)$$
ethanol   ethene

deliquescent substances take up water vapour from the air and dissolve in it. Examples are calcium chloride, potassium hydroxide and sodium hydroxide. Deliquescent substances make good drying agents in *desiccators*.

Deliquescent substances cannot be used as *primary standards* in volumetric analysis because they cannot be weighed accurately.

delocalised electrons: bonding electrons which are not fixed between two atoms in a bond but are shared between three or more atoms. Electron delocalisation accounts for the shape, stability and properties of *benzene* rings, *nitrate* ions and other oxoanions, the acidity of *carboxylic acids* and *phenols*, and the *colour* of some organic compounds.

Electron delocalisation takes places in molecules where the conventional structure shows alternating double and single bonds. Delocalisation also affects ions where an atom with a lone pair of electrons and a negative charge is separated by a single bond from a double bond.

The extreme example of delocalisation is *metallic bonding* where electrons are shared between all the atoms in a crystal. Extended delocalisation over the planes of carbon atoms explains the electrical conductivity of graphite.

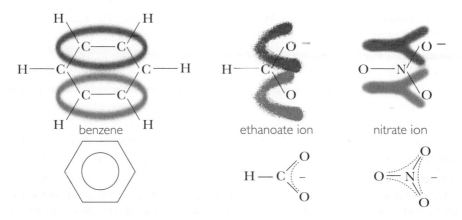

Examples of delocalisation in molecules and ions. The *bond lengths* of the bonds affected by delocalisation in each of these examples are the same – shorter than single bonds but longer than double bonds.

denaturation of proteins happens when the three dimensional shape of a *protein* is disrupted by heating, by extremes of *pH*, or by the presence of *heavy metal* ions. The chemical activity of protein molecules is linked to their shape. Changing the structure of proteins means that they lose their normal activity. *Enzymes*, for example, cease to act as *catalysts* when denatured.

density is the mass per unit volume of a material.

$$\text{density} = \frac{\text{mass}}{\text{volume}}$$

The symbol for density is ρ and the SI unit is kg m^{-3}. In chemistry densities are generally given in g cm^{-3}. The densities for most solids and liquids are in the range 0.5 to 10 g cm^{-3}.

The density of water at 0°C is 1.00 g cm^{-3} and in approximate laboratory work it is common to assume that dilute aqueous solutions have the same density as water. Solids and immiscible liquids float if they are less dense than water. The density of *ice* at the same temperature is 0.917 g cm^{-3} so ice floats on water.

At room temperature and pressure the densities of gases are about a thousand times smaller than those of solids and liquids. Measurement of gas densities can be used to calculate the *molar masses* of gases.

depression of freezing point: see *colligative properties*.

deprotonation is the removal of a *proton* from a molecule or ion. Adding aqueous hydroxide ions to a solution of *aluminium(III) ions* leads to a series of deprotonation reactions.

derivatives are used by chemists to identify unknown organic compounds. Converting a compound to a crystalline derivative produces a substance which can be purified by recrystallisation and then identified by measuring its melting point. Chemists identify carbonyl compounds by converting them to crystalline derivatives with *Brady's reagent*.

desalination is a process for obtaining pure water from seawater or other sources of water containing dissolved salts. Methods of desalination include *distillation* at low pressure, freezing, reverse *osmosis*, electrodialysis and *ion exchange*.

desiccator: a container with a *drying agent* used to remove the moisture (or other liquids) from chemical products which decompose if warmed in an oven. Chemicals which can be oven dried may also be stored in a desiccator as they cool to stop them picking up moisture from the air. A desiccator is a useful place to store chemicals which must be kept dry.

Vacuum desiccators are fitted with a tap so that air can be pumped out. The partial vacuum speeds up evaporation and diffusion of the vapour to the drying agent.

detergents clean things by removing dirt from surfaces. The chemicals which act as detergents are surface active agents or *surfactants*. Water is a polar solvent with a high *surface tension*. This means that water alone it is not good at removing dirt and grease. Detergents help to clean by:

- lowering the surface tension of water so that is spreads out and wets the surface

- separating grease and particles from the surface
- suspending the dirt in water so that it can be rinsed away.

There are two main types of detergent:

- soap detergents (usually just called *soaps*) made from animal *fats* or *vegetable oils*
- soapless detergents (usually just called detergents) which are made using chemicals from oil.

Washing powders or liquids for clothes are complex formulations which may include:

- detergents to increase wetting power, separate grease from fabrics and keep dirt in suspension,
- *sequestering agents* such as citrates or *zeolites* to soften *hard water*
- *enzymes* to break down protein stains such as blood stains
- optical brighteners to keep white fabrics looking bright and white
- oxygen *bleach* such as sodium peroxoborate(III) which only acts above 60°C
- *perfume* and colour to make the product distinctive and attractive to customers.

deuterium is the *isotope* of hydrogen with atomic number 1 but mass number 2. The symbols used for deuterium are ^2_1H or D. About 0.015% of natural hydrogen is the deuterium isotope. Deuterium oxide (heavy water) is D_2O.

The relative mass of D is twice that of H and this difference means that deuterium compounds react more slowly than their normal hydrogen equivalents. During *electrolysis* of acidified water H_2 is formed at the cathode more readily than D_2 so the concentration of deuterium increases as electrolysis continues. Eventually it is possible to make almost pure heavy water; this is used as a moderator in some nuclear power stations.

diagonal relationship: the similarities between the first member of one group in the periodic table with the second element in the next group, to the right, found particularly with these three pairs of elements: Li–Mg, Be–Al, and B–Si.

The diagonal relationship is a consequence of the relatively small size of the ions of the elements in the second short period. The *polarising power* of the positive ion of an element determines to a large extent the type of bonding between the element with non-metals such as oxygen and chlorine and hence the chemical characteristics of the compounds (see *Fajan's rules*).

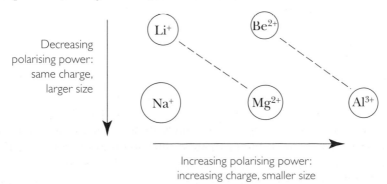

Diagonal relationships related to polarising power

These are some of the similarities between *beryllium* and *aluminium* which are not shared by the other group 2 elements:

- their bonding in compounds is mainly covalent
- their oxides and hydroxides are amphoteric
- their anhydrous chlorides vaporise easily and form dimers in the vapour (Be_2Cl_4 and Al_2Cl_6).

dialysis is used to separate dissolved ions and small molecules from *colloid* particles. Dialysis uses a *selectively-permeable membrane* which lets through to the small ions and molecules but traps larger colloid particles. Dialysis is used in medicine to treat the blood of patients with kidney failure. Dialysis removes the waste products of metabolism from blood and helps to adjust the concentration of ions. Blood cells and colloidal sized protein molecules cannot pass through the membrane.

diamond is one of the *allotropes* of carbon; it consists of a giant structure of atoms. Larger diamonds are valued as gem stones because of the mineral's high *refractive index* and dispersive power. Diamond is also a valuable industrial material because it is the hardest natural substance and the best solid conductor of heat energy. Its thermal conductivity is five times as great as that of copper. This makes diamonds very useful for cutting, grinding and polishing other hard materials. Diamond tipped tools are hard wearing and they do not overheat. Industrial diamonds are manufactured by heating graphite to 1800°C at very high pressure in the presence of a metal such as nickel or iron.

diatomic molecule: strictly a molecule with two atoms such as N_2 or HCl but the term is generally used for the molecules of elements with two identical atoms. Examples of diatomic elements are oxygen, O_2, nitrogen, N_2 and the halogens, F_2, Cl_2, Br_2 and I_2.

diazonium salts: salts formed when aryl amines, such as *phenylamine*, react with *nitrous acid* (HNO_2) below about 10°C.

Diazonium salts are unstable so they are made as needed and kept cold. Above 10°C, benzene diazonium chloride decomposes to phenol and nitrogen.

The commercial importance of diazonium salts is based on their coupling reactions to form *azo dyes*.

Diazonium salts are also useful intermediates which make it possible to form derivatives of arenes.

Diazotisation – the formation of benzene diazonium chloride

dienes are hydrocarbons with two double bonds. Buta-1,3-diene is a product of *thermal cracking* which is a useful intermediate for making other organic compounds. Synthetic *rubbers* are copolymers of butadiene, or one of its derivatives, with other unsaturated compounds such as styrene.

diffusion is a spreading out and mixing process in gases or solutions as molecules or ions mingle with each other. Molecules diffuse from a region where they are more concentrated to a region where their concentration is lower. Eventually diffusion evens out differences in concentration. The smaller the molecules or ions the faster they diffuse. Diffusion is a consequence of the rapid random motion of molecules in gases and liquids. Diffusion through membranes is important in *dialysis* and *osmosis*.

dilution is the process of adding more solvent to a solution to lower the concentration.

Quantitative dilution is an important procedure in analysis. The purpose is to make a solution with known concentration by accurately diluting a *standard solution*. Successive dilutions provide a series of solutions which can be used to *calibrate* instruments such as *colorimeters*.

The procedure is to take a measured volume of the more concentrated solution with a pipette and run it into a graduated flask. The flask is then carefully filled to the mark with purified water.

The key to calculating the volumes to use when diluting a solution is to remember that the amount in moles of the reagent in the final solution must equal the amount in moles of the sample taken from the concentrated solution. If c is the concentration in mol dm^{-3} and V is the volume in dm^3:

- the amount in moles of the reagent in the concentrated solution $= c_A V_A$
- the amount in moles of the reagent in the diluted solution $= c_B V_B$

So: $c_A V_A = c_B V_B$

Worked example:

What volume of a 1.00 mol dm^{-3} solution of copper(II) sulfate is needed to prepare 100 cm^3 of a 0.1 mol dm^{-3} solution?

Notes on the method

Start by a *conversion* to give the volume of the diluted solution in dm^3.

V_A can be calculated because all the other terms in the relationship $c_A V_A = c_B V_B$ are known.

Answer

Final volume of the diluted solution is to be: $100 \text{ cm}^3 \times \dfrac{1 \text{ dm}^3}{1000 \text{ cm}^3}$
$= 0.1 \text{ dm}^3$

$1.0 \text{ mol dm}^{-3} \times V_A = 0.1 \text{ mol dm}^{-3} \times 0.1 \text{ dm}^3$

$V_A = \dfrac{0.1 \text{ mol dm}^{-3} \times 0.1 \text{ dm}^3}{1.0 \text{ mol dm}^{-3}} = 0.01 \text{ dm}^3 = 10 \text{ cm}^3$

Pipetting 10 cm^3 of the concentrated solution into a 100 cm^3 graduated flask and making it up to the mark gives the required dilution.

---

dimerisation: two molecules linking together. Dimers may be held together by *covalent bonds* (see *aluminium chloride*) or by *hydrogen bonding.*

$$CH_3 - C \overset{\displaystyle O \cdots H - O}{\underset{\displaystyle O - H \cdots O}{\Big\langle}} C - CH_3$$

An ethanoic acid dimer in a *non-polar solvent.* In aqueous solution, ethanoic acid does not dimerise because there are so many water molecules with which they can form hydrogen bonds.

diols are alcohols with two — OH groups. An example is ethan-1,2-diol used as *antifreeze.*

$$H - \overset{\displaystyle H}{\underset{\displaystyle OH}{C}} - \overset{\displaystyle H}{\underset{\displaystyle OH}{C}} - H$$

Structure of ethane-1,2-diol

dioxins are a family of stable compounds which persist in the environment. Chemically they contain chlorine atoms bonded to benzene rings. Some dioxins are toxic and can cause skin disease, cancer and birth defects. Traces of dioxins may form during the manufacture of chlorinated phenols, the bleaching of wood pulp for paper and during the *incineration of wastes* containing chlorine compounds such as *PCBs* and the *addition polymer,* PVC.

dipole moments: see *polar molecules.*

dipole–dipole interactions are *intermolecular forces* involving *polar molecules* with permanent dipoles. The cohesive forces are stronger than *intermolecular forces* and account for the fact that compounds such as propanone are liquids at room temperature while comparable non-polar hydrocarbons are gases.

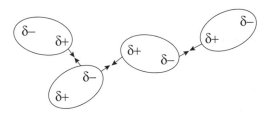

Attractions between molecules with permanent dipoles

diprotic acid: an *acid* such as sulfuric acid (H_2SO_4), which can give away two *protons* (H^+).

disaccharides: see *carbohydrates.*

disinfectants are chemicals which destroy micro-organisms. Chlorine and oxygen *bleaches* are used as disinfectants. Unlike *antiseptics,* disinfectants cannot be used on skin and other living tissues.

displacement reactions are redox reactions which can be used to compare the relative strengths of metals as *reducing agents* and non-metals as *oxidising agents*.

A more reactive metal displaces a less reactive metal from one of its salts. Zinc, for example, displaces copper from a copper(II) sulfate. The zinc atoms reduce the copper ions.

$$Zn(s) + Cu^{2+}(aq) \longrightarrow Zn^{2+}(aq) + Cu(s)$$

In *group* 7, a more reactive halogen displaces a less reaction halogen. The order of reactivity for the halogens is Cl>Br>I. The more reactive halogen oxidises the ions of a less reactive halogen.

$$Br_2(aq) + 2I^-(aq) \longrightarrow 2Br^-(aq) + I_2(s)$$

Standard electrode potentials make it possible to predict the direction of change in a displacement reaction (see *electrochemical series*).

displayed formulae show all the atoms and bonds in a molecule (see also *structural formula*).

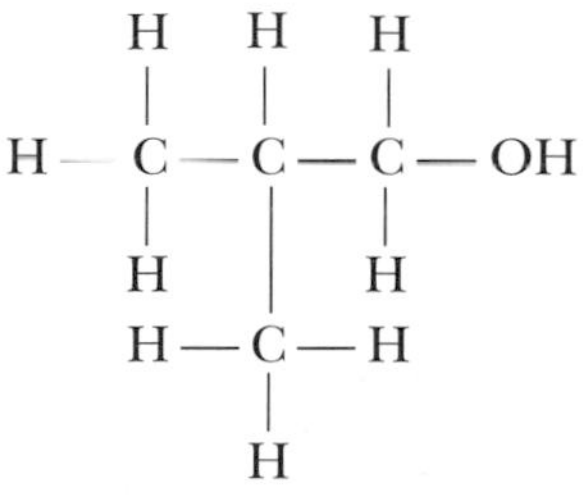

2-methylpropan-1-ol 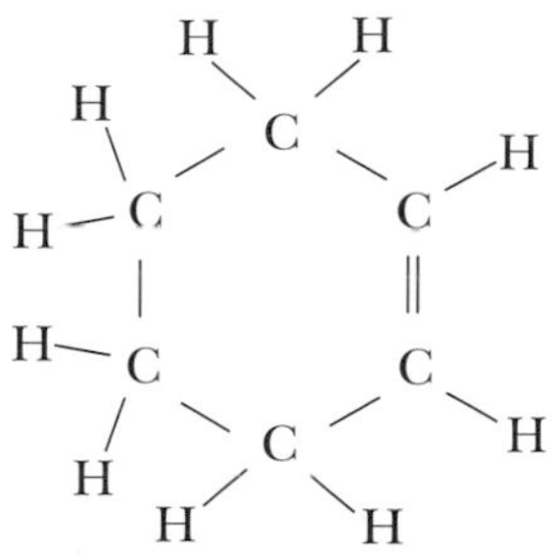   cyclohexene

Examples of displayed formulae

disproportionation reaction: a reaction in which the same element both increases and decreases its *oxidation number*. *Copper*(I) ions disproportionate in aqueous solution to a mixture of $Cu^{2+}(aq)$ ions and Cu atoms.

A series of disproportionation reactions is used to make *chlorine oxoanions*. The oxygen in *hydrogen peroxide* disproportionates when it decomposes. Note that disproportionation refers to a particular element in a compound not to the compound as a whole.

(See the *electrochemical series* for the use of standard electrode potential values to predict whether or not disproportionation will take place.)

dissociation: a reaction in which a compound splits into two or more smaller products which can recombine to form the original compound if the conditions change.

Some compounds dissociate on heating. Examples are ammonium chloride and dinitrogen tetroxide. These *thermal dissociation* reactions are *reversible*. On cooling the original compounds reform.

$$NH_4Cl(s) \underset{\text{cool}}{\overset{\text{heat}}{\rightleftharpoons}} NH_3(g) + HCl(g)$$

$$N_2O_4(g) \underset{\text{cool}}{\overset{\text{heat}}{\rightleftharpoons}} 2NO_2(g)$$

Acids dissociate into ions on solution in water. The extent to which they ionise distinguishes strong and weak acids and this is measured by the *acid dissociation constant.*

distillation: a technique for separating and purifying a liquid by heating to vaporise the liquid and then cooling to condense it. Simple distillation is used to purify water, to recover pure solvents and to separate liquid product from reaction mixtures. Substances which do not evaporate are left behind in the distillation flask.

Thermometer

Condenser

Anti-bumping granule

Apparatus for simple distillation

One of the easiest ways to determine the boiling point of a liquid is to distil a pure sample of the liquid in an apparatus with a thermometer.

(See also *fractional distillation, steam distillation* and *vacuum distillation.*)

disulfide bridges: see *cross-linking.*

DNA (deoxyribonucleic acid) is a double helix made up of two poly*nucleotide* chains. Both chains have a sugar phosphate backbone. Every sugar unit has one of four bases linked to it. The four bases are adenine (A), cytosine (C), guanine (G) and thymine (T). The links between the chains are formed by hydrogen bonding between pairs of bases: A pairs with T and C pairs with G.

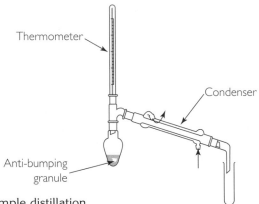

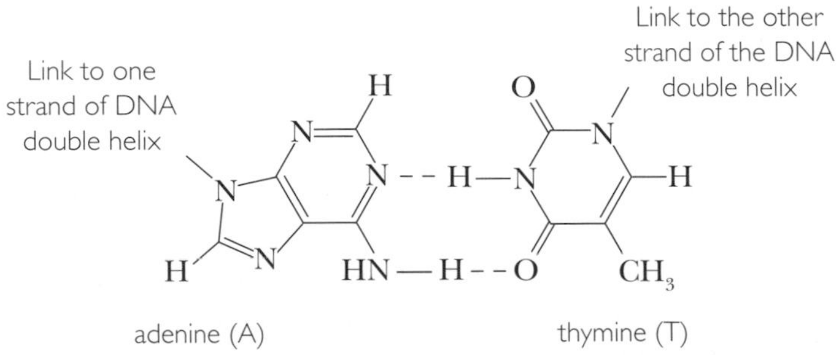

Hydrogen bonding between adenine and thymine in a DNA double helix

Genes are sections of DNA molecules found in the chromosomes of cells.

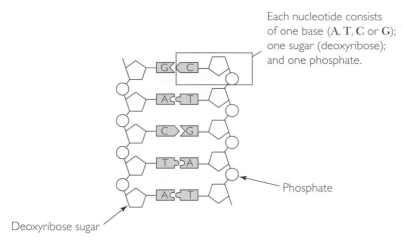

Each nucleotide consists of one base (A, T, C or G); one sugar (deoxyribose); and one phosphate.

Phosphate

Deoxyribose sugar

Representation of part of a DNA double helix

dolomite is a carbonate mineral with the composition $MgCO_3.CaCO_3$. It is a raw material for the production of magnesium and its salts.

_d_-orbitals are the five atomic orbitals in a _d_-sub-shell. In a free atom or ion the five _d_-orbitals are at the same energy level. The five orbitals are not the same shape and they split into two groups with different energies when a _d-block element_ ion is surrounded by molecules or ions in a complex ion. This helps to account for the _colour_ of complex ions.

dot and cross diagrams show the way in which the outer electrons of atoms are shared or transferred when _covalent bonds_ and _ionic bonds_ form. Dots, crosses and small circles help to keep count of the number of electrons contributed by each atom to bonding.

Dot and cross diagrams for molecules show both bonding and lone pairs of electrons; they help to predict the _shapes of molecules._

Despite the use of dots and crosses for the electrons coming from different atoms, all the electrons are the same. Once bonds form it is impossible to say which electron came from which atom.

double bond: two covalent bonds between two atoms as in _oxygen, alkenes_ and _ketones_. With two electron pairs involved in the bonding, there is a region of high electron density between the two atoms joined by a double bond.

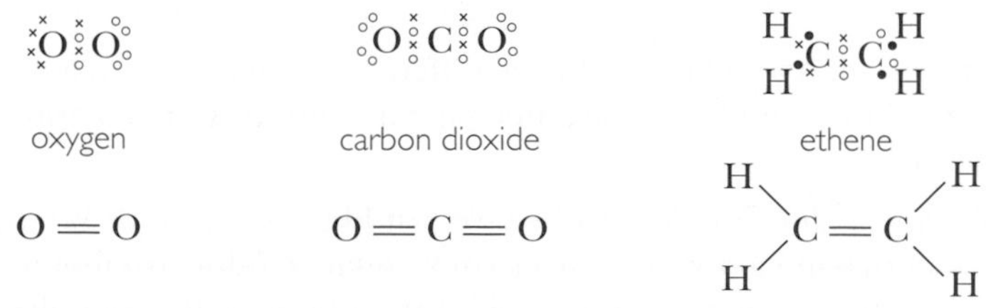

oxygen

$O = O$

carbon dioxide

$O = C = O$

ethene

$$\underset{H}{\overset{H}{\diagdown}}C = C\underset{H}{\overset{H}{\diagup}}$$

Examples of molecules with double bonds

The molecular orbital model for a double bond shows that one of the bonds is a normal *sigma(σ) bond* while the second bond is a *pi(π) bond*.

double indicator titration: a procedure for using two *indicators* successively to determine the two end points during a *titration*. This procedure can be used to find the concentrations in a solution containing a mixture of sodium carbonate and sodium hydrogencarbonate by titration with a standard solution of an acid such as hydrochloric acid.

The solution starts alkaline because carbonate and hydrogencarbonate ions are bases. Phenolphthalein is added at the start and it turns red. Acid is run in from the burette. The phenolphthalein becomes colourless at the first end point. This corresponds to converting all CO_3^{2-}(aq) ions in the mixture to HCO_3^-(aq).

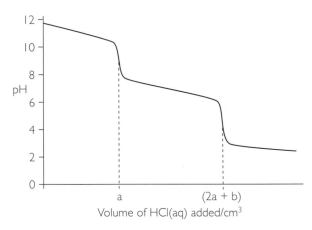

The *pH curve* during a titration of a mixture of Na_2CO_3(aq) and $NaHCO_3$(aq) with dilute 0.10 mol dm^{-3} hydrochloric acid

Next, methyl orange indicator is added and it starts yellow. More acid is run in from the burette and the indicator turns red at the second end point. The second end point corresponds to the complete neutralisation of all the HCO_3^-(aq) ions.

$$Na_2CO_3 \xrightarrow{\ a\ } NaHCO_3 \xrightarrow{\ a\ } NaCl$$
$$NaHCO_3 \xrightarrow{\ b\ } NaCl$$
present in the
original mixture

The chemical changes and the titres during a double indicator titration

double salts are solid ionic salts with two different metal cations in a lattice of negative ions. Crystalline double salts are not mixtures but distinct compounds with a definite structure.

When dissolved in water double salts behave just like a mixture of two simple *salts*. The usual way of making a double salt is to mix equal amounts (in moles) of the two single salts in solution and then to crystallise the double salt. Examples of double salts are the *alums*.

drugs are substances which affect natural chemical processes in the body.

In medicine, drugs are substances used for the treatment, relief, diagnosis or prevention of disease. A drug is what a doctor prescribes. A *medicine* is the whole formulation.

There are many classes of medical drug which include:

- drugs which act on the central nervous system, such as hypnotics, sedatives, tranquillisers, antidepressants, *narcotics, anaesthetics and analgesics*
- drugs which kill bacteria, such as *antibiotics* and sulfonamides (see *chemotherapy*)
- anti-viral drugs, such as those used to suppress HIV
- cardiovascular drugs, which help to lower blood pressure, stimulate the heart or control the heart beat
- drugs such as *cisplatin* used in cancer treatment
- drugs used for the digestive system, such as *antacids*, laxatives and drugs to treat ulcers
- *hormones* such as *insulin*, growth hormone and *steroids*.

People also take drugs for pleasure, stimulation and relaxation. Some recreational drugs such as alcohol and nicotine are legal. Others such as ecstasy, marijuana, cocaine and opiates are illegal. Drugs used to treat disease, such as steroids, can also be exploited to enhance performance in sport. Sensitive methods of chemical analysis are widely used to detect traces of drugs in blood, breath and urine (see *breathalyser*).

drug development is the research and development programme which leads to the launch of a new drug. Scientists in the *pharmaceutical industry* use the term 'new chemical entity' (NCE) to describe a newly-discovered chemical for treating disease. Discovering an NCE can be hugely profitable, so pharmaceutical companies spend large sums on research and development.

Before a new compound can become a marketable drug the manufacturer has to show that:

- the drug meets a definite need of patients and their doctors
- it is technically possible to make it on a large scale
- there is a big enough market to make it commercial
- it is safe.

The US pharmaceutical chemist, Gertrude Elion (born 1918) developed a rational process of drug design based on detailed studies of the differences between the *biochemistry* and molecular biology of human cells and the cells of micro-organisms which cause disease. Since the 1940s, the team she worked with has developed drugs to treat a range of diseases including leukaemia, malaria, gout and HIV infection.

dry ice is solid carbon dioxide. Above $-79°C$ it *sublimes* to carbon dioxide gas. This makes it a useful coolant.

drying agents remove water from moist liquids and gases. Liquid products of organic preparations are often dried with small amounts of anhydrous sodium sulfate or anhydrous magnesium sulfate before the final distillation step. Gases made with laboratory reagents are often moist and can be dried by bubbling them through concentrated sulfuric acid, or more safely by passing them though a U-tube containing *silica gel*. Anhydrous calcium chloride is frequently used in *desiccators* to absorb water, alcohols or amines.

drying oils are oils such as linseed oil which polymerise and set when exposed to air. As a drying oil linseed oil is used in the manufacture of *paints* and varnishes.

ductility is a property of materials, especially metals, which can be drawn out under tension without breaking. *Copper* is a ductile metal which can be pulled through a die (narrow hole) to make thin wire.

dyes are coloured materials used to dye fabrics and other materials such as paper, hair, leather and food. A good dye is a fast dye which means that it is not only coloured but can also attach itself strongly to a material so that it does not wash out nor should it fade in the light.

Plants were the main source of natural dyes until William Perkin chanced on the first synthetic dye, mauveine in 1856. Alizarin and indigo were plant dyes which were the basis of large scale industries. The discovery of synthetic routes to these dyes in the late nineteenth century were early triumphs for *organic chemistry*.

Examples of synthetic dyes are *azo dyes*, *vat dyes*, and *fibre reactive dyes*.

dynamic equilibrium: the state of balance in a *reversible* process when neither the forward change nor the backward change is complete; both changes are still going on at equal rates so that they cancel each other out and there is no overall change.

Examples of systems which can be in dynamic equilibrium include:

- water and ice at 0°C, or more generally the liquid and solid states of a substance at its melting point
- a liquid with its saturated *vapour* in a closed container
- a solid with a *saturated solution* of the solid
- a solute distributed between two immiscible solvents
- a *weak acid* in aqueous solution, or any other reversible reaction at constant temperature and pressure.

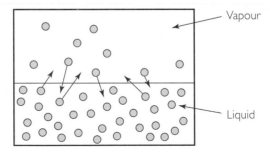

Dynamic equilibrium between a liquid and its vapour

edta is the common abbreviation for a particular ion which binds so firmly with metal ions that it holds them in solution and effectively makes them chemically inactive. A very little edta is added to salad dressing to trap traces of metal ions which otherwise would catalyse the oxidation of oils. Another practical application is where edta is included in bathroom cleaners to help remove scale by dissolving deposits of calcium carbonate left by hard water. Thanks to this ability to act as a *chelating agent*, edta can be used to treat lead poisoning.

In laboratories edta is the disodium salt of **e**thylene **d**iamine**t**etra**a**cetic **a**cid (or as it is now called 1,2-bis[bis(carboxymethyl)amino]ethane). The ion can fold itself round metal ions so that four oxygen atoms and two nitrogen atoms can present lone pairs to form co-ordinate bonds with the metal ion. It is a hexadentate *ligand*.

(See also *complex forming titration.*)

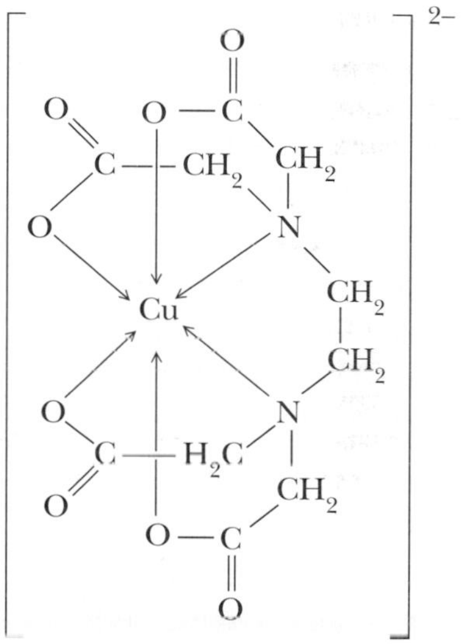

Complex ion formed by edta with a metal ion. For simplicity the disodium salt is often represented as Na$_2$Y.

efflorescence is the loss of *water of crystallisation* from a hydrated salt kept in an open container. Hydrated crystals of sodium carbonate (washing soda, Na$_2$CO$_3$.10H$_2$O) turn powdery in an unsealed container as they gradually lose most of their water.

elastomers are *polymers* which can be stretched to several times their normal length and recover their original size and shape when released. They are elastic materials. Natural rubber is an elastomer. So are synthetic rubbers such as the *copolymer* of butadiene and styrene.

Lycra is an elastic fibre with an exceptional ability to stretch and recover. It is a condensation polymer combining two polymer segments – *polyurethane* (—NH—CO—O—) and polyether (—CH$_2$—CH$_2$O—). The polyether sections of the chains are the soft stretchy parts, while the urethane sections make sure that the fibres snap back into shape after stretching. Many sporting fabrics now include Lycra fibres but a garment is never made entirely from Lycra.

electric arc furnace: a furnace for making *steel* by *recycling* scrap iron and steel. The heat of an arc from carbon electrodes melts the metal. Added lime combines with impurities to form a *slag*. Other ingredients are added as necessary. The process can be precisely controlled to give small batches of steel matched to a specified composition.

electrical conductors are solids, liquids or gases which conduct electricity because they contain charged particles which are free to move. They include:

- metals and graphite in which the charge carriers are *delocalised electrons*
- electrolytes in which the charge carriers are positive and negative ions.

electrical insulators are solids, liquids or gases which do not conduct electricity because they contain no free moving charged particles. Increasing the applied voltage may turn an insulator into a conductor if the voltage becomes high enough to ionise the atoms or molecules.

electrochemical cells produce an electric potential difference from a redox reaction. Some electrochemical cells are designed for practical use. Examples include *alkaline manganese cells* and rechargeable *nicad* (nickel–cadmium) and *lead–acid cells.*

The reaction of zinc metal with aqueous copper(II) ions is a redox reaction which can be described by two half equations. Zinc is oxidised to zinc ions as copper(II) ions are reduced to copper metal.

$$Zn(s) \longrightarrow Zn^{2+}(aq) + 2e^-$$
$$Cu^{2+}(aq) + 2e^- \longrightarrow Cu(s)$$

In an electrochemical cell the two half reactions happen in separate *half cells*. The electrons flow from one cell to the other through a wire connecting the electrodes. The electric circuit is completed by a *salt bridge* connecting the two solutions.

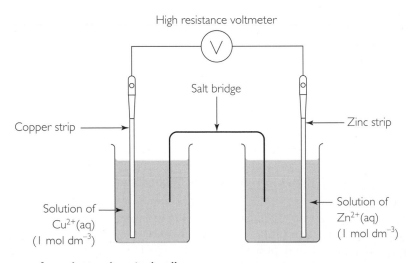

Diagram of an electrochemical cell

The tendency for the current to flow is measured with a high resistance voltmeter which measures the cell's *emf* when no current is flowing. In the example shown electrons tend to flow out of the zinc electrode (negative) round the circuit and into the copper electrode (positive). The emf of the cell is 1.14 V under standard conditions (298 K and concentrations of 1.0 mol dm^{-3}).

There is a convenient shorthand for describing cells. The standard emf of the cell, $E^{\ominus}_{cell}$ is written alongside the cell diagram. The agreed convention is that the sign of $E^{\ominus}_{cell}$ is the charge on the right-hand electrode.

$$Zn(s)|\ Zn^{2+}(aq)\ \vdots\ Cu^{2+}(aq)|Cu(s) \qquad E^{\ominus}_{cell} = +1.14\ V$$

If the cell emf is positive the reaction tends to go according to the cell diagram read from left to right. As a current flows in a circuit connecting the two electrodes, zinc atoms turn into zinc ions and dissolve, as copper ions turn into copper atoms and deposit on the copper electrode.

$$Zn(s) + Cu^{2+}(aq) \longrightarrow Zn^{2+}(aq) + Cu(s)$$

$E^{\ominus}_{cell}$ values can be worked out for any redox reaction with the help of a table of *standard electrode potentials*.

$$E^{\ominus}_{cell} = E^{\ominus}_{right\text{-}hand\ electrode} - E^{\ominus}_{left\text{-}hand\ electrode}$$

electrochemical series: a series showing half-reactions for oxidation and reduction, with the most powerful *reducing agent* at the top (or bottom) of the list and the most powerful *oxidising agent* at the bottom (or top) of the list.

The order of the half-reactions is determined by their *standard electrode potentials*. In the list below, the *half-cell* with the most negative electrode potential is at the top of the list. The most negative electrode has the greatest tendency to give up electrons, so it is powerfully reducing.

The half-cell with the most positive electrode potential is shown at the bottom of the list. The most positive electrode has the greatest tendency to gain electrons so it is the most powerfully oxidising.

(Note that standard electrode potentials may also be listed in the opposite order, in which case the most powerfully reducing electrode is at the bottom and the most powerfully oxidising electrode is at the top.)

Half cell	Half reaction	$E^{\ominus}/V$
$Na^+(aq)|Na(s)$	$Na^+(aq) + e^- \longrightarrow Na(s)$	-2.71
$Mg^{2+}(aq)|Mg(s)$	$Mg^{2+}(aq) + 2e^- \longrightarrow Mg(s)$	-2.37
$Zn^{2+}(aq)|Zn(s)$	$Zn^{2+}(aq) + 2e^- \longrightarrow Zn(s)$	-0.76
$2H^+(aq)|H_2(g)$	$2H^+(aq) + 2e^- \longrightarrow H_2(g)$	0.00 (by definition)
$Cu^{2+}(aq)|Cu(s)$	$Cu^{2+}(aq) + 2e^- \longrightarrow Cu(s)$	$+0.34$
$[I_2(aq),2I^-(aq)]|Pt$	$I_2(aq) + 2e^- \longrightarrow 2I^-(aq)$	$+0.54$
$[Br_2(aq),2Br^-(aq)]|Pt$	$Br_2(aq) + 2e^- \longrightarrow 2Br^-(aq)$	$+1.09$
$[Cl_2(aq),2Cl^-(aq)]|Pt$	$Cl_2(aq) + 2e^- \longrightarrow 2Cl^-(aq)$	$+1.51$

The top part of the above electrochemical series shows half-reactions for metal ions and metals. Sodium is the most reactive of the metals when it reacts as a reducing agent forming metal ions; copper is the least reactive. This broadly corresponds to the reactivity series for metals and the order of reaction shown by metal/metal ion *displacement reactions.*

The bottom part of the above electrochemical series shows half-reactions for halogens and halide ions. Chlorine is the most reactive of these halogens when it acts as an oxidising agent forming chloride ions. Iodine is the least reactive. This corresponds to the order of reactivity of the halogens as shown by their displacement reactions.

An electrochemical series based on electrode potentials can be used to predict the direction of change in redox reactions. First identify the two half-equations. Write them down one above the other. The more positive half-reaction tends to go from left to right taking in electrons, while the more negative half-reaction goes from right to left.

More
negative $I_2(aq) + 2e^- \rightleftharpoons 2I^-(aq) \quad E^\ominus = +0.54\,V$
electrode

More
positive $Cl_2(aq) + 2e^- \rightleftharpoons 2Cl^-(aq) \quad E^\ominus = +1.51\,V$
electrode

Using half-reactions and $E^\ominus$ values to predict the direction of change

In this way we can predict from electrode potentials whether or not a *disproportionation* reaction is likely to occur. Copper(I) ions disproportionate in aqueous solution, but iron(II) ions tend not to disproportionate.

More
negative $Cu^{2+}(aq) + e^- \rightleftharpoons Cu^+(aq) \quad E^\ominus = +0.15\,V$
electrode
                                                                    Copper(I)
                                                                    disproportionates
More
positive $Cu^+(aq) + e^- \rightleftharpoons Cu(s) \quad E^\ominus = +0.52\,V$
electrode

More
negative $Fe^{2+}(aq) + 2e^- \rightleftharpoons Fe(s) \quad E^\ominus = -0.44\,V$
electrode
                                                                    Iron(III) reacts
                                                                    with iron metal
More   to form iron(II)
positive $Fe^{3+} + e^- \rightleftharpoons Fe^{2+} \quad E^\ominus = +0.77\,V$
electrode

Using standard electrode potential values to decide whether or not ions are likely to disproportionate

Standard electrode potentials predict the direction of change but say nothing about the rate of change. A reaction which is *feasible* may not in fact happen because it is so slow. Also, the predictions apply only under standard conditions. Changing the concentrations can alter the direction of change.

electrode: see *anode* and *cathode*.

electrode potential: see *standard electrode potential.*

electrolysis is a process which uses an electric current to decompose a molten ionic compound, or a solution of ions, into elements. Ions can conduct electricity only when they are free to move, which is why electrolytes are molten compounds or solutions.

Large scale manufacturing processes which use electrolysis include *aluminium extraction* and the *electrolysis of brine. Electroplating, copper refining* and *anodising* are also electrolytic processes.

During the electrolysis of molten salts, such as lead bromide:

- metals deposits appear at the *cathode*, e.g. $Pb^{2+} + 2e^- \longrightarrow Pb$ (reduction)
- non-metals appear at the *anode*, e.g. $2Br^- \longrightarrow Br_2 + 2e^-$ (oxidation).

During the electrolysis of a solution of a salt in water the change at the *cathode* depends on the type of metal ion in the salt. If the metal is:

- low in the electrochemical series (unreactive) it forms at the cathode, e.g. $Cu^{2+}(aq) + 2e^- \longrightarrow Cu(s)$
- high in the electrochemical series (reactive) the product at the cathode is hydrogen, from hydrogen ions produced by the ionisation of water, $2H^+(aq) + 2e^- \longrightarrow H_2(g)$

During the electrolysis of a solution of a salt in water the change at the anode may depend on the type of electrode. If the anode is made of carbon or platinum the products are:

- halogen molecules if the ions in the solution are chloride, bromide or iodide ions, e.g. $2Cl^-(aq) \longrightarrow Cl_2(g) + 2e^-$
- otherwise, oxygen from the water, $4OH^-(aq) \longrightarrow O_2(g) + 2H_2O(l) + 4e^-$

The concentration of the solution can affect the product at the anode. Chlorine forms at the anode during electrolysis of a concentrated solution of sodium chloride, but as the solution gets more dilute increasing amounts of oxygen form.

If the anode is a metal such as copper or silver, the atoms in the electrode turn into ions,

e.g. $Cu(s) \longrightarrow Cu^{2+}(aq) + 2e^-$

The amount of change at an electrode depends on:

- the charge on the ions
- the amount of electric charge that flows.

electrolysis of brine is the basis of the chlor-alkali industry. *Electrolysis* of brine is used to manufacture *chlorine, hydrogen* and *sodium hydroxide*. The process is also the source of chlorine *bleach*, made by allowing the chlorine and sodium hydroxide to mix and react forming sodium chlorate(I).

Brine is a solution of sodium chloride in water. During electrolysis chlorine forms at the positive electrode (anode). Hydrogen bubbles off the negative electrode (cathode) while the solution turns into sodium hydroxide.

interpretation of the information from X-ray diffraction studies. This is possible because it is the electrons in a crystal which scatter X-rays (see *X-ray crystallography*)

Electron density maps for an ionic crystal provide evidence for electron transfer from the metal to the non-metal. The ions show up as distinct particles. Electron density maps for molecules show electrons shared between atoms in *covalent bonds.*

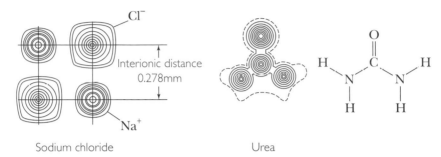

Sodium chloride	Urea

Electron density maps for an ionic crystal and for a molecule

electron-pair repulsion theory: see *shapes of molecules.*

electron transfer takes place during *redox reactions.* An atom, molecule or ion loses electrons when oxidised (**oxidation is loss**). The electrons are transferred to the atom, molecule or ion being reduced (**reduction is gain**). Hence the memory aide: OIL RIG. *Oxidation numbers* and *half-equations* for redox reactions help to keep track of electron transfer reactions.

electronegativity measures the pull of an atom of an element on the electrons in a chemical bond. The stronger the pulling power of an atom the higher its electronegativity. There are two quantitative scales of electronegativity; one devised by Linus Pauling and the other by Robert Mulliken. However the term electronegativity is generally used to compare one element with another qualitatively so it is enough to know the trends in values across and down the periodic table.

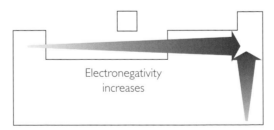

Electronegativity increases

Trends in electronegativity for *s* and *p* block elements. Values are low to the left of each period and rise towards the right, giving rise to a repeating (periodic) pattern.

The highly electronegative elements, such as fluorine and oxygen are at the top right of the periodic table. The least electronegative elements, such as caesium, are at the bottom left.

The bigger the difference in the electronegativity of the elements forming a bond, the more polar the bond. Oxygen is more electronegative than hydrogen so an

O — H bond is *polar* with a slight negative charge on the oxygen atoms and a slight positive charge on the hydrogen atom.

$$\overset{\delta+}{H} - \overset{\delta-}{Cl} \qquad \overset{H}{\underset{H}{\diagdown\diagup}}\overset{\delta+}{C} = \overset{\delta-}{O} \qquad \overset{\delta+}{H} - \overset{2\delta-}{O} \\ | \\ \underset{\delta+}{H}$$

Examples of *polar covalent bonds* between atoms with different electronegativities

The bonding in a compound becomes ionic if the difference in electronegativity is large enough for the more electronegative element to remove completely electrons from the other element. This happens in compounds such as sodium chloride, magnesium oxide or calcium fluoride.

electronic structure: the arrangement of electrons in an atom in its main energy levels and sub-levels. (See *electron configuration.*)

electrophiles are reactive ions and molecules which attack parts of molecules which are rich in electrons. They are 'electron-loving' reagents. Electrophiles form a new bond by accepting a pair of electrons from the molecule attacked during a reaction

Examples of electrophiles are the H^+ and the NO_2^+ ions. Other electrophiles are the atoms at the δ^+ end of a *polar covalent* bond. See *electrophilic addition* and *electrophilic substitution.*

electrophilic addition takes place when *alkenes* undergo *addition reactions.* The *electrophile* attacks the electron-rich region of the double bond between two carbon atoms. Electrophiles which add to alkenes include hydrogen bromide, bromine, sulfuric acid and water (in the presence of an acid catalyst).

Hydrogen bromide molecules are *polar.* The hydrogen atom, with its δ^+ charge is the electrophilic end of the molecule.

Electrophilic addition of hydrogen bromide to ethene. The reaction takes place in two steps. The intermediate has a positive charge on a carbon atom. It is a *carbocation*. Curly arrows show the movement of a pair of electrons. Bond breaking is *heterolytic* and the intermediates ions.

Bromine molecules are not polar but they become polarised as they approach the electron-rich double bond. Electrons in the double bond repel electrons in the bromine molecule. The δ^+ end of the molecule is electrophilic (see Figure on page 132.).

(See also *Markovnikov's rule.*)

Electrophilic addition of bromine to ethene (see page 131)

electrophilic substitution is the characteristic reaction of *arenes* such as benzene. The electrophile attacks the electron-rich benzene ring with its six *delocalised electrons*. When describing the mechanism of this types of reaction it is easier to keep track of what is happening to the electrons using the *Kekulé structure* for *benzene*.

Electrophilic substitution of bromine in benzene is rapid in the presence of iron. Iron reacts with some bromine to form iron(III) bromide, which is the catalyst. Iron(III) bromide is an electron pair acceptor (*Lewis acid*) which produces the electrophile, Br^+. An alternative catalyst is *aluminium chloride* (another Lewis acid). The catalysts for this reaction are sometimes described as 'halogen carriers'.

The first step of electrophilic substitution in benzene is very similar to the first step of *electrophilic addition* to alkenes. The intermediate could then complete the reaction by adding a bromide ion but in fact what happens is the loss of H^+ leading to substitution. Loss of H^+ is preferred because it keeps the stable benzene ring stabilised by six delocalised electrons.

Formation of the electrophile

The mechanism of electrophilic substitution. Note that an alternative second step (addition of a bromide ion) is not favoured because it gives a product which does not have a stable benzene ring.

Other examples of electrophilic substitution are the *nitration of benzene* and the *Friedel–Crafts reaction*.

electrophoresis is a technique for separating and identifying organic compounds. It can be used to separate compounds such as *amino acids* which become electrically charged when dissolved in water at a particular pH. Electrophoresis is used in hospitals to diagnose disease. It is important in genetic profiling to separate out the bands which form the 'fingerprint'. It is also used in research to study the structures of *proteins* and *nucleic acids*.

Electrophoresis takes place on strips of a gel soaked in a buffer solution to keep the pH constant. Small spots of several samples can be put on each gel. The charged molecules move when an electric voltage is connected between the ends of the gel and they separate into bands according to their size and charge. Most samples are colourless so, once the separation is complete, the gel is dipped into a stain to show up the bands.

electroplating uses *electrolysis* to coat one metal with another. The process is often used to coat metals with thin layers of copper, nickel, chromium, silver and zinc. The object to be plated forms the cathode in an electrolysis cell containing a solution of ions of the metal to be plated. Good results depend on careful control of the conditions, including the composition and concentration of the electrolyte, the temperature and the size of the current. Often the metal ion is present as a *complex ion*. During silver plating, for example, the silver ions are complexed with cyanide ions.

electrostatic forces are the forces between charged particles. Unlike charges attract each other. Positive ions, for example, attract negative ions in ionic compounds. Like charges repel each other. Positive ions in an electrolyte are repelled by the positive anode while being attracted to the negative cathode.

The size of the electrostatic force between two charges varies according to Coulomb's law. The bigger the charges the stronger the force – the force is proportional to the size of the charges, Q. The greater the distance, d, between the two charges the smaller the force – the force is inversely proportional to the distance squared.

$$\text{electrostatic force} \propto \frac{Q_1 \times Q_2}{d^2}$$

electrovalent bonding: see *ionic bonding.*

elements are chemically the simplest substances. All the atoms of an element have the same number of protons in the nucleus. The atomic number identifies an element and fixes its position in the *periodic table.* When an atom turns into an ion by gaining or losing electrons it becomes a charged atom of the same element. The *isotopes* of an element have different numbers of neutrons in the nucleus but they have the same chemical properties because they have the same number of protons and therefore the same *electron configuration.*

elevation of boiling point: see *colligative properties.*

elimination reaction: a reaction which splits off a simple compound from a molecule while forming a double bond.

Elimination of a *hydrogen halide* from a *halogenoalkane* produces an *alkene*. The conditions are to heat the halogenoalkane with a solution of potassium hydroxide in ethanol. Using ethanol as the solvent instead of water makes the alternative *nucleophilic substitution reaction* less likely.

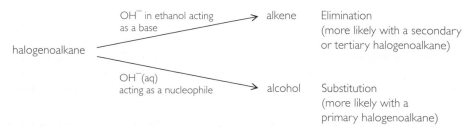

Competition between elimination and substitution

The elimination reaction is favoured if the halogen atom is in the middle of the carbon chain or at a branch in the chain.

Elimination of hydrogen bromide from 2-bromopropane

Another example of an elimination reaction is the removal of water from an *alcohol* to form an alkene. This is a useful reaction in synthesis, for introducing double bonds into molecules. The conditions for reaction are either to pass the vapour of the alcohol over a hot, solid catalyst such as aluminium oxide or to dehydrate the alcohol by heating it with concentrated sulfuric acid.

Formation of an alkene from an alcohol. The acid catalyst protonates the —OH group so that the *leaving group* is a water molecule – a better leaving group than a hydroxide ion. The proton is reformed in the second step of the reaction, so overall the acid catalyst is unchanged.

Ellingham diagrams explain the conditions for *metal extraction* from oxide ores. The diagrams are graphs showing how the *free energy changes* of formation ($\Delta G^{\ominus}$) of oxides vary with temperature.

emf (electromotive force) measures the 'voltage' of an electrochemical cell. The symbol for emf is E and the SI unit is the volt (V). The emf is the energy transferred in joules per coulomb of charge flowing through a circuit connected to a cell. The definition of the volt as a 'joule per coulomb' makes it possible to calculate the *free energy change, ΔG*, for reactions from the values of the emf for *electrochemical cells, $E^{\ominus}_{cell}$*.

$$\Delta G^{\ominus} = - zFE^{\ominus}_{cell}$$

where z is the number of electrons transferred according to the equation for the reaction and F is the *Faraday constant*.

empirical formula: the formula of a compound found by calculation from the combining masses of elements. The empirical formula shows the simplest ratio of the amounts of elements in the compound; it therefore gives the ratio of the numbers of atoms.

For molecular compounds, the *relative molecular mass* shows whether or not the empirical formula is the same as the *molecular formula*.

Worked example:

Analysis of a sample of a salt shows that it consists of 1.30 g zinc combined with 0.24 g carbon and 0.96 g oxygen. What is the formula of the salt?

Notes on the method

Look up the molar masses of the elements in a book of data.

Recall that: amount of substance/mol $= \dfrac{\text{mass of substance/g}}{\text{molar mass/g mol}^{-1}}$

Answer

	zinc	carbon	oxygen
Combining masses	1.30 g	0.24 g	0.96 g
Molar masses of elements	65 g mol^{-1}	12 g mol^{-1}	16 g mol^{-1}
Amounts combined	$\dfrac{1.30 \text{ g}}{65 \text{ g mol}^{-1}}$	$\dfrac{0.24 \text{ g}}{12 \text{ g mol}^{-1}}$	$\dfrac{0.96 \text{ g}}{16 \text{ g mol}^{-1}}$
	= 0.02 mol	= 0.02 mol	= 0.06 mol
Simplest ratio of amounts	1	1	3

The formula is $ZnCO_3$.

emulsions consist of droplets of one liquid finely dispersed in another liquid. Emulsions are *colloids*. Milk is an emulsion of fat droplets in water. Salad cream is an emulsion of vegetable oil in vinegar.

An emulsion is often 'thicker' (more *viscous*) than either liquid on its own. Many cosmetic and medical creams are emulsions. Foundation creams and brushless shaving creams are oil-in-water emulsions. Cold creams and cleansing creams are water-in-oil emulsions.

Just shaking two liquids together generally produces droplets which quickly separate back into two layers on standing. It is usually necessary to add an emulsifying agent to stabilise an emulsion. Salad dressing is an emulsion stabilised using mustard as the emulsifying agent. Mayonnaise is stabilised with egg yolk. Many creamy foods are emulsions and the lists of their ingredients include emulsifying agents such as lecithin and alginates.

Emulsifying agents are *surfactants*.

enantiomers: see *chiral compounds*.

endothermic changes take in energy from their surroundings. Melting and evaporation are endothermic *changes of state*.

For an endothermic reaction the enthalpy change, ΔH, is positive. During an endothermic reaction more energy is taken up in breaking bonds in the reactants than is given out as new bonds form in the products.

The fact that there are *spontaneous* endothermic reactions shows that it is not safe to predict that a reaction will not happen just because the *enthalpy change*, ΔH, is positive. Even if ΔH is positive, the *free energy change* (ΔG) can be negative if the *entropy change* (ΔS_{system},) is large enough and positive.

See the Figures on page 136.

Note that the enthalpy change of solution is a small difference between two large enthalpy changes. Also note that the lattice energy and the hydration enthalpies tend to be affected in the same way by changes in the sizes of the ions and their charges. The smaller the ions and the larger the charges the greater the lattice energy but also the greater the hydration enthalpies. This means that it is not easy to make use of enthalpy cycles to account for trends in solubilities of ionic compounds down group 1 or group 2 in the periodic table.

Also note that an ionic salt may well dissolve in water even if $\Delta H^{\ominus}_{solution}$ is slightly positive. Dissolving makes a significant difference in the numbers of ways that particles and energy can be arranged. The *entropy change* of the system, $\Delta S^{\ominus}_{system}$ may be positive and large enough to outweigh a negative entropy change in the surroundings.

enthalpy change of vaporisation is the *enthalpy change* when one mole of a liquid turns to a *vapour* at its boiling point. Evaporation separates the particles in a liquid so values for specific enthalpy of vaporisation give a measure of the strength of the bonding between particles in liquids. Substances with strong ionic or metallic bonding have much higher boiling points and enthalpies of vaporisation than substances consisting of molecules with weak *intermolecular forces*.

Physicists generally use specific latent heat values in studying the energy changes when liquids evaporate. Specific latent heats of vaporisation are measured in joules per kilogram rather than joules per mole.

entity: a term which can refer to any distinct particle such as an atom, molecule, ion, electron or free radical. The term is used in precise definitions such as the definition of *amount of substance* in moles. It is important to be precise when specifying a particular entity. These, for example, are four distinct chemical entities: a hydrogen atom, H; a hydrogen molecule, H_2; a hydrogen ion, H^+; and a hydride ion, H^-.

entropy (S) is a thermochemical quantity which makes it possible to predict the direction of changes (see also *standard molar entropy*). Change happens in the direction which leads to a total increase in entropy. When considering chemical reactions it is convenient to calculate the total entropy change in two parts: the entropy change of the system and the entropy change of the surroundings.

$$\Delta S^{\ominus}_{total} = \Delta S^{\ominus}_{system} + \Delta S^{\ominus}_{surroundings}$$

The entropy of a system measures the number of ways, W, of arranging the molecules and sharing out the energy between the molecules. Nothing seems to be happening to a closed flask of gas kept at a constant temperature. *Kinetic theory* tells us, however, that the molecules are in rapid motion, colliding with each other and with the walls of the container. The gas stays the same while constantly rearranging its molecules and redistributing the energy. The 'number of ways', W, is large for a gas. Gases generally have higher entropies than liquids which higher entropies than solids.

The relationship between S and W was derived by Ludwig Boltzmann (1844–1906), the Austrian physicist who was the first person to explain the laws of thermodynamics in terms of the behaviour of atoms and molecules in motion.

$$S = k \ln W$$

where S is the entropy of the system, k the Boltzmann constant and $\ln W$ the natural logarithm of the number of ways of arranging the particles and energy in the system.

The entropy change of the surroundings during a chemical reaction is determined by the size of the *enthalpy change*, ΔH, and the temperature, T. The relationship is:

$$\Delta S_{surroundings} = -\frac{\Delta H}{T}$$

The minus sign is included because the entropy change is bigger the more energy is transferred to the surroundings. For an exothermic reaction, which transfers energy to the surroundings, ΔH is negative, so $-\Delta H$ is positive.

For many exothermic reactions at about room temperature the value of $-\Delta H/T$ is much larger than ΔS_{system} which means that ΔS_{total} is positive. This explains why exothermic reactions generally tend to go (are *feasible*), but there are exceptions.

Chemists often find it more convenient to work with *free energy changes* which can be regarded as 'the total entropy changes in disguise'.

environmental chemistry is the study of chemical changes in the environment. Chemists use a range of analytical methods to estimate the amounts of elements in the environment. They apply the theory of *reaction kinetics*, the *equilibrium law*, and *thermochemistry* to explain the changes they observe.

Environmental chemists use models to summarise their findings and to make predictions. The models show how the various elements cycle through the environment (see *carbon cycle* and *nitrogen cycle*). The models show the main reservoirs for an element and the size of the flows of an element from one reservoir to another.

Knowledge of the scale of natural changes shows whether or not human activity is likely to disturb natural changes locally, regionally or globally. So environmental chemistry helps scientists to assess the seriousness of various types of pollution (see *acid rain, incineration, landfill, ozone, greenhouse effect, steady state systems* and *waste*).

environmental issues are of growing concern to chemists, the chemical industry and all who use chemicals as scientists have shown the harm which some chemicals can do to living things. Analytical chemists have contributed greatly to our knowledge of what happens to chemicals in the environment by developing very sensitive analytical techniques, such as chromatography and spectroscopy, to measure minute traces of substances which were once undetectable.

A challenge for chemists in recent years has been to redesign chemical processes to reduce both the energy required and the waste by-products formed. This is both more profitable and less damaging to the environment.

enzymes are *protein* molecules which are the *catalysts* for biochemical reactions that would otherwise be very slow. Saliva contains the enzyme amylase which aids digestion by speeding up the *hydrolysis* of starch to *sugar*. Catalase is an enzyme which protects living cells from a build up of hydrogen peroxide by speeding up its decomposition to water and oxygen. Some washing powders contains proteases – enzymes which break down the proteins such as blood on dirty clothes.

Enzymes are highly specific. Each enzyme catalyses a particular reaction. An enzyme has an *active site* which is just the right shape and size for the *substrate* molecules. Each enzymes works best at a particular temperature and pH and is less effective under other conditions. (See also *inhibitor.*)

eutectic mixture: the mixture with the lowest freezing point of all the possible mixtures. Metals can form eutectic mixtures as can solutions of salts such as sodium nitrate in water.

For pairs of metals, such as tin and lead, the eutectic mixture is the *alloy* with the lowest melting point of all the possible alloys. An alloy with the eutectic composition freezes (or melts) at one temperature like a pure metal. The melting point of the eutectic, however, is lower than the melting points of the pure metals.

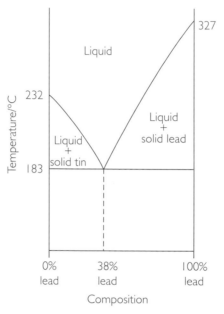

Phase diagram for mixtures of tin and lead

The eutectic mixture of tin and lead melts at 183°C. This low temperature makes the eutectic useful as solder.

eutrophication happens when the water in rivers or lakes is enriched by fertilisers from farmland or by nutrients from sewage and this makes it possible for algae to multiply rapidly. Thick layers of algae block out the light from plants growing below the surface so that they cannot produce oxygen as fast as usual. Then bacteria start to break down the mass of algae using up the remaining oxygen in the water. Other organisms such as fish die because they are starved of oxygen.

evaporation happens at the surface of a liquid as it turns to a gas. This is an example of a *change of state*. As a liquid such as water evaporates, the faster moving molecules near the surface escape from the pull of intermolecular forces and break away to become a vapour. Liquids tend to cool as they evaporate because the faster moving molecules break away so that the average kinetic energy of the remaining molecules falls.

Evaporation is an *endothermic* process. Energy must enter from the surroundings to keep a substance at a constant temperature as it evaporates. So liquids feel cold as they evaporate on the skin.

excited state: the state of an atom or molecule when one or more of its electrons is raised to a higher energy above the stable *ground state*. Heating, electricity or electromagnetic radiation can provide the energy to excite atoms or molecules. Energy is released back to the surroundings when the electrons fall back to the stable energy levels of the ground state. This happens during *flame tests*. Heat from a Bunsen flame excites the electrons in metal atoms from the sample. As the electrons drop back to the lower energy levels they give off radiation with a wavelength determined by the size of the energy jump (see *quantum theory*).

exothermic reactions give out energy to their surroundings. Freezing and condensing are exothermic *changes of state.*

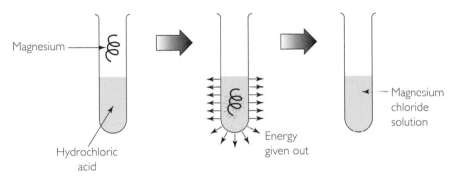

Magnesium

Hydrochloric
acid

Energy
given out

Magnesium
chloride
solution

Diagrams to illustrate an exothermic reaction

For an exothermic reaction the enthalpy change, ΔH, is negative. During an exothermic reaction more energy is given out as new bonds form in the products than is needed to break bonds in the reactants.

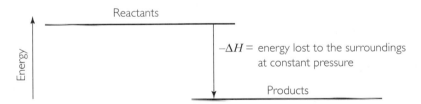

Reactants

Energy

$-\Delta H =$ energy lost to the surroundings
at constant pressure

Products

Energy level diagram for an exothermic reaction

Most *spontaneous* reactions are exothermic but there are exceptions so the *free energy change* (or total *entropy change*) must be examined to make reliable predictions about the direction and extent of change.

explosions are reactions which go with a bang. Gunpowder explodes if ignited in a confined space (see *fireworks*). The explosion is caused by the rapid build up of pressure as the reaction produces a large volume of hot gas from a small volume of solid.

The 'pop' during the test-tube test for hydrogen is a small explosion. This reaction explodes because it is a *free radical chain reaction* with a step in which one radical reacts to produce two radicals which can then both react. The branching chain reaction produces more and more reactive free radicals so the reaction goes faster and faster.

Explosives are compounds or mixtures of chemicals designed to produce explosions. Explosions are used in quarrying, mining, road building and to demolish old buildings. They are also used in warfare and as rocket propellants.

The Nobel prize was endowed with the wealth of the Swedish chemist Alfred Nobel (1833–96) who discovered that the dangerous explosive nitroglycerine (propane-1,2,3-triol trinitrate) could be used safely if absorbed in the mineral kieselguhr. He sold this product as 'dynamite'. He also developed gun cotton (cellulose nitrate) and gelignite (nitroglycerine in nitro-cellulose).

High explosives only detonate if set off by a sudden shock which is usually supplied by a detonator. Mercury fulminate can be used as a detonator because it explodes instantly when ignited.

extrusion is a process for shaping a metal, such as *aluminium*, or a *plastic*, such as pvc, by forcing it through a shaped nozzle (die). Cooks use extrusion in a similar way when decorating a cake with coloured icing sugar or when mincing meat. A plastics extruder works by melting pellets of the polymer and then using a rotating screw to push the molten material through a die.

face-centered cubic structure is one of the two *close packed* crystal structures of metals. Metals with the face-centered cubic structure are calcium, aluminium, cobalt, nickel and copper.

Fajan's rules are a guide which helps to predict the extent to which the bonding in a compound will be ionic, covalent or intermediate between ionic and covalent. Kasimir Fajan (1887–1975) was a physical chemist who noted that ionic bonding is favoured if:

- the charges on the ions are small (1+ or 2+, 1– or 2–)
- the radius of the positive ion is large and the radius of the negative ion small.

The way to apply Fajan's rules is to start by picturing ionic bonding between two atoms and then to consider the extent to which the positive metal ion will polarise the neighbouring negative ions giving rise to some degree of electron sharing (that is a degree of covalent bonding).

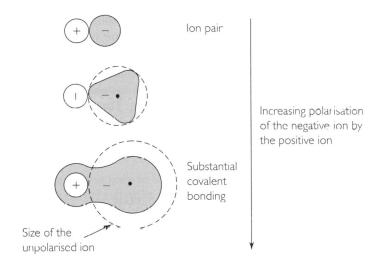

Ionic bonding with increasing degrees of electron sharing because the positive ion has polarised the neighbouring negative ion. Dotted circles show the unpolarised ions.

The smaller a positive ion and the larger its charge the greater the extent to which is tends to polarise a negative ion. So polarising power increases along the series: Na^+, Mg^{2+}, Al^{3+}, Si^{4+}. Sodium chloride is an ionic, crystalline solid. The bonding in anhydrous *aluminium chloride* is largely covalent. Silicon chloride is a covalent, molecular liquid.

The larger the negative ion and the larger its charge the more *polarisable* it becomes. So iodide ions are more polarisable than fluoride ions. Fluorine, which forms the

ferromagnetism: the property of materials which can be used to make permanent magnets. The three ferromagnetic metals are iron, cobalt and nickel. Alnico is a magnetic alloy of these three metals.

fertilisers supply plants with the mineral salts they need for growth. Plants need three major nutrients: nitrogen, phosphorus and potassium which must be available in the soil in a soluble form so that they can be taken up by roots.

Manure and compost are traditional fertilisers used in 'organic farming'. They release nutrients slowly as they rot down through the action of bacteria and fungi in the soil.

Intensive agriculture relies on fertilisers made from minerals and the air. 'Straight fertilisers' contain one of the three elements nitrogen (N), phosphorus (P) or potassium (K). 'Compound fertilisers' contain two or more of these elements.

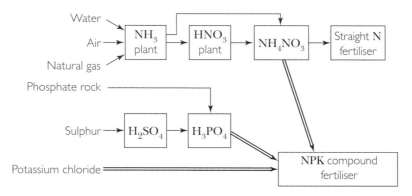

Flow diagram for manufacture of nitrogen and NPK compound fertilisers

Organic fertilisers, such as *urea,* have advantages over inorganic nitrates and ammonium salts. Organic fertilisers:

- release nitrogen more slowly by hydrolysis in the soil
- do not affect the pH of the soil to the same extent when spread on fields.

The use of manufactured fertilisers greatly increases crop yields. However the use of large amounts of fertiliser increases the risk of nutrients leaching from the soil into the rivers, lakes and groundwater. This can lead to *eutrophication.*

fibres are long thin threads which make up materials such as asbestos, wool and polyester fabrics. Most of the fibres used for textiles consist of natural or synthetic polymers. Wool, silk and hair are natural fibres made of *protein.* Cotton, linen and kapok are plant fibres made of the *carbohydrate* cellulose.

Semi-synthetic fibres are made by modifying cellulose. Heating cellulose from wood with sodium hydroxide and then carbon disulfide produces a solution called 'viscose'. Forcing this solution through very small holes in a spinneret produces fine filaments. The filaments turn back into cellulose fibres as they flow though a tank of sulfuric acid. Cloth made from these fibres is known as rayon.

Treating cellulose from wood or cotton with ethanoic anhydride produces cellulose acetate. Cellulose acetate fibres are made by spinning a solution of the polymer in propanone. Fibres form as the solvent evaporates

Examples of synthetic polymers used to make fibres are *polyester, polyamides* and the *addition polymer,* polypropylene. Synthetic polymers are spun into fibres by forcing the hot molten polymer through a spinneret. Fibres form as the liquid cools and solidifies.

Fibres of inorganic materials are increasingly important for making *composite* materials. Examples are the fibres of glass and graphite.

fibre reactive dyes are strongly bound to the fabrics they dye because they react with the molecules in the fibres, forming covalent bonds. They are fast *dyes* which do not fade during washing.

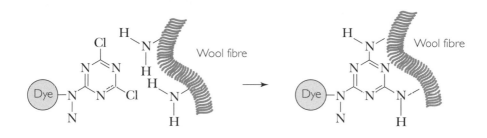

Structure of a fibre reactive dye which forms covalent bonds with amino groups in the *protein* molecules of wool

fillers are materials mixed with *plastics* or *rubbers* to modify their properties. Fillers can make plastics stronger, tougher, harder or easier to mould. Fillers are also used to cut costs by bulking out plastics with something cheaper.

filtration separates an insoluble solid from a liquid. The liquid which passes through the filter is the filtrate. The solid retained on the paper is the residue.

Filtering through a funnel with a filter paper folded into a cone can be very slow. For efficient filtration it is important that the paper is folded carefully and fitted accurately into the funnel. The paper is then wetted with water, or other solvent, before starting to filter. The funnel should never be filled to the top of the paper.

One way of filtering faster is to fold the circle of paper more times to produce a fluted filter paper so that the whole surface is available for filtration.

An even quicker method is to use a *Buchner flask and funnel.*

The use of a small plug of mineral wool or cotton wool in a small funnel cuts losses when filtering off a drying agent from an organic liquid before final distillation.

fire extinguishers put out fires by excluding air or by cooling to slow down burning. The choice of fire extinguisher depends on the type of fire. Water is the cheapest way to put out burning wood, paper or rubbish. Water cools the fire as it evaporates and produces steam which helps to exclude air. Water is dangerous on electrical fires and useless on burning hydrocarbons or metals. Burning oils float on water and continues burning so water just spreads the fire. Burning metals react vigorously with water. Alternative fire extinguishers use carbon dioxide gas or non flammable liquids such as halons and dry powders.

fireworks make *redox reactions* entertaining. The main active ingredient of any firework is gunpowder. Gunpowder is a mixture of two *fuels*, carbon and sulfur, with an *oxidant*, potassium nitrate. Gunpowder burns rapidly because the fuels are finely powdered and well mixed with the source of oxygen. Gunpowder explodes with a bang when confined because, as the fuels burn they produce a large volume of the gases carbon dioxide and sulfur dioxide (see *explosions*). One cubic centimetre of gunpowder burns fast to make about 300 cm³ of hot gas. The blue fuse which sets off fireworks is paper impregnated with potassium nitrate so that it will burn even when it is windy or damp.

The colours in fireworks come from tiny pellets of salts coated with gunpowder. The colours are the same as in flame tests. Sodium for yellow, copper for blue-green and strontium for red.

first law of thermodynamics: see *thermodynmamics (laws of)*.

first order reaction: a reaction is first order with respect to a reactant if the rate of reaction is proportional to the concentration of that reactant. The concentration term for this reactant is raised to the power one in the *rate equation*.

$$\text{Rate} = k\,[X]^1 = k\,[X]$$

Variation of concentration of a reactant plotted against time for a first order reaction The gradient of this graph at any point measures the rate of reaction. The *half-life* for a first order reaction is a constant so it is the same wherever it is read off the curve. It is independent of the initial concentration.

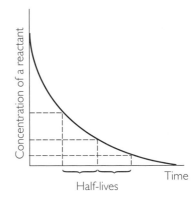

Variation of reaction rate with concentration for a first order reaction. The graph is a straight line through the origin showing that the rate is proportional to the concentration of the reactant.

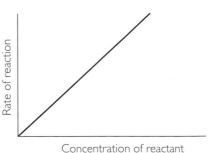

The rate of reaction of 2-bromo-2-methylpropane with hydroxide ions is first order with respect to the halogenoalkane but *zero order* with respect to hydroxide ions.

$$\text{Rate} = k[(CH_3)_3CBr]$$

fission means splitting. The word is used in two contexts in chemistry: bond breaking (*homolytic* and *heterolytic fission*) and the splitting of atomic nuclei (nuclear fission).

flame tests help to detect some metal ions in salts. An element gives a coloured flame if its *atomic spectrum* includes lines in the visible region of the spectrum.

Flame tests are particularly useful in qualitative analysis to distinguish metal ions which otherwise behave in a similar way. Both magnesium and calcium ions, for example, are precipitated by sodium hydroxide to give white precipitates which are insoluble in excess of the alkali. Calcium compounds give an orange-red flame but magnesium compounds do not colour a flame.

Metal ion	Colour
lithium	bright red
sodium	bright yellow
potassium	pale mauve
calcium	orange–red
strontium	scarlet
barium	yellowish–green
copper(II)	blue–green

flammable substances are hazardous because they easily catch fire in air. Be warned that the word 'inflammable' is still quite commonly used for flammable substances. Inflammable does not mean non-flammable. All flammable substances carry the same hazard warning sign but they are classified according to their degree of flammability:

- **flammable** – a liquid with a *flash point* which is equal or below 55°C but above 21°C.
- **extremely flammable** – a liquid with a flash point below 0°C and a boiling point less than or equal to 35°C.

Substances are labelled as **highly flammable** if they:

- may spontaneously catch fire in air
- are liquids with a flash point below 21°C but above 0°C
- are solids which may catch fire and keep burning after brief contact with a flame
- are gases which burn in air if ignited at normal pressure
- react to form flammable gases (such as hydrogen) in contact with water or water vapour.

flash point: the lowest temperature at which a small flame can ignite the *vapour* from a volatile liquid or solid.

flavourings are mixtures of chemicals which give food, confectionary, toothpaste and other products their flavours. Many natural fruit flavours are subtle blends of *esters* and other chemicals. Chemists can imitate fruit flavours by mixing esters.

The *stereochemistry* of a molecule can affect its flavour.

float glass process: the industrial process for manufacturing large, uniform and smooth sheets of *glass*. Molten glass from a furnace at 1500°C pours onto a bath of molten tin where it stays liquid long enough for both surfaces to become flat and parallel. The glass cools and solidifies as it moves over the bath of tin. Once the glass is solid at about 600°C it can move onto rollers and gradually travel through an oven where it cools slowly to room temperature. This is *annealing*.

fluids are materials which flow. Liquids or gases are fluids because the atoms or molecules are not held in fixed positions but are free to move around.

fluoridation involves adding traces of fluoride compounds to water supplies to prevent tooth decay. The benefits were discovered in areas where water supplies naturally contain fluoride ions. Adding fluoride ions to drinking water is controversial and opposed by people who think that any kind of enforced treatment is wrong. Others point to possible harmful effects because fluoride ions are *toxic* at higher concentrations.

fluorine (F) is a pale yellow gas made up of F_2 molecules. It is the most reactive of the halogens in *group 7* of the periodic table. Its electron configuration is [He] $3s^23p^5$.

Fluorine is the most *electronegative* of all elements so it forms ionic compounds with metals. Fluorine is the most powerful *oxidising agent* and its oxidation state is –1 in all its compounds. It oxidises other elements to their highest positive oxidation state. Sulfur, for example, forms SF_6.

Uses of fluorine include the manufacture of a wide range of compounds consisting of only carbon and fluorine (see *fluorocarbons*). The most familiar of these is the very slippery, non-stick *addition polymer* poly(tetrafluorethene), better known as ptfe.

fluorite structure: the cubic crystal structure of the ionic compound calcium fluoride, CaF_2. Each positive ion is surrounded by eight nearest neighbours at the corners of a cube and each negative ion is surrounded by four positive ions. So the *co-ordination number*s are 8 for the positive ion and 4 for the negative ion.

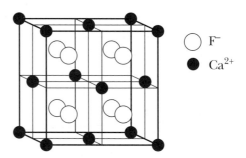

Structure of calcium fluoride showing 8:4 co-ordination. The structure consists of a face-centered cubic array of positive ions with negative ions in the tetrahedral holes

○ F^-

● Ca^{2+}

Other compounds with this structure are the fluorides of Sr, Ba, Cd, Pb and Hg.

fluorescent substances can absorb energy from ultraviolet (or other) radiation and immediately re-emit the energy as visible light. The inside surfaces of fluorescent lamp bulbs or tubes are coated with compounds such as magnesium tungstate and zinc silicate. These substances fluoresce when irradiated by the ultraviolet light from the mercury vapour in the lamp.

The optical brighteners in *detergents* are fluorescent. They absorb ultraviolet light from the Sun and re-emit the energy as blue light. This compensates for any yellowing of the fabric with age.

Fluorescence is an example of *luminescence.*

fluorocarbons are compounds of fluorine and carbon only. They are very *inert*, thermally stable and non-flammable. They are also non-toxic. They are electrical and thermal insulators. They are used during the manufacture and testing of electronic components and also as refrigerants, coolants and lubricants. One fluorocarbon is a good solvent for oxygen and, in emergencies, can act as 'artificial blood'.

foams consist of gases finely dispersed in liquids or solids. Foams are *colloids*. The mass of bubbles in a liquid foam holds the liquid and gas in place. This is useful in foam fire extinguishers which can blanket a fire with carbon dioxide held by the foam. It is also useful in shaving foam which retains the liquid soap on the skin.

Solid foams can be rigid but have a low density. Expanded polystyrene and foamed polyurethanes are examples. These materials are excellent thermal insulators because they trap a large volume of gas and gases are very poor thermal conductors.

Foams improve the texture of food including bread, meringues and cakes. In bread the gas bubbles form by *fermentation*, in cakes they are produced by heating baking powder and in meringues by whipping air into egg white before cooking.

food additives are natural or synthetic chemicals added to food, which can act as:

- colours to make food look more attractive
- preservatives to prevent the growth of micro-organisms
- anti-oxidants to stop oxygen from the air making food unfit to eat – oxidation, for example, turns fats and oils rancid
- *emulsifiers* and stabilisers to control food texture
- sweeteners to improve the taste (see *sweetness*).

forensic science is the application of scientific techniques to gather evidence for a court of law. Forensic scientists make increasing use of sophisticated analytical techniques such as *mass spectrometry*, *chromatography* and various forms of *spectroscopy*.

formula (plural formulae): used by chemists to represent the composition and structure of elements and compounds. (See *empirical formula, molecular formula, structural formula, skeletal formulae* and *displayed formula*.)

formula mass: see *formula unit*.

formula unit: the symbols used in equations and calculations to represent elements and compounds with *giant structures*. Examples are:

- sodium chloride, a giant structure of ions, formula unit NaCl
- silicon dioxide, a giant covalent structure, formula unit SiO_2.

The molar mass of a substance with a giant structure is the relative mass of its formula unit. For silicon dioxide (SiO_2) the molar mass = $[28 + (2 \times 16)]$ = 60 g mol^{-1}. This is sometimes called the formula mass.

fossil fuels are non-renewable energy resources which we now burn to utilise the energy from the Sun which was stored up millions of years ago by *photosynthesis*. The fossil fuels are coal, oil and natural gas.

Fossil fuels contribute to the formation of *acid rain* when they burn. The fuels contain varying amounts of sulfur compounds. The proportion of sulfur compounds in crude oil, for example, varies from less than 1% up to 7%, depending on its origin. When crude oil is distilled, sulfur compounds tend to be concentrated in the heavy fractions such as fuel oils. Oil refining removes much of this *sulfur* and there is very little in petrol. Sulfur compounds burn to produce sulfur dioxide. Nitrogen oxides form by direct combination of nitrogen with oxygen in the air as fuels burn at high temperatures in furnaces or engine cylinders.

fractional distillation is a method for separating mixtures of liquids with different boiling points. On a laboratory scale, the process takes place with a distillation apparatus fitted with a glass fractionating column fitted between the flask and the still-head. Separation is improved if the column is packed with inert glass beads or rings to increase the surface area where rising vapour can mix with condensed liquid running back to the flask. The column is hotter at the bottom and cooler at the top. The thermometer reads the boiling temperature of the compound passing over into the condenser.

Suppose the flask contains a mixture of two liquids. The boiling liquid in the flask produces a vapour which is richer in the more volatile of the liquids (the one with the lower boiling point).

Most of the vapour condenses in the column and runs back. As it does so it meets more of the rising vapour. Some of the vapour condenses. Some of the liquid evaporates. In this way the mixture repeatedly evaporates and condenses as it rises up the column. It is like carrying out a series of simple distillations. This can be represented as a series of steps in the boiling-point composition diagram.

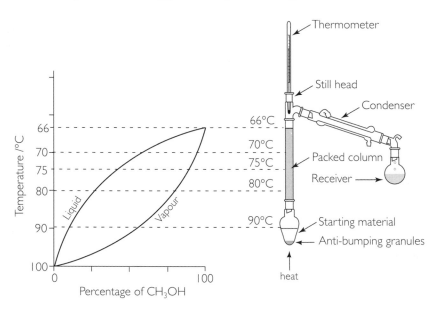

Boiling-point composition diagram for mixtures of methanol and water. Methanol boils at a lower temperature so it the more *volatile substance*. Boiling any mixture of the liquids produces a vapour which is richer in methanol.

fractional distillation of oil is a large-scale, *continuous process* for separating crude oil into fractions. A furnace heats the oil which then flows into a fractionating tower containing about 40 'trays' pierced with small holes. Condensed vapour flows over the trays and runs down into the tray below. Rising vapour mixes with liquid on a tray as it bubbles up through the holes.

The column is hotter at the bottom and cooler at the top. Rising vapour condenses when it reaches the tray with liquid at a temperature below its boiling point. Condensing vapour releases energy which heats the liquid in the tray which in turn evaporates the more volatile compounds in the mixture.

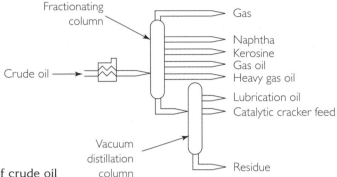

Fractional distillation of crude oil

With a series of trays the outcome is that the *hydrocarbons* with small molecules rise to the top of the column while larger molecules stay at the bottom. Fractions are drawn off from the column at various levels.

Some components of crude oil have boiling points too high for them to vaporise at atmospheric pressure. Lowering the pressure in a *vacuum distillation* column reduces the boiling points of the hydrocarbons and makes it possible to separate them.

fragmentation patterns: see *mass spectrometry*.

free energy change, ΔG: the thermochemical quantity used by chemists to decide whether a reaction tends to go and how far it will go. ΔG is the test for the *feasibility* of a reaction. If ΔG is negative the reaction is feasible.

The advantage of ΔG values for chemists is that tables of standard free energies of formation can be used to calculate the standard free energy change for any reaction. The calculations follow exactly the same steps as the calculations to calculate standard *enthalpy changes for reactions* from standard *enthalpy changes of formation*.

The idea of *entropy* underlies the quantity 'free energy'. Any change tends to happen if the total entropy change is positive. Working with entropy values is generally less convenient because of the need to consider the entropy changes in the surroundings as well as in the reaction mixture. This was the reason why Willard Gibbs suggested the concept 'free energy', which he defined by: $\Delta G = -T\Delta S_{total}$

$$\Delta S_{total} = \Delta S_{system} + \Delta S_{surroundings}$$

For a change at constant temperature and pressure:

$$\Delta S_{surroundings} = \frac{-\Delta H}{T}$$

and so: $\Delta S_{total} = \Delta S_{system} + \dfrac{-\Delta H}{T}$

Hence $-T\Delta S_{total} = -T\Delta S_{system} + \Delta H$

From Gibb's definition this becomes: $\Delta G = \Delta H - T\Delta S_{system}$

Often the $T\Delta S_{system}$ term is relatively small compared to the enthalpy change so that:

$\Delta G \approx \Delta H$.

This is the reason that chemists often use ΔH values to decide whether or not a reaction will tend to go. The approximation becomes less justified at higher temperatures when T is bigger and so $T\Delta S_{system}$ is bigger.

free radicals are reactive particles with unpaired electrons. Free radicals form when covalent bonds break in the way which leaves one electron on each of the atoms joined by the bond. This is *homolytic fission*. The symbol for a free radical generally shows the unpaired electron as a dot. Other paired electrons in the outer shells are generally not shown.

Formation of free radicals
$$Br \overset{\frown}{\underset{\smile}{:}} Br \longrightarrow Br^{\bullet} + {}^{\bullet}Br$$

Free radicals are *intermediates in reactions* taking place:

- in the gas phase at high temperature or in ultraviolet light
- in a non-polar solvent, either when irradiated by ultraviolet light or with an initiator.

Examples of free radical processes include the *cracking* of hydrocarbons, burning of *petrol* in an engine cylinder and the formation and destruction of the *ozone* layer. (See also *free radical chain reactions*.)

free-radical chain reactions involve three stages:

- initiation – the step which produces free radicals
- propagation – steps giving products and more free radicals
- termination – steps which remove free radicals by turning them into molecules.

Free-radical substitution – the reaction of an alkane with bromine in sunlight in a free-radical chain reaction. The main products are bromomethane and hydrogen bromide. The presence of some ethane in the mixture of products is evidence for the termination step.

Initiation: $\quad Br-Br \longrightarrow Br^{\bullet} + Br^{\bullet}$

Propagation: $\quad CH_4 + Br^{\bullet} \longrightarrow CH_3^{\bullet} + HBr$

$\qquad\qquad\quad CH_3^{\bullet} + Br_2 \longrightarrow CH_3Br + Br^{\bullet}$

Termination: $\quad CH_3^{\bullet} + CH_3^{\bullet} \longrightarrow CH_3CH_3$

$\qquad\qquad\quad CH_3^{\bullet} + Br^{\bullet} \longrightarrow CH_3Br$

Free-radical addition polymerisation – free-radical chains reactions can also be used to make *addition polymers* from alkenes and from epoxy compounds.

The initiator can be an organic *peroxide* such as benzoyl peroxide, which acts as a source of free radicals, $R-O^{\bullet}$.

$$R-O \overset{\frown}{\underset{\smile}{\Big/}} O-R \longrightarrow 2R-O^{\bullet}$$

This leads to the propagation steps which are repeated many times, producing a very long chain molecule.

$$2R — O\bullet \quad \underset{\substack{H \\ H}}{\overset{\substack{H \\ H}}{C = C}} \quad \longrightarrow \quad R — O — \overset{\substack{H \\ | \\ H}}{\underset{}{C}} — \overset{\substack{H \\ | \\ H}}{\underset{}{C}}\bullet$$

Termination takes place when two free radicals combine to make a molecule.

$$R — O — (CH_2)_p — CH_2 — CH_2\bullet \quad \bullet CH_2 — CH_2 — (CH_2)_q — O — R$$

$$\downarrow$$

$$R — O — (CH_2)_p — CH_2 — CH_2 — CH_2 — CH_2 — (CH_2)_q — O — R$$

freezing point: the temperature at which a liquid turns to a solid. A pure liquid has a sharp freezing point which is the same as its melting point. A *phase* diagram for the liquid shows how the freezing point varies with pressure.

Dissolving a solute in a liquid lowers its freezing point. Adding antifreeze to the water in an engine lowers the freezing point and so prevents the coolant freezing in winter. When there is a threat of ice, the highway authorities scatter salt on roads because a mixture of salt and water freezes at temperatures well below 0°C

frequency of electromagnetic radiation: the number of complete waves passing any point per second. The SI unit of frequency, v, is the hertz (Hz). Frequencies of *electromagnetic radiation* vary from about 10^5 Hz for radiowaves up to 10^{20} Hz for gamma rays.

All electromagnetic radiation travels at the same speed, c, in a vacuum. The *frequency, wavelength, λ,* and speed are simply related by: $c = v\lambda$.

Quantum theory shows that the higher the frequency the higher the energy of the quanta (photons) of radiation.

Friedel–Crafts reaction: a method for forming $C — C$ bonds to build up the carbon skeleton by adding a side chain to an *arene* such as benzene. In a Friedel–Crafts reaction a *halogenoalkane* or an *acyl chloride* undergoes an *electrophilic substitution* reaction with an arene. The reaction takes place in the presence of a catalyst such as aluminium chloride (a *Lewis acid*).

One example is the substitution of a group with three carbon atoms in place of a hydrogen atom in the benzene ring, using 2-chloropropane.

$$\bigcirc \quad + \quad \underset{\substack{| \\ Cl}}{CH_3CHCH_3} \quad \xrightarrow[\text{catalyst}]{AlCl_3} \quad \bigcirc\overset{\substack{CH_3 \\ | \\ CHCH_3}}{} \quad + \quad HCl$$

Friedel–Crafts alkylation with a chloroalkane. Benzene reacting with
2-chloropropane.

The reaction with an acyl chloride produces a *ketone*.

Friedel–Crafts acylation with an acyl chloride. Benzene reacting with ethanoyl chloride.

This important type of reaction was discovered and developed jointly by the French organic chemist Charles Friedel (1832–1899) and the American, James Crafts (1839–1917). The reaction is used both on a laboratory and on an industrial scale. A Friedel–Crafts reaction to make ethyl benzene is the first step in the production of the *addition polymer* poly(phenylethene), more often called polystyrene.

The purpose of the of the aluminium chloride is to produce the electrophile which attacks the benzene ring.

The mechanism of Friedel–Crafts alkylation

froth flotation is a process for separating valuable mineral from a rock made up of several minerals. The rock is finely crushed to separate the minerals and then stirred with a solution of *surfactants* as a stream of air bubbles rises up through the mixture. The surfactants are selected to make sure that the particles of metal *ore* are caught up by the bubbles and rise to the surface where they can be skimmed off. The unwanted minerals (or gangue) sink and flow away as a stream of waste.

fuel cell: an electrochemical cell which is continuously supplied with fuel and oxidising agent. A fuel cell produces electric power from a fuel, directly, without having to burn it and then use the energy to drive a turbine and spin a generator. In a power station about 70% of the energy released when fuels burn is wasted. A fuel cell running on hydrogen or methanol and oxygen provides a much more efficient source of electricity.

Hydrogen–oxygen fuel cells with strong alkali as the electrolyte are used in the space shuttle. In proton-exchange membrane cells the electrolyte is a polymer which can conduct hydrogen ions in one direction from one side to the other. In this type of cell the electrodes are made of conducting carbon cloth impregnated with platinum, which acts as the catalyst.

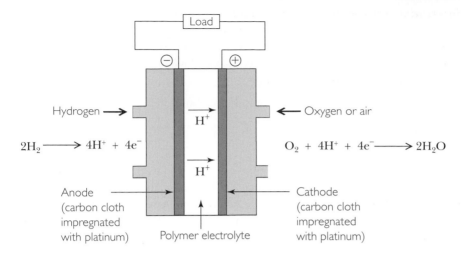

Schematic diagram of a hydrogen–oxygen fuel cell. The cell continues to supply electric power so long as it is supplied with the hydrogen fuel and with oxygen.

fuels burn in air or oxygen to release energy. Power stations often use the *combustion* of *fossil fuels* to raise steam and generate electricity. Most of the fuels for transport are made from fractions produced by the *fractional distillation of crude oil*. There is growing interest in the use of *biofuels*.

Petrol is a blend of *hydrocarbons* based on the gasoline fraction (hydrocarbons with 5–10 carbon atoms). Jet fuel is produced from the kerosene, or paraffin, fraction (hydrocarbons with 10–16 carbon atoms). Fuel for diesel engines is made from diesel oil (hydrocarbons with 13–25 carbon atoms).

Fuels have to be refined to remove components which would harm engines or cause air pollution when they burn.

The energy density of a fuel is the amount of energy available from a kilogram of fuel. This is especially important for fuels used for transport. The higher the energy density of fuels the smaller the mass of the fuel in the tanks of motor vehicles or aeroplanes.

The energy density of a fuel can be calculated given its standard enthalpy change of combustion. The energy density of hexane (a hydrocarbon in petrol) is 48 400 kJ kg^{-1}.

Ethanol is a biofuel used on its own, or in gasohol. $\Delta H_c^{\ominus}$[ethanol] = –1367 kJ mol^{-1}.

The molar mass of ethanol, C_2H_5OH = 46 g mol^{-1}.

So burning 46 g (= 4.6×10^{-2} kg) ethanol gives out 1367 kJ energy.

The energy density of ethanol = 1367 kJ ÷ 4.6×10^{-2} kg = 29 700 kJ kg^{-1}.

full equation: see *chemical equations*.

fullerenes are molecular *allotropes* of carbon. The best known example is the 'bucky ball', *buckminsterfullerene*, C_{60}. Other fullerenes have formulae C_{28}, C_{32}, C_{50}, and C_{70}.

functional group: the atoms and bonding which give a series of organic compounds its characteristic properties and is responsible for most of the reactions. The hydrocarbon chain which makes up the rest of any organic molecule is generally inert to most common reagents such as acids and alkalis.

Examples of functional groups include the:

- two carbon atoms joined by a double bond in *alkenes*
- — Cl in chloroalkanes
- — OH in *alcohols*
- — NH_2 in *amines*
- — CO_2H in *carboxylic acids.*

fundamental particles are the particles that make up an *atom*. In chemistry the fundamental particles are protons, neutrons and electrons. Physicists with high-energy particle accelerators have shown that protons and neutrons are not fundamental but in turn made up of quarks.

	Relative mass	Charge
proton	1	+1
neutron	1	0
electron	$\frac{1}{1870}$ (negligible)	−1

fusion means either melting or joining. (See *enthalpy change of melting* and *nuclear fusion.*)

gamma radiation (γ) is high-energy *electromagnetic radiation* given off during *radioactive decay*. The new nucleus formed when a radioactive atom emits an *alpha* or *beta particle* is often in an excited state; it gives off gamma rays as it loses energy. Emission of gamma rays does not cause a change in the structure of the nucleus.

The wavelengths of gamma rays are shorter than those of *X-rays*. Gamma radiation is *ionising radiation* which is more penetrating than alpha or beta radiation and is stopped only by several centimetres of lead.

gas: see *kinetic theory of gases*.

gas constant: the constant R in the *ideal gas* equation $PV = nRT$. The value of the constant depends on the units used for pressure and volume. If all quantities are in SI units with the pressure in N m^{-2} (pascals) and volume in cubic metres (m^3), then $R = 8.314\,\text{J K}^{-1}\,\text{mol}^{-1}$.

gas laws: the laws which describe the behaviour of *ideal gases* and are summarised by the ideal gas equation. The gas laws include *Boyle's law*, *Charles' law* and *Avogadro's law*.

gas–liquid chromatography (glc) is a sensitive analytical technique for analysing mixtures of liquids. In a modern gas–liquid apparatus, the stationary phase is a thin film of liquid *adsorbed* on the inside surface of a coiled capillary tube about 30 metres long inside an oven. A sample is injected into the hot column from a syringe. The mobile phase, which is a gas, carries the vapours through the column. The components in the mixture separate as they pass through the column. They are detected as they emerge and the signal from the detector is fed to a chart recorder.

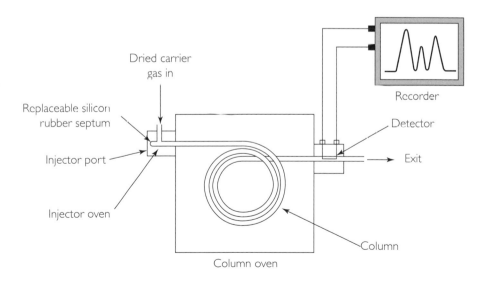

The main features of gas–liquid chromatography

1 mol zinc produces 1 mol hydrogen.

Volume of hydrogen $= 0.03$ mol $\times 24\ 000$ cm^3 mol^{-1} $= 720$ cm^3

Gay-Lussac's law of combining volumes states that when gases are involved in reactions, the ratios of the volumes of gases are simple whole numbers so long as all measurements are taken at the same temperature and pressure. For example 50 cm^3 hydrogen react with 50 cm^3 chlorine to form 100 cm^3 of hydrogen chloride. The ratios are 1:1:2. The law is named after the French chemist Joseph Gay-Lussac (1778–1850).

Avogadro's law accounts for Gay-Lussac's observations. If equal volumes of gases contains equal numbers of molecules, under the same conditions, it follows that the ratios of the volumes are also the ratios of the number of molecules as the equation shows.

$$H_2(g) + Cl_2(g) \longrightarrow 2HCl(g)$$

gel: a *colloid* in which a liquid is finely dispersed in a solid. Table jelly is a typical gel with water loosely held in a network of large gelatine molecules. Warming or even shaking can break up the solid network so that the gel liquefies.

geometrical isomerism: molecules with the same molecular and structural formulae but different shapes (geometries). *Alkenes* and other compounds with $C = C$ *double bonds* may have geometrical isomers because there is no rotation about the double bond. The isomers are labelled *cis* and *trans*. In the *cis* isomer, similar functional groups are on the same side of the double bond. In the *trans* isomer similar functional groups are on opposite sides of the double bond.

cis-but-2-ene   trans-but-2-ene

Geometric isomers of but-2-ene. They are distinct compounds with different melting points, boiling points and densities.

Some inorganic *complexes* also have geometrical isomers.

cis isomer   trans isomer

Cis and trans isomers of the planar complex [Ni(NH$_3$)$_2$Cl$_2$]. The cis form is blue–violet. The trans form is green.

The anti-cancer drug, *cisplatin* is the *cis* isomer of a flat (planar) complex.

germanium (Ge) is an element with some metallic and some non-metallic features. It is a metalloid; shiny like a metal but hard and brittle like a non-metal. It is the element below silicon but above tin in *group 4* of the periodic table with the *electron configuration* $[Ar]3d^{10}4s^24p^2$. Germanium is a *semi-conductor*.

giant structures: crystal structures in which all the atoms or ions are strongly linked by a network of bonds extending throughout the crystal. Substances with giant structures generally have high melting and boiling points.

All *metals* consist of giant structures of atoms held together by *metallic bonding*. Metal giant structures are good conductors of electricity because of the delocalised bonding electrons. The layers of atoms in a metal can slide over each other so that metals are *malleable* and *ductile*.

Ionic compounds have giant structures held together by *ionic bonding*. Ionic compounds do not conduct electricity when solid but they do conduct when the ions are free to move on melting the compound or dissolving it in water.

A few *non-metal* elements consist of giant structures of atoms held together by *covalent bonding*. Examples are carbon (as *graphite* or *diamond*) and silicon. Typically these elements do not conduct electricity because the electrons in covalent bonds are localised between pairs of atoms. Graphite is the exception with electrons *delocalised* over the layers of carbon atoms.

Some compounds of non-metals with non-metals, such as silicon dioxide and boron nitride, have giant structures with covalent bonding. These compounds are non-conductors because each bonding pair of electrons is localised between two atoms.

Many silicates are based on networks of silicon and oxygen atoms as in the *clay minerals*. Chains, sheets and three dimensional arrays of silicate ions combined with metal ions give rise to many different minerals.

Gibbs free energy: see *free energy change*. The concept is named after the US physical chemist, Josiah Gibbs (1839–1903). Hence the symbol ΔG.

glasses are *ceramic* materials which are rigid like solids but which are not crystalline. Glass is made by melting one or more oxides in a furnace. The liquid glass is cooled until thick enough to mould and then shaped and cooled further until it sets solid.

It is possible to make glass from pure silicon dioxide (SiO_2) but it melts at 1700°C. Molten silicon dioxide is a thick, sticky liquid (*viscous*). As the liquid cools back to its melting point, the atoms cannot move freely enough to return to the ordered arrangement of crystalline silica. They become 'frozen' into a disordered state.

Most glass is made from silica mixed with other oxides which melt at a lower temperature than pure silica. Windows, bottles and drinking glasses are made of soda lime glass. Lead glass is used for decorative cut glassware. *Borosilicate glass* is used to make ovenware and laboratory glassware.

glass electrode: an electrode used in combination with a *reference electrode* to make a pH probe for a pH meter. A glass electrode has a thin walled bulb made of special glass which responds to changes in pH. A salt bridge connects the glass electrode to a reference electrode to make a complete electrochemical cell.

global warming is the gradual warming of the Earth's atmosphere which many scientists believe is leading to climate change. The theory is that global warming is the result of an increased concentration of gases such as carbon dioxide in the air which is enhancing the *greenhouse effect*.

Graham's law of diffusion says that the rate of *diffusion* of a gas is *inversely proportional* to the square root of its *molar mass* at constant temperature and pressure.

$$\text{rate of diffusion} \propto \frac{1}{\sqrt{M_r}}$$

This means that a gas with low molar mass such as hydrogen ($M_r = 2$) diffuses faster than a gas with a greater molar mass such as oxygen ($M_r = 32$). According to Graham's law, hydrogen molecules diffuse four times faster than oxygen molecules. A balloon filled with hydrogen deflates much more rapidly than a balloon filled with oxygen.

graphical formula: an alternative name for *displayed formula*.

graphite is one of the *allotropes* of carbon, consisting of a giant structure of atoms. Carbon has by far the highest melting point (over 3800 K) of any element. Graphite has an almost ideal combination of physical properties for use as a high temperature *ceramic* except that it has to be used in a non-oxidising environment. Graphite bonded with clay is used to make crucibles for melting metals. Large blocks of graphite are used as *refractories* to line furnaces.

gravimetric analysis is a method of quantitative analysis for finding the composition and formulae of compounds based on accurate weighing of reactants and products. A familiar example, in school, is to find the formula of magnesium oxide by weighing a piece of magnesium in a crucible and then weighing the magnesium oxide produced after strong heating in air. Chemists have developed sophisticated and precise methods of gravimetric analysis which include *combustion analysis* to find the formulae of organic compounds.

greenhouse effect: the effect of some gases in the air which keeps the surface of the Earth about 30°C warmer that it would be if there were no *atmosphere*. Without the greenhouse effect there would be no life on Earth.

When radiation from the Sun reaches the Earth's atmosphere about 30% is reflected into space, 20% is absorbed by gases in the air and about half reaches the surface of the Earth.

The warm surface radiates energy back into space but at longer, infra-red wavelengths. Some of the infra-red radiation is absorbed and warms up the atmosphere. This is the greenhouse effect.

The gases in the air which absorb infra-red radiation are called greenhouse gases. Nitrogen and oxygen make up most of the air but they are not greenhouse gases. The main natural greenhouse gases are carbon dioxide, methane, dinitrogen oxide and water vapour. *CFCs* are also greenhouse gases.

The concentrations of greenhouse gases in the air are rising because of human activity such as burning *fossil fuels* and agriculture. This is enhancing the greenhouse effect. There is growing evidence that this is responsible for *global warming*.

Grignard reagents are reactive organic compounds containing magnesium atoms which are used to form C — C bonds in organic synthesis. They are examples of *organometallic compounds*. The reagents were discovered by the French organic chemist Victor Grignard (1871–1935).

Grignard showed that *halogenoalkanes* react with magnesium in dry ether (ethoxyethane).

$$RBr + Mg \longrightarrow RMgBr,$$ where R represents an alkyl group.

The C—Mg bond in a Grignard reagent is polarised with the metal atom at the positive end of the dipole and the carbon atom at the negative end. As a result, the carbon atom attached to magnesium is nucleophilic and C—C bonds form when Grignard regents take part in *nucleophilic substitution* or *nucleophilic addition* reactions. This is what makes the reagents so useful in synthesis.

Summary of the reactions of Grignard reagents

ground state: the stable state of an atom or molecule with the electrons in the lowest available energy levels.

group: a vertical column of elements in the *periodic table*. Elements in the same group have similar chemical properties because they have the same outer *electron configuration*.

There are trends in properties down a group as the number of full inner electron shells increases and the atoms get larger. Generally the metallic characteristics of elements increase down a group. In *groups 1 and 2* the metals get more reactive down the group.

Non-metal characteristics decrease down a group. Down *group 7* the halogens get less reactive. *Group 4* shows a trend from non-metals at the top to metals at the bottom with germanium, a *metalloid*, in the middle.

IUPAC now recommends that the groups should be numbered from 1 to 18. Groups 1 and 2 are the same as before. Groups 3 to 12 are the vertical families of *d-block elements*, the groups traditionally numbered 3 to 8 then become groups 13 to 18.

group I elements belong to the family of *alkali metals* with similar chemical properties because they all have one electron in an outer *s-orbital*. These elements are more similar to each other than the elements in any other group. Even so, because of the increasing number of full, inner shells, there are trends in properties down the group from lithium to caesium. The element in period 7, francium, is very rare and all its isotopes are radioactive.

lithium, Li $[He]2s^1$ rubidium, Rb $[Kr]5s^1$

sodium, Na $[Ne]3s^1$ caesium, Cs $[Xe]6s^1$

potassium, K $[Ar]4s^1$

The metals are powerful *reducing agents.* They react by losing the outer *s* electron to form M^+ ions. The first ionisation energies decrease down the group as the increasing number of full shells means that the outer electrons gets further away from the same effective nuclear charge (see *shielding*).

Atomic and ionic radii increase down the group. For each element the 1+ ion is smaller than the atom because of the loss of the outer shell of electrons. The tendency to react and form ions increases down the group.

The small size of the lithium ion means that it has a relatively high *polarising power* and is heavily *hydrated* in solution. As a result lithium is in some ways not typical of the group as a whole. It resembles magnesium in some ways. This is an example of the *diagonal relationship.*

The metals are soft and easily cut with a knife. They are shiny when freshly cut but quickly dull in air as they react with moisture and oxygen. The metals are stored under oil.

All the metals burn brightly in oxygen on heating to form ionic oxides. Lithium forms a simple oxide, Li_2O. Sodium forms mainly the *peroxide*, Na_2O_2. Potassium forms a superoxide, KO_2. The oxides are basic; they react with acids to form salts.

$$Li_2O(s) + 2HCl(aq) \longrightarrow 2LiCl(aq) + H_2O(l)$$

All the metals react with water to form hydroxides and hydrogen. The rate and violence of the reaction increases down the group. Lithium reacts steadily with cold water. Caesium reacts explosively.

All the metals react vigorously with chlorine to form colourless, ionic chlorides, M^+Cl^-, which are soluble in water. The crystal structures depend on the size of the metal ion (see *caesium chloride structure* and *sodium chloride structure*).

The hydroxides are:

- similar in that they all have the formula MOH, and are soluble in water forming alkaline solutions (they are *strong bases*)
- different in that their solubility increases down the group.

The carbonates are similar in that they all have the formula M_2CO_3, and, with the exception of lithium, do not decompose on heating (see *thermal stability of carbonates*). The carbonates of sodium and potassium are soluble, forming alkaline solutions because the carbonate ion is a *base.*

The nitrates are:

- similar in that they all have the formula MNO_3, are colourless crystalline solids, are very soluble in water and decompose on heating,
- different in that they become more difficult to decompose down the group. Lithium nitrate, like magnesium nitrate, decomposes on heating to the oxide, nitrogen dioxide and oxygen. The nitrates of sodium and potassium need strong heating to decompose and they form the nitrite:

$$2KNO_3(s) \longrightarrow 2KNO_2(s) + O_2(g)$$

group 2 elements belong to the family of *alkaline earth metals* with similar chemical properties because they all have two electrons in an outer *s-orbital.*

beryllium, Be	[He]$2s^2$	strontium, Sr	[Kr]$5s^2$
magnesium, Mg	[Ne]$3s^2$	*barium,* Ba	[Xe]$6s^2$
calcium, Ca	[Ar]$4s^2$		

The first member of the group, beryllium, is not a typical member of group 2. Because of the small size of the Be^{2+} ion, its chemistry is in many ways more like the chemistry of aluminium than of magnesium. This is an example of the *diagonal relationship.*

The following similarities and differences refer to the elements Mg, Ca, Sr and Ba. The symbol M is here used to represent any one of these elements.

The metals are *reducing agents* which react by losing their two s electrons to form M^{2+} ions. The first and second *ionisation energies* decrease down the group as the increasing number of full shells means that the outer electrons are further away from the same effective nuclear charge (see *shielding*).

Atomic radii and *ionic radii* increase down the group. For each element the 2+ ion is smaller than the atom because of the loss of the outer shell of electrons. The tendency to react and form ions increases down the group.

The metals are harder and denser than group 1 metals and have higher melting points. In air the surface of the metals is covered with a layer of oxide

All the metals burn brightly in oxygen on heating to form white, ionic oxide, $M^{2+}O^{2-}$. The oxides are basic. They react with acids forming salts.

$$BaO(s) + 2HNO_3(aq) \longrightarrow Ba(NO_3)_2(aq) + H_2O(l)$$

All the metals react with water to form hydroxides and hydrogen or acids to form salts and hydrogen. Magnesium reacts only slowly with hot water. Barium reacts quite fast even with cold water. Barium is so reactive with air and moisture that it is generally stored under oil like the alkali metals.

All the metals react vigorously with chlorine to form colourless, ionic chlorides, MCl_2. Unlike the group 1 chlorides, the chlorides of this group are usually *hydrated.* They are soluble in water.

The hydroxides are:

- similar in that they all have the formula $M(OH)_2$ and are to some degree soluble in water, forming alkaline solutions
- different in that their solubility increases down the group.

The carbonates are:

- similar in that they all have the formula MCO_3, are insoluble in water and decompose on heating: e.g. $CaCO_3(s) \longrightarrow CaO(s) + CO_2(g)$
- different in that they become more difficult to decompose down the group (see the *thermal stability of carbonates*).

The nitrates are:

- similar in that they all have the formula $M(NO_3)_2$, are colourless crystalline solids, are very soluble in water and decompose to the oxide on heating:
  $$2Mg(NO_3)_2 \longrightarrow 2MgO(s) + 4NO_2(g) + O_2(g)$$

Haber process: the process for *ammonia manufacture* developed by the German physical chemist Fritz Haber (1868–1934).

haemoglobin is the red globular *protein* in blood which carries oxygen from the lungs to the cells in body tissues. A haemoglobin molecule consists of four poly*peptide* chains, each with a nitrogen atom forming a dative bond to an Fe^{2+} ion in a haem group. So there are four haem groups in each haemoglobin molecule.

Nitrogen atoms in the porphyrin ring form four more dative bonds to the Fe^{2+} ion, leaving one remaining site on the metal ion which can accept a pair of electrons from an oxygen molecule (in oxyhaemoglobin which is bright red) or a water molecule (in deoxyhaemoglobin which is dull red).

A haem group

The reactions between haem groups and oxygen or water are reversible, allowing haemoglobin to pick up and release oxygen. The reaction with *carbon monoxide* is irreversible, which explains why the gas is dangerously toxic.

half-cell: an electrode dipping into a solution of ions. Two half-cells connected by a salt bridge make up an *electrochemical cell.*

half-equation: an *ionic equation* used to describe either the gain or the loss of electrons during a redox process. Half-equations help to show what is happening during a *redox reaction.* Two half-equations combine to give an overall balanced equation for a redox reaction.

Iron(III) ions can oxidise iodide ions to iodine. This can be shown as two half-equations:

electron gain (reduction): $Fe^{3+}(aq) + e^- \longrightarrow Fe^{2+}(aq)$

electron loss (oxidation): $2I^-(aq) \longrightarrow I_2(s) + 2e^-$

The number of electrons gained must equal the number lost. So the first half-equation must be doubled to arrive at the overall balanced equation:

$$2Fe^{3+}(aq) + 2e^- \longrightarrow 2Fe^{2+}(aq)$$
$$2I^-(aq) \longrightarrow I_2(s) + 2e^-$$
$$\overline{2Fe^{3+}(aq) + 2I^-(aq) \longrightarrow 2Fe^{2+}(aq) + I_2(s)}$$

half-life (radioactive): the time for half the atoms in a sample of a radioactive isotope to decay away. It is also the time for the count rate for alpha or beta particles from a sample to fall by half. Half-lives for radioactive isotopes can be as short as a fraction of a second or as long as millions of years. The half-lives of radioactive isotopes are unaffected by changes in temperature or pressure or the presence of catalysts. The half-life remains the same whether the atoms are in the elemental state or in one of its compounds.

half-life (chemical): chemists measure the half-lives of chemical reactions. The half-life of a reaction is the time for the concentration of one of the reactants to fall by half. Half-lives help to identify first order reactions. At a constant temperature, the half-life of a *first order reaction* is the same wherever it is measured on a concentration/time graph. So the half-life is independent of the initial concentration.

halide ions are the ions of the halogen elements. They include the fluoride, F^-, chloride, Cl^-, bromide, Br^- and iodide, I^-, ions. (See also *silver halides*.)

Warming sodium chloride with concentrated sulfuric acid produces clouds of hydrogen chloride gas. This *acid–base* reaction can be used to make hydrogen chloride.

$$NaCl(s) + H_2SO_4(l) \longrightarrow HCl(g) + NaHSO_4(s)$$

Both sulfuric acid and hydrogen chloride are strong acids. The reaction goes from left to right because the hydrogen chloride is a gas and escapes from the reaction mixture, so the reverse reaction cannot happen.

This type of reaction cannot be used to make hydrogen bromide or hydrogen iodide, because bromide and iodide ions are strong enough reducing agents to reduce sulfur from the +6 state to lower *oxidation states*.

The reactions of halide ions with sulfuric acid show that there is a trend in the strength of the halide ions as *reducing agents*:

- chloride ions do not reduce sulfuric acid at all, so the only gaseous product is hydrogen chloride gas
- bromide ions turn to orange bromine molecules as they reduce H_2SO_4 to SO_2 (mixed with some hydrogen bromide gas)
- iodide ions are the strongest reducing agents, they turn into iodine molecules as they reduce H_2SO_4 to S and H_2S; scarcely any hydrogen iodide forms.

So the trend as reducing agents is: $I^- > Br^- > Cl^-$. In *group* 7, iodine is the weakest oxidising agents so it has the least tendency to form negative ions. Conversely iodide ions are the ones that most readily give up electrons and turn back into iodine molecules. (See also *displacement reactions*.)

hallucinogens: drugs which powerfully affect the brain and cause people to experience hallucinations. Some hallucinogens occur naturally in plants such as the cactus, peyote, which produces mescaline. Other hallucinogens are synthetic chemicals such as LSD and MDMA (the *amphetamine* called ecstasy). There is much that is not understood about the effects of drugs on the brain but it seems that chemicals such as LSD and MDMA affect the nerve cells which release a chemical transmitter called serotonin. Part of the scientific case against the use of MDMA is that there is evidence from studies on animals that it can damage and destroy the nerve fibres of cells that produce serotonin in the brain.

halogenation: see *chlorination, bromination* and *iodination*.

halogens is the family name of the *group 7* elements. The halogens are *fluorine, chlorine, bromine, iodine* and *astatine*. The name halogen means 'salt-former' and is based on the fact that the elements combine with most metals to form salts (*halides*).

halogenoalkanes (or haloalkanes) are compounds formed by replacing one or more of the hydrogen atoms in *alkanes* with halogen atoms.

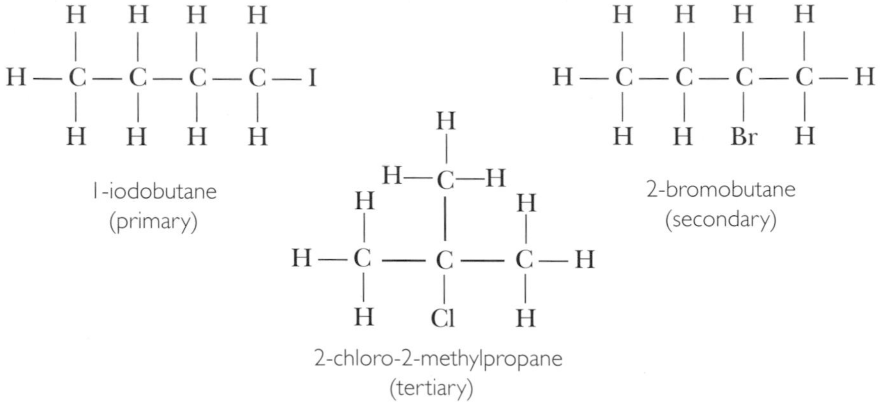

1-iodobutane
(primary)

2-chloro-2-methylpropane
(tertiary)

2-bromobutane
(secondary)

Names and structures of some halogenoalkanes. A primary halogenoalkane has the halogen atom at the end of the chain. A secondary compound has the halogen atom somewhere along the chain but not at the ends. A tertiary halogenoalkane has the halogen atom at a branch in the chain.

Chloromethane, bromomethane and chloroethane are gases at room temperature. Most other halogenoalkanes are colourless liquids which do not mix with water.

Halogen atoms are more electronegative than carbon atoms so a carbon—halogen bond is polar. The characteristic reactions of haloalkanes are *nucleophilic substitution* reactions.

$CH_3CH_2CH_2CH_2Br$

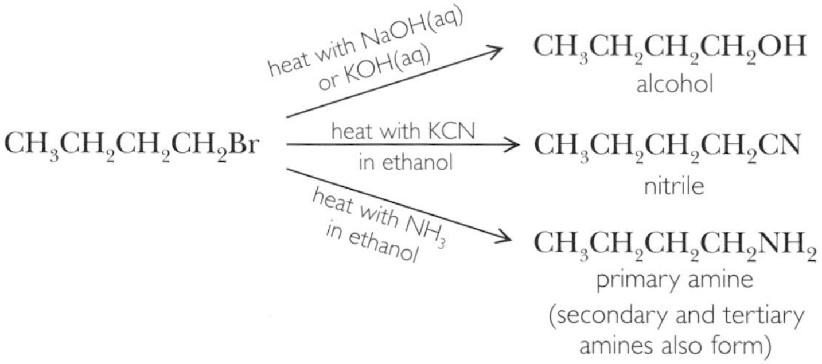

heat with NaOH(aq) or KOH(aq) → $CH_3CH_2CH_2CH_2OH$
alcohol

heat with KCN in ethanol → $CH_3CH_2CH_2CH_2CN$
nitrile

heat with NH$_3$ in ethanol → $CH_3CH_2CH_2CH_2NH_2$
primary amine
(secondary and tertiary amines also form)

Nucleophilic substitution reactions of 1-bromobutane

The *hydrolysis* of halogenoalkanes makes it possible to distinguish chloro-, bromo- and iodo- compounds. Heating the compound with alkali releases halide ions.

Acidifying with nitric acid and then adding silver nitrate produces a precipitate of the *silver halide*.

The rates of reaction of halogenoalkanes are in the order: RI > RBr > RCl where R represents an alkyl group.

The C—I bond is the longest and the weakest (as measured by the mean *bond energy*). The C—Cl bond is the shortest and the strongest. Bond polarity does not appear to be a factor in determining the rates because chlorine is the most electronegative of the elements so the C—Cl bond is the most polar.

An *elimination reaction* happens on warming a halogenoalkane with a solution of potassium hydroxide in ethanol. The reaction goes more readily in compounds with a halogen atom at a branch in the carbon chain.

Halogenoalkanes are important intermediates in the manufacture of other chemicals. They are also used as:

- solvents (e.g. dichloromethane)
- refrigerants (e.g. hydrochlorofluorocarbons, such as $CHClF_2$, which are replacing *CFCs*)
- pesticides (e.g. bromomethane)
- fire extinguishers (e.g. CBr_2ClF).

There are growing restrictions on the uses of many halogenoalkanes because of concern about their hazards to health, their persistence in the environment and their effect on the *ozone layer*.

halogenoarenes are compounds formed by replacing one or more of the hydrogen atoms in an *arenes* with halogen atoms. An example is chlorobenzene which is a colourless liquid at room temperature.

Structure of chlorobenzene

hardness is a property of materials which shows how easy they are to dent or scratch (see *Mohs' scale*). The term is also used in the expression *hard water*.

hard water does not easily lather with *soap* but instead forms a greasy scum. Water is hard if it contains calcium or magnesium ions. Scum is a *precipitate* formed when soap mixes with water containing these ions.

$$Ca^{2+}(aq) + 2C_{17}H_{35}CO_2^{-}(aq) \longrightarrow Ca(C_{17}H_{35}CO_2)_2(s)$$

octadecanoate   insoluble precipitate (scum)
(stearate) ions in soap

Water is temporarily hard if it contains the hydrogencarbonates of calcium or magnesium. This type of hardness is removed by boiling. Boiling reverses the reaction which produced the hard water as rain containing carbon dioxide trickled through limestone, chalk or dolomite.

$$Ca(HCO_3)_2(aq) \underset{\text{formation of hard water}}{\overset{\text{boiling}}{\rightleftharpoons}} CaCO_3(s) + H_2O(l) + CO_2(g)$$

The solid calcium carbonate precipitates as scale which coats heating elements and gradually blocks the pipes in heating systems.

Permanent hardness is not removed by boiling. The mineral gypsum, $CaSO_4$, is slightly soluble in water and makes water permanently hard.

Ion exchange resins and other methods of *water treatment* soften water by removing calcium and magnesium ions.

hazard warning signs identify chemicals which can harm people. The signs have been agreed internationally and they have very specific meanings. Words beside or below the signs indicate the degree of hazard. Very toxic and *toxic substances*, for example, carry the same sign but are distinguished by the words alongside.

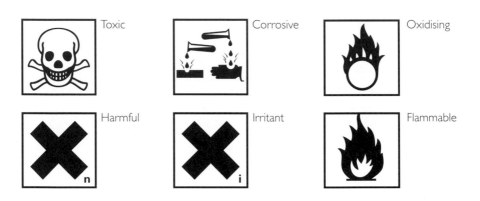

Hazard warning signs for chemicals

heat exchangers are widely used in the chemical industry to transfer energy from a hot liquid or gas to a cooler liquid or gas.

A laboratory water condenser is a simple example of a heat exchanger. The water flowing in the jacket round the inner tube cools the hot vapour from a *distillation* column. The water flows in the opposite direction to the vapour.

An industrial heat exchanger is normally made of steel rather than glass and has many tubes carrying the hotter *fluid* through the surrounding container of cooler fluid.

Heat exchangers are important to the economics of industry. In *sulfuric acid manufacture* by the contact process the reaction of sulfur dioxide with oxygen to make sulfur trioxide is highly exothermic. The gases must be cooled after each pass through the *catalyst* to maximise the yield. Instead of letting the energy go to waste, the hot gases are cooled in heat exchangers in which water turns to steam. The steam is then used to generate electricity. Typically a sulfuric acid plant has no fuel bills because it can generate all the electricity it needs. This helps to make the process commercially viable.

heavy metals are metals such as cadmium, mercury and lead which have relatively high relative atomic masses. The term does not have a precise chemical meaning but crops up in descriptions of pollution caused by *toxic* heavy metal ions.

heavy water is the common name for deuterium oxide, D_2O. This is water in which both the atoms linked to oxygen are the *deuterium* isotope of hydrogen.

helium (He) is the first member of the group of noble gases, *group 8*. Helium is the second most common element in the universe after hydrogen. The energy of stars such as the Sun comes from *nuclear fusion* which turns hydrogen nuclei into helium nuclei. Helium was detected on the Sun by *spectroscopy* during a total eclipse in 1868 over 20 years before it was isolated and identified on Earth.

On Earth one of the main sources of helium is *natural gas*. Some natural gas wells produce up to 7% helium. The gas boils at 4.2 K and is the coldest liquid available to study the properties of materials at low temperatures.

Helium is used to provide an inert atmosphere for welding and for growing crystals of pure *semiconductors* such as germanium and silicon.

Divers breathe a mixture of helium with oxygen. Helium is less soluble in blood than nitrogen and so this mixture reduces the risk of divers suffering from the 'bends' when they come to the surface.

Helium is less dense than air under the same conditions of temperature and pressure. It is a much safer gas than hydrogen for filling airships and weather balloons.

Henderson–Hasselbalch equation: an equation which helps chemists to explain the behaviour of *buffer solutions* and indicators. The equation is derived by rearranging the expression for K_a of a weak acid and then putting both sides of the equation into logarithmic form. The advantage of taking logs is to produce an equation which can be easily used to calculate *pH* values, which are themselves logarithmic.

For a weak acid HA:

$$HA(aq) + H_2O(l) \rightleftharpoons H_3O^+(aq) + A^-(aq)$$

$$K_a = \frac{[H_3O^+(aq)][A^-(aq)]}{[HA(aq)]}$$

This rearranges to give:

$$[H_3O^+(aq)] = \frac{K_a[HA(aq)]}{[A^-(aq)]}$$

Taking logs and substituting pH for $-\lg [H_3O^+(aq)]$ and pK_a for $-\lg K_a$, gives:

$$pH = pK_a + \lg \frac{[A^-(aq)]}{[HA(aq)]} \text{ because } -\lg \frac{[HA(aq)]}{[A^-(aq)]} = +\lg \frac{[A^-(aq)]}{[HA(aq)]}$$

In a mixture of a weak acid and its salt, the weak acid is only slightly ionised while the salt is fully ionised, so it is often accurate enough to assume that all the anions come from the salt present and all the unionised molecules from the acid.

Hence: $pH = pK_a + \lg \frac{[salt]}{[acid]}$, which is the Henderson–Hasselbalch equation.

So in a buffer solution with [acid] = [salt], $pH = pK_a + \lg 1 = pK_a$, since $\lg 1 = 0$.

Diluting a buffer solution does not change [salt]/[acid] so the pH does not change. The equation also shows why it takes a big change in the ratio [salt]:[acid] in a buffer solution to make a significant change in pH. Since $\log 1 = 0$, $\log 10 = 1$ and $\log 100 = 2$, the ratio has to change by a factor of ten to make the pH change by one unit.

The equation also explains why *acid–base indicators* change colour roughly in the range $pH = pK_{in} \pm 1$. When $pH = pK_a$, $[HIn] = [In^-]$ and the two different colours of the indicator are present in equal amounts the indicator is mid-way through its colour change.

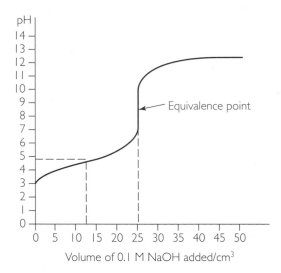

Curve showing the change of pH when a 25 cm³ 0.1 mol dm⁻³ ethanoic acid, a weak acid, is titrated with a strong base. Half way to the end-point half the acid has been converted to its salt, so [acid] = [salt] and pH = pK_a = 4.8

Add a few drops of acid and the pH falls. The characteristic acid colour of the indicator is distinct when $[HIn] \approx 10 \times [In^-]$. At this point pH = p$K_a$ + log 0.1 = pK_a – 1, since log 0.1 = –1

Add a few drops of alkali and the pH rises. The characteristic alkaline colour of the indicator is distinct when $[In^-] \approx 10 \times [HIn]$. At this point pH = p$K_a$ + log 10 = pK_a + 1, since log 10 = +1.

Henry's law describes the effect of changing the pressure on the solubility of a gas in a liquid. The law applies only to gases which are slightly soluble. It does not apply to the solubility in water of gases such as ammonia or hydrogen chloride which react as they dissolve.

The law states that the mass of gas that dissolves in a certain volume of a liquid at constant temperature varies with the *partial pressure* of the gas above the liquid.

The law applies to the solubility of gases such as nitrogen, oxygen and *helium* in water and can be used to explain the dangers faced by divers breathing gas mixtures under pressure.

Henry's law is a special case of the *equilibrium law* and can be described in the following form since the partial pressure of a gas is a measure of its concentration:

$$\frac{\text{concentration of gas in solution at equilibrium}}{\text{concentration of gas above the liquid}} = K \text{ (a constant)}$$

herbicides: see *pesticides*.

Hess's law makes it possible to calculated enthalpy changes which cannot be measured. The law states that the enthalpy change for a reaction is the same whether it takes place in one step or a series of steps. So long as the reactants and the products are the same the overall enthalpy change will be the same whether the reactants are converted to products directly or through two or more intermediate reactions.

Hess's law is used to calculate:

- *enthalpy changes of formation* from enthalpy changes of combustion
- *enthalpy changes of reaction* from enthalpy changes of formation
- lattice energies in *Born–Haber cycles*.

heterocyclic compounds are compounds with rings made of carbon atoms with atoms of at least one other element such as nitrogen, sulfur or oxygen. The bases in *DNA* are heterocyclic compounds. Adenine is an example. Many *alkaloids* are also heterocyclic compounds.

The structure of the heterocyclic compound nicotine or 3-(1-methyl-2-pyrrolidyl)pyridine. Nicotine is present in the leaves of tobacco plants. Pure nicotine, a colourless liquid, is highly toxic. Nicotine from tobacco plants is used as an insecticide but the compound is better known as the addictive drug in tobacco smoke.

nicotine

heterogeneous catalyst: a *catalyst* which is in a different *phase* from the reactants. Generally a heterogeneous catalyst is a solid while the reactants are gases or in solution in a solvent.

The advantage of a heterogeneous catalyst is that it can be held in a reaction vessel as the reactants flow in and the products flow out. There no difficulty in separating the products from the catalyst.

Heterogeneous catalysts are used in large scale continuous processes such as *catalytic cracking*. In the Haber process for *ammonia manufacture* the nitrogen and hydrogen gases flow through a reactor containing small lumps of iron.

Platinum metal alloyed with other metals such as rhodium is an important catalyst. It is used to oxidise ammonia during *nitric acid manufacture*. It is also used in *catalytic converters*. In a catalytic converter the expensive catalyst is finely divided and supported on the surface of a ceramic to increase the surface area in contact with the exhaust gases. Similarly, the expensive silver catalyst used to manufacture *epoxyethane* is finely spread over the surface of alumina.

Impurities in the reactants can poison catalysts so that they become less effective. Carbon monoxide poisons the iron catalyst used in the Haber process. Lead compounds in the exhaust from a car engine poison catalytic converters so lead-free petrol must be used.

Heterogeneous catalysts work by adsorbing reactants at *active sites* on the surface of the solid. Nickel acts as a catalyst for the addition of hydrogen to $C = C$ in unsaturated compounds by adsorbing hydrogen molecules which probably split up into single atoms held on the surface of the metal crystals.

If a metal is to be a good catalyst for a *hydrogenation* reaction it must not adsorb the hydrogen so strongly that the hydrogen atoms become unreactive. This happens with tungsten. Equally, if adsorption is too weak there will not be enough adsorbed atoms tor the reaction to go at a useful rate, as is the case with silver. The strength of adsorption must have a suitable intermediate value. Suitable metals for hydrogenation are nickel, platinum and palladium.

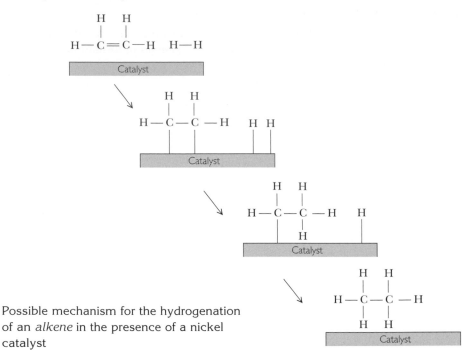

Possible mechanism for the hydrogenation of an *alkene* in the presence of a nickel catalyst

heterogeneous equilibrium: an equilibrium system in which the substances involved are not all in the same *phase*. An example is the equilibrium state involving two solids and a gas, formed on heating calcium carbonate in a closed container.

$$CaCO_3(s) \rightleftharpoons CaO(s) + CO_2(g)$$

The concentrations of solids do not appear in the expression for the equilibrium constant. Pure solids have a constant 'concentration'.

$$K_c = [CO_2(g)] \text{ or } K_p = p_{CO_2}$$

The same applies to the equilibrium between a solid and its saturated solution in water (see *solubility product constant*).

heterolytic bond breaking (fission): the type of bond breaking which produces ionic *intermediates* in reactions. In heterolytic fission, a *covalent bond* breaks so that one of the atoms joined by the bond takes both of the shared pair of electrons while the other is left with none. This type of bond breaking happens during *electrophilic* and *nucleophilic substitution* and *addition reactions* and is favoured when reactions take place in *polar solvents* such as water.

$$H\!:\!\overset{\times\times}{\underset{\times\times}{Br}}\!\!{\times} \longrightarrow H^+ \qquad \overset{\times\times}{\underset{\times\times}{:}Br}{\times}^-$$

$$H \frown Br \longrightarrow H^+ + Br^-$$

Heterolytic bond breaking

hexadentate ligand: a molecule or ion which forms six *co-ordinate* (dative) *bonds* with the central metal ion in a complex. The common example is *edta*. (See also *chelates*.)

hexagonal close packed structure (hcp): see *close packed structures*.

high-performance liquid chromatography (hplc) is a sophisticated version of *liquid chromatography*. The technique is able to separate components in a mixture which are very similar to each other. The mobile phase is a solvent of very high purity. The stationary phase consists of very small particles of a solid such a silica packed into a long steel tube. The use of fine particles increases the surface area helping to make the separation efficient. A pump provides the very high pressure needed to force the solvent through the tightly packed column. One application of hplc is to study what happens to drugs as they are metabolised in the body.

Hofmann degradation is a reaction which turns an *amide* into a primary *amine* while removing a carbon atom from the molecule. The reaction shortens the carbon chain, hence the term 'degradation'. The reaction was discovered by August von Hofmann (1818–92). He was a pioneering German organic chemist who was head of the Royal College of Chemistry in London for 20 years from 1845.

Amides react in this way when treated with bromine and concentrated sodium hydroxide solution.

$$CH_3CH_2CH_2C\!\!\begin{array}{c} O \\ \diagup\diagup \\ \diagdown \\ NH_2 \end{array} \quad \xrightarrow[\text{warm}]{Br_2/KOH} \quad CH_3CH_2CH_2NH_2$$

amide   amine

Equation for the Hofmann degradation of an amide to a primary amine with one less carbon atom

homogeneous catalyst: a *catalyst* which is in the same phase as the reactants. Typically the reactants and the catalyst are dissolved in the same solution.

Transition metal ions can be effective homogeneous catalysts because they can gain and lose electrons as they change from one *oxidation state* to another. The oxidation of iodide ions by persulfate ions, for example, is very slow.

$$S_2O_8^{2-}(aq) + 2I^-(aq) \longrightarrow 2SO_4^{2-}(aq) + I_2(aq)$$

The reaction is catalysed by iron(III) ions in the solution. A possible mechanism is that iron(III) is reduced to iron(II) as it oxidises iodide ions to iodine. The $S_2O_8^{2-}(aq)$ ions oxidise the iron(II) back to iron(III) ready to oxidise some more of the iodide ions and so on.

Acid catalysis and *base catalysis* are other examples of homogeneous catalysis.

homogeneous equilibrium: an equilibrium in which all the substances involved are in the same *phase*.

homologous series: a series of closely related organic compounds. The compounds in a homologous series have the same functional group and can be described by a general formula. The formula of one member of the series differs from the next member by CH_2. Primary *alcohols* are an example of a homologous series with the general formula $C_nH_{(2n+2)}O$. Physical properties, such as the boiling point, show a steady trend in values along a homologous series.

homolytic bond breaking (fission): the type of bond breaking which produces *free radicals* with unpaired electrons. In homolytic fission, a *covalent bond* breaks so that the atoms joined by the bond separate, each taking one of the shared pair of electrons.

$$:\overset{\bullet\bullet}{\underset{\bullet\bullet}{Br}}{:}\overset{\times\times}{\underset{\times\times}{Br}}{\times} \longrightarrow :\overset{\bullet\bullet}{\underset{\bullet\bullet}{Br}}{\bullet} \quad \times\overset{\times\times}{\underset{\times\times}{Br}}{\times}$$

Homolytic bond breaking

$$Br \overset{\frown\frown}{-} Br \longrightarrow Br^\bullet + {}^\bullet Br$$

hormones are chemical messengers carried round the body in the bloodstream. Hormones produce a response from particular target cells. Hormones are produced by glands and pass directly into the blood. Examples of hormones are *insulin* from the islet cells in the pancreas, adrenaline from the adrenal glands near the kidneys and sex hormones produced by the testes in men and the ovaries or other organs in women.

Hund's rule is one of the rules which help to predict electron configurations of atoms (see *aufbau principle*). The rule states that the most stable arrangement of electrons in an atoms is the one with as many unpaired electrons as possible all with parallel *spins*. For example, in a nitrogen atom, the three *p*-electrons have parallel spins and each occupy a separate *p* energy level.

Electron configuration of a nitrogen atom   N   1s ⇅   2s ⇅   2p ↑ ↑ ↑

hydration takes places when water molecules bond to ions or add to molecules. Water molecules are *polar* and so they are attracted to both positive and negative ions.

It is hard to see why the charged ions in a crystal of sodium chloride separate and go into solution in water with only a small energy change. Where does the energy come from to overcome the attraction between the ions? The explanation is that the ions are so strongly hydrated by the polar water molecules that the sum of the *enthalpy changes of hydration* nearly balances the *lattice energy*.

Some metal ions are hydrated in crystals as well as solution. In many hydrated salts the metal is present as a complex ion. This is true of copper(II) ions in blue copper sulfate (see diagram on page 189). Cobalt(II) ions also bond to water molecules when blue (anhydrous) cobalt(II) chloride turns pink with water.

Hydrogen ions are hydrated in solution. They do not float around freely, but bond to water molecules forming *oxonium ions.*

The term hydration is also used to describe the addition of water to molecules such as *alkenes*. Ethene, for example, is hydrated when mixed with steam and passed over a phosphoric acid catalyst (see diagram on page 189).

$$CH_2{=}CH_2(g) + H_2O(g) \xrightarrow[\text{heat under pressure}]{\text{phosphoric acid catalyst}} CH_3CH_2OH(l)$$

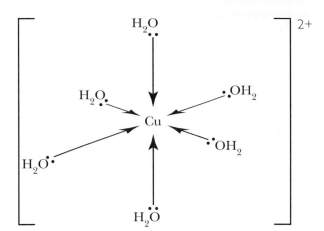

Hydration Hydrated copper(II) ion. Here co-ordinate (dative covalent) bonds hold the water molecules to the metal ions (see page 188).

hydrazine (N₂H₄) is a colourless fuming liquid which is manufactured by oxidising ammonia with sodium chlorate(I) in the presence of gelatin. Hydrazine, like ammonia, is a base and a reducing agent. It has a *lone pair of electrons* on each nitrogen atom.

Hydrazine with one hydrogen replaced by a methyl group can be used as a rocket fuel in space; it catches fire on contact with liquid N_2O_4.

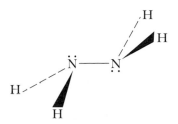

Structure and shape of a hydrazine molecule

hydrides are compounds of elements with hydrogen. The hydrides of the group 1 and 2 metals are ionic and contain the hydride ion, H^-.

$$2Na(s) + H_2(g) \longrightarrow 2Na^+H^-(s)$$

Ionic hydrides are ionic crystals. They are rapidly hydrolysed by water because the hydride ion is a strong base.

$$H^-(s) + H_2O(l) \longrightarrow H_2(g) + OH^-(aq)$$

The hydrides of non-metals are covalently bonded molecular liquids and gases. They include the *hydrogen halides*, *water*, hydrogen sulfide and *ammonia*. There are very many hydrides of carbon including the *alkanes*, *alkenes* and *arenes*.

The properties of the hydrides of the three highly *electronegative* elements nitrogen, oxygen and fluorine are affected by *hydrogen bonding*.

Hydrogen also forms *interstitial hydrides* with some *d-block elements*.

hydrocarbons are compounds which consist of just carbon and hydrogen. Important classes of hydrocarbons are the *alkanes*, *alkenes*, *alkynes* and *arenes*.

hydrochloric acid is a solution of hydrogen chloride gas in water. Hydrogen chloride is a strong acid so the solution in water is fully ionised into aqueous hydrogen ions (*oxonium ions*) and chloride ions. The concentration of commercial concentrated hydrochloric acid is about 12 mol dm^{-3}. Dilute hydrochloric acid (about 2 mol dm^{-3}) is commonly used as a laboratory reagent. Hydrochloric acid has the advantage of being cheaper and safer to dilute than sulfuric acid. It is a non-oxidising acid unlike dilute nitric acid.

hydrogen is the commonest element in the universe. Stars, like the Sun, consist mainly of hydrogen and get their energy from *nuclear fusion* which turns hydrogen into *helium.*

Hydrogen atoms are the smallest of all atoms consisting, normally, of one proton and one electron in the 1*s orbital.* There are two other *isotopes: deuterium* and *tritium.*

Life on Earth depends on hydrogen. Most *organic compounds* contain hydrogen and life requires water – hydrogen oxide.

Hydrogen is an important industrial chemical. It is produced from steam and natural gas for *ammonia manufacture.* It is also one of the products of the *electrolysis of brine.* Hydrogen is used to *hydrogenate* vegetable oils for margarine and similar spreads.

Hydrogen forms a wide range of *hydrides* with other elements.

hydrogen bonding is a type of attraction between molecules which is much stronger than other types of *intermolecular force,* but much weaker than covalent bonding.

Hydrogen bonding affects molecules in which hydrogen is covalently bonded to one of the three highly *electronegative* elements nitrogen, oxygen and fluorine.

The hydrogen atoms lie between two highly electronegative atoms. They are hydrogen bonded to one of them and covalently bonded to the other. The *covalent bond* is highly polar. The small hydrogen atom ($\delta+$) can get close to the other electronegative atom ($\delta-$) to which it is strongly attracted.

The three atoms associated with a hydrogen bond are always in a straight line.

Hydrogen bonding in hydrogen fluoride

Hydrogen bonding accounts for:

- the relatively high boiling points of ammonia, water and hydrogen fluoride which are out of line for the trends in the properties of the other hydrides in groups 5, 6 and 7
- the open structure of *ice*
- the tertiary structure of *proteins*
- the pairing of bases in a *DNA* double helix
- the pairing up (*dimerisation*) of carboxylic acids in a non-aqueous solvent.

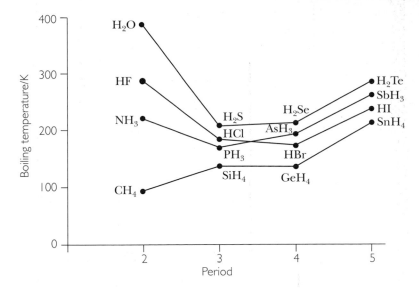

Boiling points for the hydrides of the elements in groups 4, 5, 6 and 7

hydrogen electrode: a *half-cell* of great theoretical importance but of limited practical significance. By definition the *standard electrode potential* of the hydrogen electrode is zero, $E^{\ominus}(\frac{1}{2}H_2 | H^+) = 0.00$ V.

A standard hydrogen electrode sets up an equilibrium between hydrogen ions in solution (1 mol dm^{-3}) and hydrogen gas (at 1 bar pressure) all at 298 K on the surface of a platinum electrode coated with platinum black.

By convention, when a standard hydrogen electrode is the left-hand electrode in an electrochemical cell, the cell e.m.f is the electrode potential of the right hand electrode.

Hydrogen electrodes are inconvenient and so most measurements are taken with alternative *reference electrodes* acting as secondary standards.

$$Pt[H_2(g)] \mid 2H^+(aq) \mathbin{\vdots} Zn^{2+}(aq) \mid Zn(s)$$

Conventional diagram to represent the cell which defines the standard electrode potential of the Zn^{2+}(aq)|Zn(s) electrode

hydrogen emission spectrum: the spectrum of radiation from excited hydrogen gas. There is a bluish glow when a high voltage is applied across the electrodes at each ends of a glass tube containing hydrogen at low pressure. A spectroscope produces a line spectrum which can be recorded photographically.

Each line in the *atomic spectrum* corresponds to electrons dropping from a higher energy level to a lower level. The series of lines are named after the people who first discovered or studied them.

The pattern of energy levels in a hydrogen atom with just one electron is simpler than in other atoms with more than one electron. In a hydrogen atom all the orbitals in the same shell have the same energy. So the energy levels correspond to the main shells.

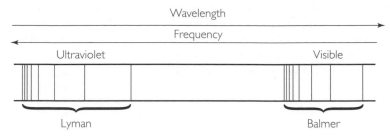

The series of lines in the hydrogen emission spectrum

The lines of the Balmer series are in the visible region. The lines of the Lyman series are in the ultraviolet. The bigger the energy jumps the higher the *frequency* of *electromagnetic radiation* emitted.

The energy gaps between energy levels get smaller as electrons recede from the nucleus. As a result, in each series, the differences between the energy jumps get smaller and smaller until they converge. They converge at the high frequency (big jump) end. The biggest jump in the Lyman series is for an electron dropping back from the very edge of an atom to the lowest energy level. The size of this jump is the ionisation energy for a hydrogen atom. So the ionisation energy of hydrogen can be calculated from the frequency of the convergence limit of the Lyman series since (according to *quantum theory*) $\Delta E = h\nu$.

hydrogen halides are compounds of hydrogen with the *halogens*. They are all colourless, molecular compounds with the formula HX where X stands for F, Cl, Br or I.

The bonds between hydrogen and the halogens are *polar*. The H — F bond is so polar that the properties of the compound are affected by *hydrogen bonding*.

Hydrogen chloride, hydrogen bromide and hydrogen iodide are similar in that they are:

- colourless gases at room temperature which fume in moist air
- very soluble in water, forming acid solutions (hydrochloric, hydrobromic and hydriodic acids)
- *strong acids* so they ionise completely in water.

Hydrogen chloride, hydrogen bromide and hydrogen iodide show some trends down *group* 7 in that they:

- become less thermally stable – heating does not decompose hydrogen chloride but a hot wire will decompose hydrogen iodide into hydrogen and iodine
- become easier to oxidise to the halogen – hydrogen iodide is a strong *reducing agent*.

Hydrogen fluoride is rather different from the other hydrogen halides:

- it is a liquid at room temperature (boiling point 19.9°C) because of hydrogen bonding
- it is a *weak acid*
- its solution in water (hydrofluoric acid) reacts with glass and can be used for etching.

hydrogen ions (H$^+$) are hydrogen atoms which have lost an electron. Since a hydrogen atom consists of one proton and one electron this means that a hydrogen ion is just a *proton*. In water, hydrogen ions do not float around freely, they become attached to water molecules forming oxonium ions, H$_3$O$^+$. A lone pair on the oxygen atom forms a dative bond with the hydrogen ion.

Bonding in an oxonium ion

Hydrogen ions (protons) transfer from an acid to a base during an *acid–base reaction*. This gives rise to the *Brønsted–Lowry* definition of an acid as a proton donor and a base as a proton acceptor.

hydrogen peroxide (H$_2$O$_2$), when pure is a pale blue liquid. Like water it is a *hydride* of oxygen and it is a *hydrogen bonded*, molecular liquid.

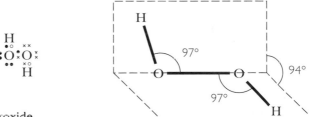

Bonding and shape of a
molecule of hydrogen peroxide

Hydrogen peroxide is usually supplied in solution in water. Solutions are sometimes labelled with the 'volume strength'. For example, '20 volume' hydrogen peroxide decomposes to give 20 cm^3 oxygen from 1 cm^3 of solution (measured at 0°C and 1 atmosphere pressure, *stp*).

Only in peroxides does oxygen have the oxidation state −1, in between 0 and −2. So hydrogen peroxide can act as both an *oxidising agent* and as a *reducing agent*.

Hydrogen peroxide decomposes slowly on standing in the dark, but much more rapidly in the present of a catalyst such a powdered manganese(IV) oxide or the *enzyme* catalase. This is a *disproportionation* reaction with oxygen starting in the −1 state and ending up in the −2 and 0 states. Hydrogen peroxide, as it were, oxidises and reduces itself.

$$2H_2O_2(aq) \longrightarrow 2H_2O(l) + O_2(g)$$

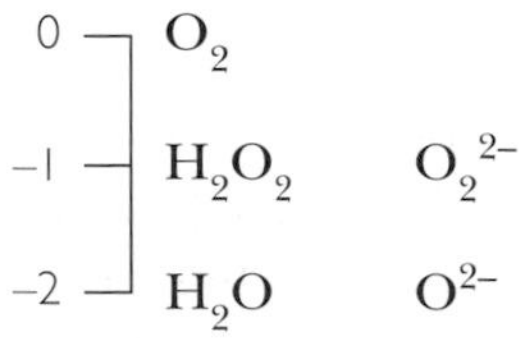

0	O$_2$	
−1	H$_2$O$_2$	O$_2^{2-}$
−2	H$_2$O	O^{2-}

Oxidation states of oxygen

Hydrogen peroxide oxidises cobalt(II) to cobalt(III) in the presence of ammonia, and chromium(III) to chromium(VI) under alkaline conditions. As an oxidising agent hydrogen peroxide is also used to bleach paper and textiles. It is used to manufacture the bleaches used in washing powders. In water treatment it destroys micro-organisms. The great advantage of hydrogen peroxide as a bleach is that it turns into water when it reacts, so it does not contaminate or pollute.

$$H_2O_2(aq) + 2H^+(aq) + 2e^- \longrightarrow 2H_2O(l)$$

Hydrogen peroxide is not a powerful reducing agent but it will decolourise an acid solution of manganate(VII) ions. A *potassium manganate(VII)* titration is one way of determining the concentration of a solution of hydrogen peroxide.

$$H_2O_2(aq) \longrightarrow O_2(g) + 2H^+(aq) + 2e^-$$

hydrogenation is a reaction which adds hydrogen to a compound. Hydrogenation converts liquid vegetable oils to the solid fats used as ingredients of *margarine*. In these reactions hydrogen adds to double bonds in the hydrocarbon chains of unsaturated *fatty acids*. Hydrogen adds to $C = C$ double bonds at room temperature in the presence of a platinum or palladium catalyst or on heating in the presence of a nickel catalyst. (See also *heterogeneous catalyst*.)

hydrolysis: a reaction in which a compound is split apart in a reaction involving water. Hydrolysis reactions are often catalysed by acids or alkalis.

Hydrolysis of the ester ethyl ethanoate catalysed by alkali

Hydrolysis is used to make *soap* from fats and oils. This is an example of *ester* hydrolysis.

Inorganic examples of hydrolysis include *hydrolysis of non-metal halides* and the *hydrolysis of salts*.

hydrolysis of non-metal halides splits the compounds into an *oxoacid* (or hydrated oxide of the non-metal) and hydrogen halide. Non-metal chlorides are typically molecular liquids at room temperature. Examples are CCl_4, $SiCl_4$ and PCl_3. Most of these compounds do not mix with water but react with it. For example:

$$SiCl_4(l) + 2H_2O(l) \longrightarrow SiO_2(s) + 4HCl(g)$$

The exception is CCl_4 which does not react with water. This is an example of *kinetic inertness* (stability). The four large chlorine atoms prevent water molecules from using their *lone pairs of electrons* to attack the small central carbon atom.

The hydrolysis of phosphorus bromide is one of the laboratory methods for making hydrogen bromide gas:

$$PBr_3(l) + 3H_2O \longrightarrow H_3PO_3(s) + 3HBr(g)$$

In a similar way, hydrogen iodide can be made by adding water to hydrolyse the PI_3 in a mixture of red phosphorus and iodine.

hydrolysis of salts changes the pH of a solution of a salt by altering the concentrations of H_3O^+ and OH^- ions.

Solutions of the salts of *aluminium(III)*, iron(III) and chromium(III) are acidic because of hydrolysis. The hydrated aluminium(III) ion, for example, is an acid. Al^{3+}, because of its *polarising power*, draws electrons towards itself, so that the water molecules can more easily give away protons than free water molecules.

$$[Al(H_2O)_6]^{3+}(aq) + H_2O(l) \longrightarrow [Al(H_2O)_5OH]^{2+}(aq) + H_3O^+(aq)$$

The hydrated ions in M^{2+} salts are much less acidic because of the smaller polarising power of these metal ions.

Solutions of the salts of weak bases are acidic. This includes the salts of ammonia and amines. The NH_4^+ ion in an ammonium salt is an acid because ammonia (the weak base) does not have a strong hold on the protons it accepts.

$$NH_4^+(aq) + H_2O(l) \rightleftharpoons NH_3(aq) + H_3O^+(aq)$$

Solutions of the salts of weak acids are alkaline. This includes the salts of:

- carbonic acid (carbonates)
- sulfurous acid (sulfites)
- ethanoic acid (ethanoates)
- hydrogen sulfide (sulfides).

The negative ions from weak acids are strong bases; they take hydrogen ions from water molecules and turn back into the acid.

$$CH_3CO_2^-(aq) + H_2O(l) \rightleftharpoons CH_3CO_2H(aq) + OH^-(aq)$$

hydrosphere: the watery part of the Earth's surface. This includes the oceans, rivers, lakes, groundwater and glaciers. Over 80% of the hydrosphere is in the oceans and shallow seas. Almost all the rest of the water is trapped in sediments.

The oceans are linked to the rest of the hydrosphere by the water cycle: water evaporates from the sea into the atmosphere, falls as rain or snow onto land and then returns to the oceans via rivers and lakes.

hygroscopic substances absorb water from the air. The chemicals used as drying agents in *desiccators* are hygroscopic, they include anhydrous calcium chloride and calcium oxide. It is difficult to weigh hygroscopic substances accurately which makes them unsuitable as *primary standards* for titrations.

ibuprofen is a drug used in *medicines* formulated to treat pain, inflammation and fever.

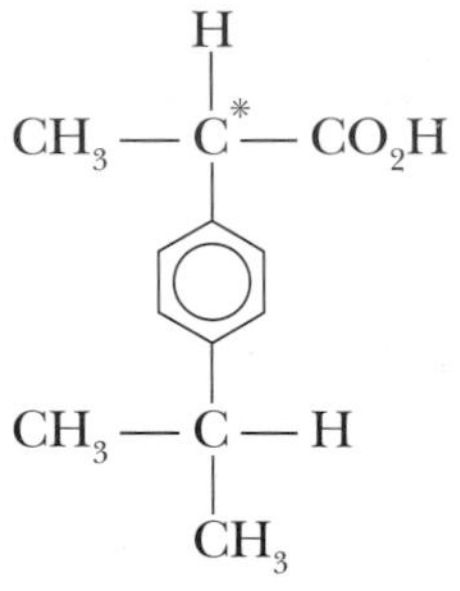

$$CH_3 - \overset{\text{H}}{\underset{|}{\overset{|}{C^{*}}}} - CO_2H$$

$$CH_3 - \underset{|}{\overset{|}{C}} - H$$
$$CH_3$$

The structure of ibuprofen. Note the asymmetric carbon atom marked with an asterisk. Ibuprofen is a chiral compound.

Ibuprofen was developed to reduce inflammation in the joints of people suffering from rheumatoid arthritis. Testing showed that the drug was also an effective pain reliever and a rival to aspirin. The research and development programme took 30 years. Ibuprofen was issued as a 'prescription only' drug in 1969 but is now available over the counter in retail pharmacies.

ice: the solid state of water which is remarkable because, at the freezing point, it is less dense than liquid water and so floats. Ice owes its open cage-like structure to *hydrogen bonding.* Each water molecule is hydrogen bonded to four others. The hydrogen bonds are 0.177 nm long which is much longer than the length of a covalent $O - H$ bond (0.094 nm). The distance between the oxygen atoms is equal to the sum of the lengths of a hydrogen bond and a normal $O - H$ bond. Note that in every instance the three atoms $O - H \cdots O$ are in a straight line.

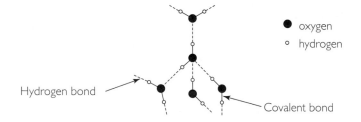

- ● oxygen
- ○ hydrogen

Hydrogen bond

Covalent bond

Structure of ice

The view from some angles shows the oxygen atoms in ice arranged in puckered hexagons – a pattern which is reflected in the hexagonal shapes of snowflakes.

When ice melts the open structure partly collapses as some of the hydrogen bonds break, so at 0°C the liquid is denser than the solid.

ideal gases: gases which obey the ideal gas equation $pV = nRT$, where p is the pressure in $N\,m^{-2}$ (pascals), V is the volume in m^3, T is the temperature on the *Kelvin* scale in K, n the *amount* in moles and R the *gas constant.* It is important to convert all the quantities to *SI units* before substituting values in the equation.

The ideal gas equation incorporates all the *gas laws*. For example, with a fixed amount of gas n is constant, so at constant temperature pV = constant, which is *Boyle's law*.

In practice *real gases* deviate from ideal gas behaviour. Under laboratory conditions the gases which are close to behaving like ideal gases are the ones which are well above their boiling points such as the *noble gases*, nitrogen, oxygen and hydrogen. For these gases at room temperature and pressure the assumptions of *kinetic theory* are nearly true. The attractive forces between these non-polar molecules are small enough to be ignored and the volume of the molecules is insignificant compared to the total volume of gas.

The ideal gas equation can be used to determine the *molar mass* of gases and of other substances which evaporate easily.

ideal mixture: a mixture of liquids which obeys *Raoult's law*. Only very similar compounds form ideal mixtures, such as the two *alkanes* hexane and heptane or the two *alcohols*, propan-1-ol and propan-2-ol.

A mixture of two liquids A and B is only ideal if the *intermolecular forces* in pure A, in pure B and in the mixture of A and B are all the same strength. A mixture of liquids deviates from Raoult's law if the intermolecular forces in the mixture are stronger or weaker than in the pure liquids.

ignition temperature: the lowest temperature at which a substance will spontaneously catch fire in air. There is no need for a flame to ignite a gas or *vapour* at or above its ignition point. Examples of ignition temperatures in air are: hexane, 487°C, and ethanol, 558°C.

immiscible liquids are liquids which do not mix. As a rule *non-polar* liquids such as hexane and other hydrocarbons do not mix with polar liquids such as water.

Oily hydrocarbon liquids are usually less dense than water so they form a separate, colourless layer on top of water. Liquid organic halogen compounds such as dichloromethane are often denser than water so that they give rise to a separate liquid layer underneath water.

Liquids which are immiscible are useful in *solvent extraction*.

immobilised enzymes are *enzymes* bound to a solid so that they can be easily recovered from a reaction mixture by filtering or centrifuging. Alternatively the solid holding the enzyme can be contained in a column and the raw material (*substrate*) passed through the column turning into products in the presence of the trapped enzyme *catalyst*. Three methods of linking an enzyme to a solid support are:

- ionic bonding between charged amino side-chains on the protein molecules of the enzyme and charged groups on the support
- trapping the enzyme in a tangle of polymer molecules
- using covalent bonds to attach the enzyme to the insoluble solid.

Immobilised enzymes are used to convert starch from maize into glucose and then to convert much of the glucose to fructose which is sweeter than glucose or sucrose (see *sweetness*). Using immobilised enzymes has meant a big fall in the cost of high fructose syrup used to make colas and other soft drinks.

incineration is one of the methods available for disposing of *waste*. Burning plastic waste or domestic refuse in a modern incinerator releases energy to generate electricity. Burning all the domestic waste from homes in Europe could generate 5% of European energy needs.

Just dumping rubbish in *landfill* is expensive and a waste of resources. Incineration greatly reduces the volume of waste to be dumped in landfill sites. It also replaces *fossil fuels*, thus cutting the overall emission of greenhouse gases.

There has been strong opposition to burning waste. Old incinerators were inefficient, operated at too low a temperature and often were not designed to generate electricity. Incomplete combustion produces *carbon monoxide* which is toxic. Another worry is that if waste contains chlorine compounds such as PVC an incinerator will give off corrosive or toxic chemicals such as hydrogen chloride or *dioxins*. High temperature combustion at 800–1250°C, however, helps to achieve complete combustion and limit the emission of toxic chemicals. Modern incinerators have elaborate gas cleaning systems to control the emission of harmful gases and dust particles.

indicators are used to detect the end-point during titrations. *Acid–base indicators* change colour over a range of pH values. Starch is used as an indicator in *iodine-thiosulfate* redox titrations. Other indicators are available for redox titrations and for *complex-forming titrations*.

induced dipoles: see *intermolecular forces*.

inductive effect: a term to describe the extent to which electrons are pulled away from, or pushed towards, an atom by the atoms or group to which they are bonded. The word crops up sometimes in accounts of reaction mechanisms.

More *electronegative* atoms pull electrons away from a carbon atom, so the carbon carries a slight positive charge. This means that that carbon atom is open to attack by nucleophiles which happens when carbon is bonded to oxygen, chlorine or bromine.

Other groups, especially *alkyl groups*, have a slight tendency to push electrons towards the carbon atom to which they are bonded. One of the effects of this kind of inductive effect is that a secondary *carbocation* is more stable than a primary carbocation. This helps to account for the products formed during *electrophilic addition* of compounds such as HBr to unsymmetrical alkenes (see *Markovnikov's rule*).

Primary alkyl *amines* are stronger bases than ammonia because of the inductive effect. The push of electrons from the alkyl group to the nitrogen atom helps to stabilise the ion formed when the amine accepts a proton.

methylamine   methylammonium ion

inert chemicals: a chemical is inert if it has no tendency to react under given circumstances. The lighter noble gases, helium and neon live up to the original name for group 8. They are inert towards all other reagents.

Nitrogen is a relatively unreactive gas which can be used to create an 'inert atmosphere' free of oxygen which is much more reactive. Nitrogen is not inert in all circumstances. It reacts, for example, with hydrogen in the *Haber process* and with oxygen at high temperatures to form *nitrogen oxides*.

Sometimes there is no tendency for a reaction to go because the reactants are *stable*. This is so if the *free energy change* for the reaction is positive.

Sometimes there is no reaction even though thermochemistry suggests that it should go. The free energy change is negative so the change is *feasible*. A high activation energy means that the rate of reaction is very, very slow. Methane, for example, does not burn in oxygen at room temperature even though it would be energetically favourable for it to do so. This is an example of a substance being kinetically inert. The term 'kinetic stability' is sometimes used but it helps to make a clear distinction between 'thermochemical stability' and 'kinetic inertness'.

inert gases: the older name for the group of *noble gases* which was used until the discovery of compounds formed by xenon.

inert pair effect: describes the observation that, for groups 3, 4 and 5 of the periodic table, the oxidation state which is 2 below the highest state becomes increasingly stable down the group. In group 4, for example, the main oxidation state for carbon and silicon is +4. The +2 oxidation state becomes important in the chemistry of tin and lead. In the chemistry of lead the +2 state is even more stable relative to the +4 state than it is in the chemistry of tin. What this means is that two of the four electrons in the outer shell become less available for bonding down the group, hence the term 'inert pair effect'. This effect is best regarded as a reminder of a trend rather than an explanation.

infra-red (IR) spectroscopy is an analytical technique used to identify functional groups in organic molecules. Most compounds absorb IR radiation. The wavelengths they *absorb* correspond to the natural frequencies at which vibrating bonds in the molecules bend and stretch. It is *polar covalent bonds* such as $O-H$, $C-O$ and $C=O$ which absorb strongly as they vibrate.

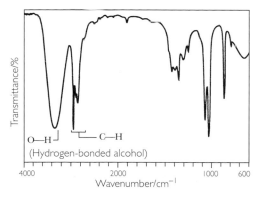

Ethanol

(Cont'd)

Answer

From experiments 1, 2 and 3: doubling $[A]_{initial}$ increases the rate by a factor of 4 (2^2). Tripling $[A]_{initial}$ increases the rate by a factor of 9 (3^2). So rate $\propto [A]^2$.

From experiments 2, 4 and 5: doubling $[B]_{initial}$ does not change the rate. So rate $\propto [B]^0 = 1$.

The reaction is second order with respect to A and zero order with respect to B. The rate equation is:

rate $= k\,[A]^2$

Rearranging this equation, and substituting values from experiment 2:

$$k = \frac{\text{rate}}{[A]^2} = \frac{8 \times 10^{-4}\ \text{mol dm}^{-3}\ \text{s}^{-1}}{(0.2\ \text{mol dm}^{-3})^2}$$

$k = 0.02\ \text{mol}^{-1}\ \text{dm}^3\ \text{s}^{-1}$

---

inorganic chemistry is the study of the chemical elements and their compounds. In an advanced level course the study of the elements concentrates on the reactions of selected elements and of their simpler compounds such as oxides, chlorides and hydrides. Inorganic chemistry includes the chemistry of the element carbon and a few of its compounds such as carbon monoxide, carbon dioxide, tetrachloromethane and the carbonates. The chemistry of most other carbon compounds belongs to *organic chemistry*.

Most of the reactions of inorganic chemicals belong to one of these four types:

- *acid–base*
- *redox*
- *ionic precipitation*
- *complex forming* (or *ligand substitution*).

insulin is the *hormone*, produced by special cells in the pancreas, which helps to control the level of glucose in the blood. Digestion of food after a meal raises the level of glucose in blood. In response the pancreas secretes insulin which circulates round the body and speeds up the rate at which cells take up glucose.

Insulin is a *protein* consisting of two folded chains of amino acids which are *cross-linked* by disulfide links.

interhalogen compound: a compound formed when one halogen combines with another. For example, passing a stream of chlorine gas over solid iodine produces a red–brown liquid, iodine monochloride, ICl. With excess chlorine this turns to a yellow solid, iodine trichloride, ICl_3, which is unstable and easily decomposes back to ICl.

intermediate bonding is bonding which is neither purely ionic nor purely covalent. In most compounds the bonding between the atoms of different elements is to some extent intermediate between the two extremes.

The *electronegativity* values for two elements are a guide to the extent to which the bonds between them will be covalent, ionic or intermediate between the two. Electronegativity values are particularly useful for discussing *polar covalent bonds*.

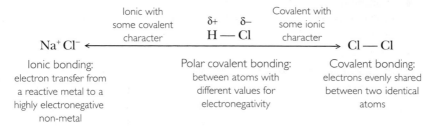

A spectrum from purely ionic to purely covalent bonding

Where the bonding is largely ionic, *Fajan's rules* are a guide to the extent to which the metal (positive) ion will distort neighbouring negative ions, giving rise to a degree of electron sharing.

intermediates in reactions are atoms, molecules, ions or *free radicals* which do not appear in the balanced equation but which are formed during one step of a reaction, then used up in the next step. Chemists can use spectroscopy to detect intermediates which only exist for a short time during a reaction.

Examples of reaction intermediates are the free radicals formed during a *free radical chain reaction* or the carbocations formed during *nucleophilic substitution* by the S_N1 mechanism and during *electrophilic addition* to alkenes.

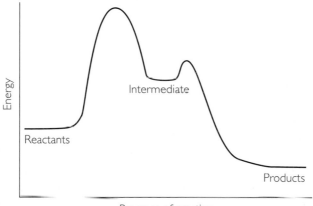

Energy changes in the course of a reaction. Note the energy dip as the intermediate forms. The dip is not deep enough for the intermediate to exist as a stable product. Contrast this with the energy profile for a reaction with just a *transition state* between reactants and products.

intermolecular forces are weak attractive forces between molecules. Without intermolecular forces there could be no molecular liquids or solids. Also, *real gases* would behave more like *ideal gases*.

Weak intermolecular forces arise from electrostatic attractions between *dipoles*, including attractions between:

- molecules with *permanent dipoles* such as hydrogen chloride

- a permanent dipole in one molecule and a dipole induced in a neighbouring molecule, such as the attraction between hydrogen bromide and an ethene molecule
- temporary dipoles created fleetingly in non-polar atoms or molecules.

Examples of substances with non-polar atoms and molecules include the noble gases, iodine and alkanes. When non-polar atoms or molecules meet there are fleeting repulsions and attractions between the nuclei of the atoms and the surrounding clouds of electrons. Temporary displacements of the electrons lead to temporary dipoles. It is the attractions between these transient dipoles which give rise to the tendency for the molecules to cohere.

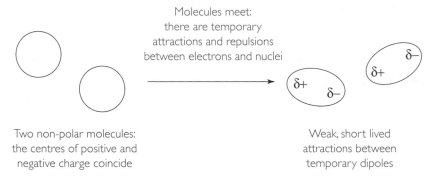

Molecules meet:
there are temporary
attractions and repulsions
between electrons and nuclei

Two non-polar molecules:
the centres of positive and
negative charge coincide

Weak, short lived
attractions between
temporary dipoles

An illustration of the origins and effect of temporary induced dipoles

The greater the number of electrons the greater the polarisability of the molecule and the greater the possibility for temporary, induced dipoles. This explains why the boiling points rise down *group 7* (the halogens) and *group 8* (the noble gases). For the same reason the boiling points in *alkanes* increase with the increasing number of carbon atoms in these hydrocarbon molecules.

In polymers the intermolecular forces become larger as molecules get longer so that there is a larger area of contact over which intermolecular forces can operate. The strength of *thermoplastic polymers* depends on the total strength of intermolecular forces between tangled, long chain molecules.

Weak intermolecular forces due to attractions between temporary and permanent dipoles are often called *van der Waals forces*. Forces of this kind are roughly a hundred times weaker than covalent bonds.

The properties of some molecular compounds are affected by stronger intermolecular forces called *hydrogen bonding*. Hydrogen bonds are about ten times stronger than van der Waals forces.

interstitial hydrides are compounds formed by *d-block elements* with hydrogen. The hydrogen atoms fit into the spaces between the atoms in the metal crystal. The holes are the interstices of the crystal, hence the name.

The amounts of hydrogen in the lattice vary with the temperature and pressure. The compounds do not have a definite composition. The hydride of titanium has the variable formula TiH_x, where x is less than 2.

Palladium reacts particularly well with hydrogen and can absorb nearly a thousand times its own volume of the gas at room temperature.

Interstitial hydrides release hydrogen gas on heating so they have been investigated as a possible safe way of storing hydrogen.

intramolecular forces are the strong *covalent bonds* between atoms in a molecule. They are about one hundred times stronger than the weak *intermolecular forces* between molecules. Substances consisting of small molecules melt and boil at relatively low temperatures because of the weakness of the forces between the molecules. The strong intramolecular forces in small molecules do not normally break during melting and boiling.

iodination is a reaction in which an iodine atom replaces a hydrogen atom in an organic molecule. One example is the iodination of propanone – a reaction which crops up in the study of reaction rates. The reaction is *acid catalysed*.

$$CH_3COCH_3(aq) + I_2(aq) \longrightarrow CH_2ICOCH_3(aq) + H^+(aq) + I^-(aq)$$

iodine (I) is a lustrous grey-black solid at room temperature, formula I_2, which *sublimes* when gently warmed to give an purple vapour. Iodine is a *halogen*, the fourth element in *group 7*, with the *electron configuration* $[Kr]4d^{10}5s^25p^5$.

Iodine consists of diatomic molecules with pairs of atoms held together by single *covalent bonds*. The molecules are non-polar so the *intermolecular forces* are relatively weak, but stronger than in bromine because the atoms are larger and have more electrons. This makes iodine molecules more *polarisable* than bromine molecules. (See *crystal structures of non-metals*.)

Iodine, like the other halogens, is an *oxidising agent* but a less powerful oxidising agent than bromine.

Iodine reacts with metals to form iodides. Because of the polarisability of the large iodide ion, the iodides formed with small cations or highly charged cations are essentially covalent (see *Fajan's rules*). Examples are lithium iodide, magnesium iodide and aluminium iodide.

Iron(III) iodide and copper(II) iodide do not exist because iodide ions reduce the metals ions to their lower oxidation state. The only iodides of these metals are iron(II) iodide and copper(I) iodide.

Iodine oxidises hydrogen on heating forming hydrogen iodide. Unlike the reactions of chlorine and bromine, this is a reversible reaction which has been studied in detail to establish the *equilibrium law*.

$$H_2(g) + I_2(g) \rightleftharpoons 2HI(g)$$

Hydrogen iodide is a fuming, acidic gas. Like the other *hydrogen halides* it is very soluble in water and a strong acid.

Iodine is only very slightly soluble in water. It is much more soluble in a solution of potassium iodide because of the formation of the *triiodide ion*. Iodine solution in aqueous potassium iodide is a brownish-yellow colour. Iodine dissolves freely in non-polar solvents such as hexane forming a solution with the same colour as iodine vapour.

As a weaker oxidising agent, iodine converts thiosulfate ions to tetrathionate ions. This is a quantitative reaction used in *iodine-thiosulfate titrations*.

Iodine and its compounds are used to make pharmaceuticals, photographic chemicals and dyes. Iodine is needed in the diet so that the thyroid gland in the neck can make the *hormone* thyroxine which regulates growth and metabolism. In many regions sodium iodide is added to table salt to supplement the iodine in the diet and drinking water and so prevent goitre.

iodine–thiosulfate titrations: a useful analytical method for measuring *amounts* of *oxidising agents*. The method is based on the fact that oxidising agents convert iodide ions to iodine quantitatively. Among the oxidising agents which do this are iron(III) ions, copper(II) ions, chlorine molecules and chlorate(I) ions in *bleach*, dichromate(VI) ions with acid, iodate(V) ions with acid and manganate(VII) ions with acid.

$$2I^-(aq) \longrightarrow I_2(aq) + 2e^-$$

The iodine stays in solution in excess potassium iodide turning a yellow–brown (see *triiodide ion*).

The iodine produced is then titrated with a *standard solution* of sodium thiosulfate which reduces iodine molecules back to iodide ions. This too happens quantitatively exactly as in the equation.

$$I_2(aq) + 2S_2O_3^{2-}(aq) \longrightarrow 2I^-(aq) + S_4O_6^{2-}(aq)$$

The greater the amount of oxidising agent added, the more the iodine formed and so the more thiosulfate needed to react with it. When thiosulfate is added from a burette the colour of the iodine gets paler. Near the end-point the solution is a very pale yellow. Adding a little soluble starch solution as an indicator near the end point gives a sharp colour change from blue–black to colourless.

Worked example:

Calculate the concentration of a solution of sodium thiosulfate standardised by this method. A 0.0642 g sample of potassium iodate(v), KIO_3, was dissolved in water in a conical flask. Excess of potassium iodide was dissolved in the solution which was then acidified with dilute sulfuric acid. The iodine formed was titrated with the solution of sodium thiosulfate from a burette. The volume of sodium thiosulfate solution needed to decolourise the blue iodine–starch colour at the end-point was 24.50 cm^3.

Notes on the method

Write the equations and work out the amount in moles of $S_2O_3^{2-}$ equivalent to 1 mol IO_3^-.

There is then no need to consider the amounts of iodine in the calculations.

Look up the *molar mass* of potassium iodate: $M_r(KIO_3) = 214.0$ g mol^{-1}.

Only use your calculator at the last step of the calculation to avoid repeated rounding errors.

Answer

The equations for producing iodine:

$$IO_3^-(aq) + 5I^-(aq) + 6H^+(aq) \longrightarrow 3I_2(aq) + 3H_2O$$

The equation for the reaction during the titration:

$$I_2(aq) + 2S_2O_3^{2-}(aq) \longrightarrow 2I^-(aq) + S_4O_6^{2-}(aq)$$

So 1 mol IO_3^- produces 3 mol I_2 which then reacts with 6 mol $S_2O_3^{2-}$.

The amount of KIO_3 at the start $= \dfrac{0.0642\ g}{214.0\ g\ mol^{-1}}$

6 mol $S_2O_3^{2-}$ react with the iodine formed by 1 mol IO_3^-.

So the amount of thiosulfate in 24.5 cm³ (= 0.0245 dm³) solution

$= 6 \times \dfrac{0.0642\ g}{214.0\ g\ mol^{-1}}$

So the concentration of the sodium thiosulfate solution

$= 6 \times \dfrac{0.0642\ g}{214.0\ g\ mol^{-1}} \times \dfrac{1}{0.0245\ dm^3}$

$= 0.0735\ mol\ dm^{-3}$

---

iodoform reaction: see *triiodomethane reaction.*

ion exchange is a process for swapping ions in a solution and is used to soften and deionise water. An ion exchange resin consists of small beads of a *polymer* which has been modified so that along the chains there are ionic groups. In a cation exchange resin the polymer has negatively charged groups which attract positive ions.

A water softener contains a cation exchanger in which all the negatively charged sites are attracting sodium ions. As hard water flows over the resin beads the calcium ions in the *hard water* change places with the sodium ions. The softened water can then be used for washing without forming insoluble precipitates with *soaps* and other *detergents.*

Ion exchange is a reversible process. Once all the sodium ions in a bed of resin have been used up it can be recharged by running a concentrated solution of sodium chloride through it to replace calcium ions with sodium ions, ready to soften more water. The salt added to a dishwasher is needed to make sure that the water softener in the machine is regenerated in this way.

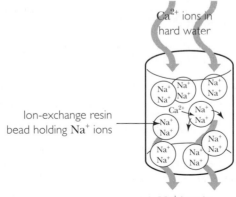

Ca²⁺ ions in hard water

Ion-exchange resin bead holding Na⁺ ions

Na⁺ ions in softened water

Using ion exchange to soften water

It takes two ion exchange resins to deionise water: a cation resin holding hydrogen ions and an anion resin with hydroxide ions. As water runs through the two resins the cation exchangers holds onto positive metal ions and replaces them with hydrogen ions. The anion exchanger holds on to the negative ions and replaces them with hydroxide ions. Then H^+ (aq) $+ OH^-$(aq) $\longrightarrow H_2O$(l), so no ions are left in solution.

ionic bonding is one of three types of strong chemical bonding. Ionic bonding occurs in compounds of metals with non-metals. Ionic compounds form particularly when the reactive *s-block* metals combine with the halogens of group 7 or oxygen in group 6. The *crystal structures of ionic compounds* are three dimensional lattices of ions.

Ionic bonding is the result of *electrostatic forces* of attraction between positive metal ions and negative non-metal ions. The larger the charges on the ions the bigger the attractive force. The smaller the ions the closer the charges get to each other and the stronger the force of attraction.

Electron *dot and cross diagrams* provide a balance sheet for keeping track of the electrons when ionic compounds form. These diagrams do not describe the mechanisms of the reactions.

$$Na^\bullet \quad + \quad {}_\times^\times Cl_\times^\times \quad \longrightarrow \quad Na^+ \quad + \quad {:}Cl_\times^\times{}^-$$

sodium atom	chlorine atom	sodium ion	chloride ion
(2.8.1)	(2.8.7)	(2.8)	(2.8.8)

$$Ca{:} \quad + \quad {}_\times^\times F_\times \quad {}_\times^\times F_\times \quad \longrightarrow \quad Ca^{2+} \quad + \quad {:}F_\times^-{} \quad {:}F_\times^-{}$$

calcium atom	two fluorine atoms	calcium ion	two fluoride ions
(2.8.8.2)	(2.7)	(2.8.8)	(2.8)

Dot and cross diagrams for the formation of sodium chloride and calcium fluoride

Energy is needed to remove electrons from metal ions (see *ionisation energy*). The energy changes on adding electrons to non metal atoms are quite small (see *electron affinity*). Ionic crystals are stable because of the large release of energy as the ions form a crystal lattice (see *lattice energy*). The *Born–Haber cycle* helps to analyse the stability of ionic crystals. This energy cycle also helps to explain why it is generally true that the atoms of s- and *p*-block elements gain or lose electrons to attain the electron configuration of the nearest *noble gas*.

ionic character of bonds: a phrase used to describe the extent to which bonds between the atoms of two different elements are polar. The greater the difference in electronegativity between the elements the more ionic the bonding. (See also *polar bonds, intermediate bonding* and *Fajan's rules*.)

ionic equations describe chemical changes by showing only the reacting ions in solution. These equations leave out the *spectator ions* which remain in solution unchanged. For example, an ionic equation shows that the use of a barium salt to test for sulfate ions is essentially the same reaction whether the reagent is barium chloride or barium nitrate and whatever the source of the sulfate ions.

$$Ba^{2+}(aq) + SO_4^{2-}(aq) \longrightarrow BaSO_4(s)$$

Note that ionic equations are balanced both for atoms and charges.

An ionic equation also shows that the neutralisation of any fully ionised *strong acid* by a fully ionised *strong base* is essentially the same process:

$$H^+(aq) + OH^-(aq) \longrightarrow H_2O(l)$$

ionic precipitation is a reaction which produces a solid precipitate on mixing two solutions. This type of reaction can be used to make an insoluble salts from two soluble salts. Silver bromide forms as a precipitate on mixing solutions of silver nitrate and potassium bromide. This is best represented by an *ionic equation* leaving out the *spectator ions*.

$$Ag^+(aq) + Br^-(aq \longrightarrow AgBr(s)$$

This is the reaction used to make silver bromide for *photography*. Silver metal is dissolved in nitric acid to make silver nitrate, $Ag^+NO_3^-$. Then the solution of silver nitrate is mixed with potassium bromide (K^+Br^-) in a solution of gelatin. This produces a very fine precipitate of silver bromide (AgBr) suitable for making a photographic film.

Ionic precipitation reactions play a big part in *anion tests* and *cation tests*.

ionic product of water, K_w: a constant for the equilibrium produced by the ionisation of water. There are oxonium and hydroxide ions even in pure water because of a transfer of hydrogen ions between water molecules. This only happens to a very slight extent. At equilibrium in pure water: $[H_3O^+(aq)] = 10^{-7}$ mol dm^{-3}.

$$H_2O(l) + H_2O(l) \rightleftharpoons H_3O^+(aq) + OH^-(aq)$$

The equilibrium constant $K_c = \dfrac{[H_3O^+(aq)]\,[OH^-(aq)]}{[H_2O(l)]^2}$

There is such a large excess of water that $[H_2O(l)]$ is a constant, so the relationship simplifies to: $K_w = [H_3O^+(aq)][OH^-(aq)]$

where K_w is the ionic product of water.

In pure water $[H_3O^+(aq)] = [OH^-(aq)] = 10^{-7}$ mol dm^{-3}.

Hence $K_w = 10^{-14}$ mol^2 dm^{-6}

K_w is a constant in all aqueous solutions at 298 K. This makes it possible to calculate the pH of alkalis.

Worked example:

What is the pH of a 0.02 mol dm^{-3} solution of sodium hydroxide?

Notes on the method

Sodium hydroxide is fully ionised in solution. So in this solution

$[OH^-(aq)] = 0.02$ mol dm^{-3}

$pH = -lg\,[H_3O^+(aq)]$

Answer

For this solution:

$K_w = [H_3O^+(aq)] \times 0.02$ mol dm$^{-3} = 10^{-14}$ mol^2 dm^{-6}

So $[H_3O^+(aq)] = \dfrac{10^{-14} \text{ mol}^2 \text{ dm}^{-6}}{0.02 \text{ mol dm}^{-3}} = 5 \times 10^{-13}$ mol dm^{-3}

Hence $pH = -lg\,(5 \times 10^{-13}) = 12.3$

---

ionic radius is the radius of an ion in a crystal. Ionic radii are determined by X-ray diffraction methods. The radius of the positive ion of an element is smaller than its atomic radius. The radius of the negative ion of an element is larger than its atomic radius.

Na $r_{atom} = 0.191$ nm	Na$^+$ $r_{ion} = 0.102$ nm	F $r_{atom} = 0.071$ nm	F$^-$ $r_{ion} = 0.133$ nm
Mg $r_{atom} = 0.160$ nm	Mg^{2+} $r_{ion} = 0.072$ nm	O $r_{atom} = 0.073$ nm	O^{2-} $r_{ion} = 0.140$ nm

Comparison of the radii of atoms and ions

Trends in ionic radius help to account for patterns of bonding and properties in the periodic table.

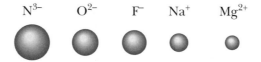

$$N^{3-} \quad O^{2-} \quad F^- \quad Na^+ \quad Mg^{2+}$$

Ionic radii for selected elements. All the ions shown have the same electron configuration. They are *isoelectronic*.

Note these patterns:

- down any group the ionic radii increase as the number of inner full shells increases
- across a period, the radii of ions with the same electron configuration decrease as the nuclear charge increases.

Ions in solution are *hydrated*. The radius of the ion plus the hydrating water molecules is larger than the ion in a crystal. A smaller ion is likely to be more heavily hydrated than a larger ion with the same charge.

Ion	Radius in a crystal/nm	Radius when hydrated/nm
Li$^+$	0.074	1.00
Na$^+$	0.102	0.79

ionisation energy (enthalpy): a measure of the energy needed to remove an electron from a gaseous atom or ion. Ionisation energies give evidence for the arrangement of electrons in atoms in shells and subshells. Ionisation energies can help to explain which ions an element can form. Values for ionisation energies are used in energy cycles, such as the *Born–Haber cycle*, to investigate bonding in compounds.

The first ionisation energy for an element is the energy needed to remove one mole of electrons from one mole of gaseous atoms. Successive ionisation energies for the same element measure the energy needed to remove a second, third, fourth electron and so on.

Ionisation energies can be measured using a *mass spectrometer* or by studying the emission spectra of atoms (see *hydrogen emission spectrum*). In these ways it is possible to measure energy changes involving ions which do not normally appear in chemical reactions.

$$Na(g) \longrightarrow Na^+(g) + e^- \quad \text{first ionisation energy} = 496 \text{ kJ mol}^{-1}$$

$$Na^+(g) \longrightarrow Na^{2+}(g) + e^- \quad \text{second ionisation energy} = 4563 \text{ kJ mol}^{-1}$$

$$Na^{2+}(g) \longrightarrow Na^{3+}(g) + e^- \quad \text{third ionisation energy} = 6913 \text{ kJ mol}^{-1}$$

There are 11 electrons in a sodium atom so there are 11 successive ionisation energies for this element. The electron configuration of the element is $1s^2 2s^2 2p^6 3s^1$. There is 1 electron in the outer shell which is furthest from the nucleus and shielded by ten inner electrons. There are eight electrons in the second shell and these are closer to the nucleus and only have two inner *shielding* electrons. The two inner electrons feel the full attraction of the nuclear charge and are closest to the nucleus. They are hardest to remove.

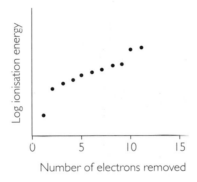

Plot of log (ionisation energy) against number of electrons removed for sodium

So the successive ionisation energies for an element rise and there are big jumps in value each time electrons start to be removed from the next shell in towards the nucleus.

In the periodic table, ionisation energies tend to rise from left to right across a period because from one element to the next the added electron goes into the same main shell as the charge on the nucleus increases by one (see the diagram on page 212). The rise in values is not smooth but shows a 2–3–3 pattern, corresponding to the way that the *s*- and *p-orbitals* fill up.

The $2s$ orbital is full at Be, there is then a slight dip as the next electron goes into a *p* orbital with a slightly higher energy. The three *p* orbitals each have one electron in a nitrogen atom. The dip at oxygen happens as the next electron has to pair up with another electron. Nevertheless, the main factor is the charge on the nucleus which increases steadily across the period. This pattern repeats from Na to Ar.

Ionisation energies decrease down a group. Down a group the number of full shells increases. The increased shielding effect balances out the increasing charge on the nucleus. The outer electrons get further and further away from the same effective nuclear charge, so they are easier to remove.

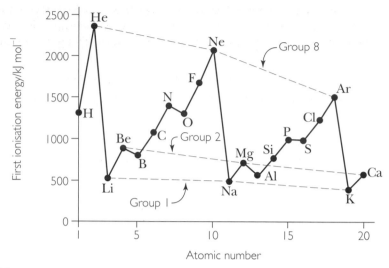

Plot of first ionisation energy against proton number for the elements H to Ca. Note that the trend is for the values of the first ionisation energies to decrease down any group.

ionisation of water: see *ionic product of water.*

ionising radiation includes all the types of radiation with enough energy per *photon* to ionise atoms. Ionising radiation includes *alpha* and *beta particles, gamma rays* and *X-rays.* Living cells are very sensitive to ionising radiation. High doses of ionising radiation can kill living cells and even small doses can cause mutation. Exposure to ionising radiation can, in time, give rise to cancer.

ions: see *anions, cations* and *ionic bonding.*

iron (Fe) is a *d-block* metal which shows the characteristic behaviour of the *transition metals.* Its *electron configuration* is: $[Ar]3d^64s^2$. The pure metal is soft and easily worked. It is strongly *ferromagnetic.* In most of its uses iron is mixed with other elements to make *steel.* Iron is so widely used that *corrosion of the metal* is a serious economic problem.

Iron can exist in more than one *oxidation state.* Iron dissolves in dilute acids to form iron(II) salts. The aqueous iron(II) ion is pale green. Oxidising agents including oxygen in the air, chlorine or potassium manganate(VII), all oxidise iron(II) to iron(III). The aqueous iron(III) ion is yellow.

$$Fe^{2+}(aq) \longrightarrow Fe^{3+}(aq) + e^-$$

Adding aqueous sodium hydroxide to solutions of iron salts helps to distinguish iron(II) from iron(III) compounds (see *cation tests*). A greenish precipitate of iron(II) hydroxide turns browny-red when exposed to the air, as it is oxidised to iron(III) hydroxide.

Iron ions form a variety of complexes in both the +2 and the +3 states. The hexaaquo complex of the iron(III) ion, $Fe(H_2O)_6^{3+}$, is acidic in solution for the same reason that the *aluminium(III) ion* is acidic.

A very sensitive test for iron (III) is to add a dilute solution of thiocyanate ions, SCN^-. The solution turns to a deep blood-red colour due to the formation of iron complexes such as $[Fe(H_2O)_5(SCN)]^{2+}$. The SCN^- *ligand* takes the place of a water molecule in the aquo complex.

An example of an iron (III) complex with a *bidentate ligand* is $[Fe(C_2O_4)_3]^{3-}$, in which the ligand is the ethanedioate ion.

Iron and its compounds act as *homogeneous* and *heterogeneous catalysts*.

iron extraction produces the metal from its oxide ores in large blast furnaces.

Coke burning in air heats the furnace.

$$C(s) + O_2(g) \longrightarrow CO_2(g)$$

Coke also produces the *reducing agent* by reacting with carbon dioxide further up the furnace to make carbon monoxide.

$$C(s) + CO_2(g) \longrightarrow 2CO(g)$$

The carbon monoxide reduces the oxide ore to iron.

$$Fe_2O_3(s) + 3CO(g) \longrightarrow 2Fe(l) + 3CO_2(g)$$

Where the furnace is hot enough, carbon too can act as the reducing agent.

Limestone ($CaCO_3$) decomposes to calcium oxide (CaO), which combines with silicon dioxide and other impurities to make a liquid *slag*. For example:

$$CaO(s) + SiO_2(s) \longrightarrow CaSiO_3(l)$$

The molten metal and slag run to the bottom of the furnace, where the slag floats on the metal so that it can be tapped off separately. Most of the iron is then turned into *steel*.

irritant chemicals do not destroy tissues, so they are less than corrosive. However skin or eyes may become inflamed or affected by sores if exposed to the chemical. This may happen quite quickly or after extended or repeated contact with the chemical.

isoelectric point: the pH at which the overall charge on an *amino acid* or *protein* is zero. At this pH amino acids or proteins do not move towards one or other of the electrodes during *electrophoresis*.

At low pH an amino acid has a positive charge because there is an extra hydrogen ion on the basic amino group. At high pH an amino acid is negatively charged because the carboxylic acid group has given away a hydrogen ion. At the pH of the isoelectric point the amino acid is present in solution as a *zwitterion* and the total charge is zero.

in acid solution — zwitterion — in alkaline solution

The charges on alanine at different pH values

Isoelectric points vary from one amino acid to another because of differences in their side chains.

A large protein molecule is at its least soluble in water at its isoelectric point when it is uncharged.

isoelectronic atoms and ions have the same number and arrangement of electrons. The following particles are all isoelectronic: N^{3-}, O^{2-}, F^-, Ne, Na^+, Mg^{2+} and Al^{3+}. They all have the electron configuration $1s^2 2s^2 2p^6$.

Methane and the ammonium ion are isoelectronic. Both have ten electrons, two in the first shell of the central atom and eight forming four covalent bonds with hydrogen atoms. They both have the same shape (see *shapes of molecules*).

isomerisation is a process used in oil refining to convert straight chain *alkanes*, such as pentane, into branched *isomers* such as 2-methylbutane. The value of the process is that branched alkanes increase the octane number of *petrol*. Isomerisation happens when the hot hydrocarbon vapour passes over a *catalyst*.

isomers are compounds with the same molecular formula but different structures. Structural isomerism may arise because:

- the hydrocarbon chain is branched in different ways

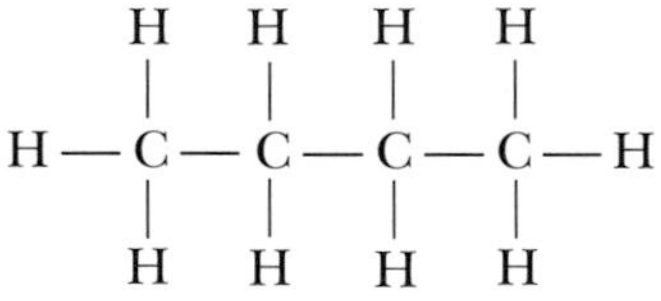

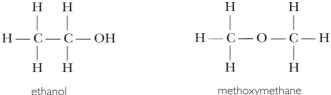

butane

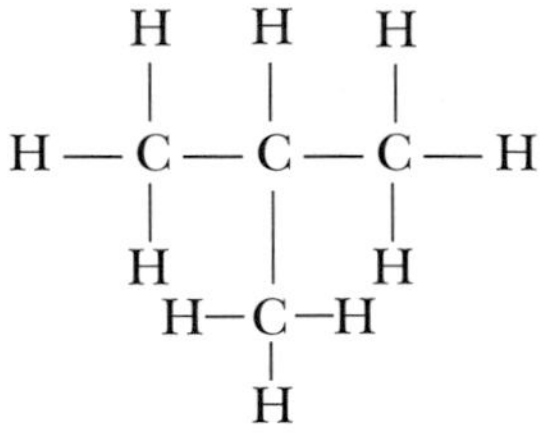

2-methylpropane

Chain isomers of C_4H_{10}

- the functional group is in a different position

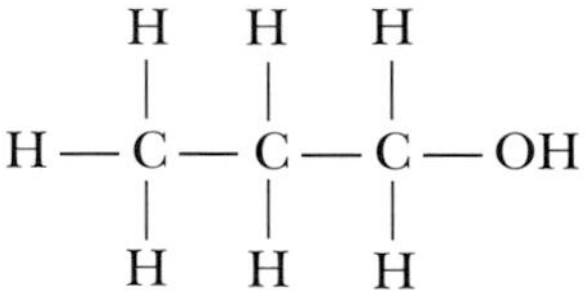

Position isomers
of C_3H_8O

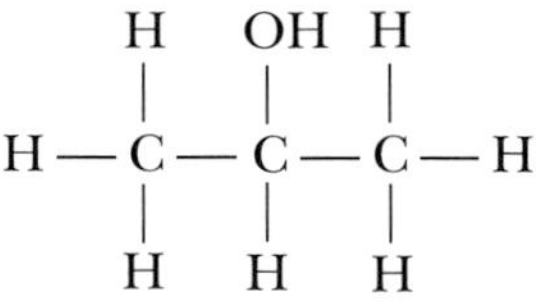

propan-1-ol

propan-2-ol

- the functional groups are different.

Functional group
isomers of C_2H_6O

ethanol

methoxymethane

(See also *geometrical isomerism* and *optical isomerism*.)

isomorphous compounds have the same type of chemical formula and crystallise with the same shape and structure. *Alums* are isomorphous. They form octahedral crystals.

isotactic polymer: an *addition polymer* with a regular structure. In isotactic poly(propene), for example, all the methyl side chains are on the same side of the carbon chain. The molecules coil into a regular helical shape and pack together to form a highly crystalline polymer which is very strong. This is the form of the polymer which is useful for hardwearing fibres, tough mouldings in motor vehicles and containers which can hold boiling water.

$$CH_3 \quad CH_3 \quad CH_3 \quad CH_3 \quad CH_3$$

Structure of isotactic poly(propene) with all the methyl groups on the same side of the carbon chain

isotopes are atoms of the same element which have the same number of protons in the nucleus but a different number of neutrons. In other words the atoms of isotopes have the same *atomic number* but different *mass numbers*. The isotopes have the same chemical properties because they have the same number and arrangement of electrons (*electron configuration*).

The isotopes of an element can be separated and detected with a *mass spectrometer*. Peaks in a mass spectrum are often multiple because of the existence of isotopes.

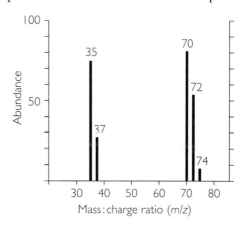

Mass spectrum of chlorine

isotopic abundance: the proportions of the different *isotopes* of an element. Chlorine, for example, has two isotopes: chlorine-35 and chlorine-37. Naturally occurring chlorine contains 75% of chlorine-35 and 25% of chlorine-37.

$$\text{So the mean } relative \text{ atomic mass} = \frac{(3 \times 35) + 37}{4} = 35.5$$

Most relative atomic mass values are not whole numbers because most elements have several naturally occurring isotopes.

isotopic labelling uses *isotopes* as markers to trace the movement of chemicals or to investigate what happens to particular atoms during chemical changes (see also *tracers*). Atoms of the normally abundant isotope of an element in a molecule are replaced by a different isotope.

Isotopes have the same chemical properties so it is possible to use them to follow what happens during a change without altering the normal course of the process. Radioactive isotopes can be tracked by detecting their radiation. The fate of non-radioactive isotopes can be followed by analysing samples with a *mass spectrometer*.

Isotopic labelling has been used to investigate the mechanism of ester hydrolysis. Using water labelled with oxygen-18 (instead of the normal isotope oxygen-16) it was shown that it is the $C-O$ bond in the ester which breaks and not the $O-C_2H_5$ bond. Mass spectrometry showed that the heavier oxygen atoms end up in the acid and not in the alcohol.

$$CH_3C\overset{\displaystyle O}{\underset{OC_2H_5}{\big\langle}} + H_2{}^{18}O \longrightarrow CH_3C\overset{\displaystyle O}{\underset{{}^{18}OH}{\big\langle}} + C_2H_5OH$$

This bond breaks

Use of labelling to investigate bond breaking during ester hydrolysis

isotropic solids have properties which are the same in all directions. Crystals of sodium chloride are isotropic because the ions are spherical and the crystal structure is cubic. Most substances are more or less *anisotropic*.

IUPAC (International Union of Pure and Applied Chemistry): the recognised authority for the names of chemical compounds, for chemical symbols and terms and the values of data such as relative atomic masses. IUPAC names are systematic names based on a set of rules which make it possible to work out the chemical structure of a compound from its name. Chemists increasingly use approved IUPAC names for simpler compounds but stick to traditional names when the systematic name is complex. The systematic name 2-hydroxypropane-1,2,3-tricarboxylic acid describes the structure but is cumbersome compared to the traditional name citric acid based on its occurrence in citrus fruits. (See *names of carbon compounds*, *names of complex ions* and *names of inorganic compounds*.)

joule (symbol J) is the *SI unit* of energy. The quantities of energy transferred during chemical reactions are relatively large, so chemists generally measure energy changes in kilojoules (kJ). 1 kJ = 1000 J.

K

K_a is the symbol for an *acid dissociation constant* (or ionisation constant).

K_c is the symbol for an equilibrium constant, with the concentrations measured in moles per litre (see *equilibrium law*).

The expression for K_c makes it possible to explain the effect of changing the concentration of one of the chemicals in an equilibrium mixture. There is an equilibrium, for example, in a solution of bromine in water:

$$Br_2(aq) + H_2O(l) \rightleftharpoons HOBr(aq) + Br^-(aq) + H^+(aq)$$

orange   colourless

At equilibrium: $K_c = \dfrac{[HOBr(aq)][Br^-(aq)][H^+(aq)]}{[Br_2(aq)]}$ where these are equilibrium concentrations

In dilute solution $[H_2O(l)]$ is constant.

Adding alkali removes $H^+(aq)$ from the right-hand side:

$$H^+(aq) + OH^-(aq) \longrightarrow H_2O(l).$$

This briefly upsets the equilibrium. For an instant after adding alkali:

$$\frac{[HOBr(aq)][Br^-(aq)][H^+(aq)]}{[Br_2(aq)]} < K_c$$

The system restores equilibrium as bromine reacts with water to produce more of the products while lowering the concentration of $Br_2(aq)$. There is very soon a new equilibrium. Once again:

$$\frac{[HOBr(aq)][Br\ (aq)][H^+(aq)]}{[Br_2(aq)]} = K_c$$

but now with new values for the various concentrations.

Speaking roughly, chemists say that adding alkali makes the 'position of equilibrium shift to the right'. The effect is visible because the orange colour of the solution fades. This is as *Le Chatelier's principle* predicts. The advantage of using K_c is that it makes quantitative predictions possible.

K_p is the symbol for an equilibrium constant for an equilibrium involving gases with the concentrations measured by *partial pressures*.

Equilibrium	K_p	Units of K_p (using the SI unit of pressure, Pa)
$H_2(g) + I_2(g) \rightleftharpoons 2HI(g)$	$K_p = \dfrac{(p_{NH_3})^2}{(p_{H_2})(p_{I_2})}$	no units
$N_2(g) + 3H_2(g) \rightleftharpoons 2NH_3(g)$	$K_p = \dfrac{(p_{NH_3})^2}{(p_{N_2})(p_{H_2})^3}$	Pa^{-2}

(cont'd)

$$N_2O_4(g) \rightleftharpoons 2NO_2(g) \qquad K_p = \frac{(p_{NO_2})^2}{(p_{N_2O_4})} \qquad Pa$$

Worked example:

When a mixture of 10.0 mol sulfur dioxide and 5.0 mol oxygen comes to equilibrium at 450°C and 200 kPa pressure, the amount of sulfur trioxide formed is 9.0 mol. Calculate K_p for the reaction.

Notes on the method

First write the equation for the reaction and work out the amount in moles of each gas at equilibrium. Note that 2 mol $SO_2(g)$ reacts with 1 mol $O_2(g)$ forming 2 mol $SO_3(g)$. 9.0 mol $SO_2(g)$ must react to form 9.0 mol $SO_3(g)$.

Hence calculate the *mole fraction* of each gas. This leads to the *partial pressures* for the gases at equilibrium. Substitute these values in the expression for K_p and then decide on the units.

Answer

The equation:	$2SO_2(g)$	+	$O_2(g)$ $\rightleftharpoons$	$2SO_3(g)$
Initial amounts	10.0 mol		5.0 mol	0.0 mol
Equilibrium amount	1.0 mol		0.5 mol	9.0 mol
Equilibrium mole fractions	$\dfrac{1.0}{10.5}$		$\dfrac{0.5}{10.5}$	$\dfrac{9.0}{10.5}$
Equilibrium partial pressures	$\dfrac{1.0}{10.5} \times 200$ kPa		$\dfrac{0.5}{10.5} \times 200$ kPa	$\dfrac{0.5}{10.5} \times 200$ kPa
	= 19.0 kPa		= 9.52 kPa	= 171.4 kPa

$$\text{Hence } K_p = \frac{(p_{SO_3})^2}{(p_{SO_2})^2 (p_{O_2})} = \frac{(171.4 \text{ kPa})^2}{(19.0 \text{ kPa})^2 (9.52 \text{ kPa})} = 8.55 \text{ kPa}^{-1}$$

K_w is the symbol for the *ionic product* of water.

Kekulé structure: a ring structure for benzene proposed by the German chemist Friedrich Kekulé in 1865. According to Kekulé the idea of a ring structure came to him while he was day dreaming in front of a fire.

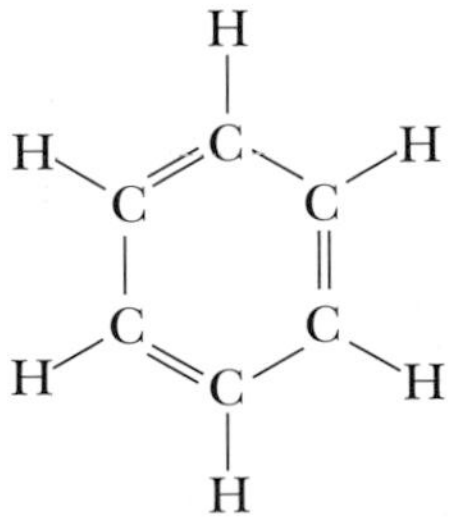

Representations of the Kekulé structure for benzene

The problem with this structure is that it suggests that there should be two isomers of 1,2-dichlorobenzene.

Isomers of 1,2-dichlorobenzene suggested by the Kekulé structure. Isomers with this formula have never been separated. There is only one form of the compound.

In practice it has never been possible to separate isomers of disubstituted benzenes. To get around this problem Kekulé suggested that benzene molecules rapidly alternate between the two possible structures. This is not the modern explanation. Today, chemists use the idea of *delocalisation of electrons* to account for the benzene structure.

kelvin (symbol K) is the *SI unit* of temperature. Temperatures above *absolute zero* are measured in kelvins. Temperatures on the Celsius scale use the same size units but zero is set at the freezing point of water (273.15 K). Temperature differences measured on either scale are the same and given as kelvins (K).

kerosine is a mixture of hydrocarbons produced by the *fractional distillation of oil*. Part of the kerosine fraction is refined for use as paraffin, while some is cracked to make *petrol*. Most of the kerosine fraction, however, is purified for use as the fuel for jet engines in aircraft.

ketones are *carbonyl compounds* in which the carbonyl group is attached to two *alkyl groups*. The carbonyl group is the *functional group* which gives ketones (and *aldehydes*) their characteristic properties. Ketones are named after the alkane with the same carbon skeleton by changing the ending 'e' to 'one'. Where necessary a number in the name shows the position of the carbonyl group.

Oxidation of secondary alcohols with acidified potassium dichromate(VI) produces ketones which, unlike aldehydes, are not easily oxidised further.

$$CH_3 - \underset{\underset{H}{|}}{\overset{\overset{OH}{|}}{C}} - CH_3 \xrightarrow[\substack{\text{sodium dichromate(VI)} \\ \text{and sulfuric acid}}]{\substack{\text{heat under reflux} \\ \text{with a mixture of}}} CH_3 - \overset{\overset{O}{||}}{C} - CH_3$$

propanone

Oxidation of propan-2-ol to propanone

Fehling's solution and *Tollen's reagent* (or ammoniacal silver nitrate solution) cannot oxidise ketones. There is no change when testing ketones with these reagents. Ketones with a methyl group next to the carbonyl group, such as propanone, give a positive result with the *triiodomethane reaction*.

Chemists can identify ketones with the help of instrumental techniques such as *mass spectrometry* and *infra-red spectroscopy*. Traditionally, chemists characterised these compounds by preparing crystalline derivatives with *Brady's reagent*.

Sodium tetrahydridoborate(III) reduces ketones to secondary alcohols, as does *lithium tetrahydridoaluminate(III)*.

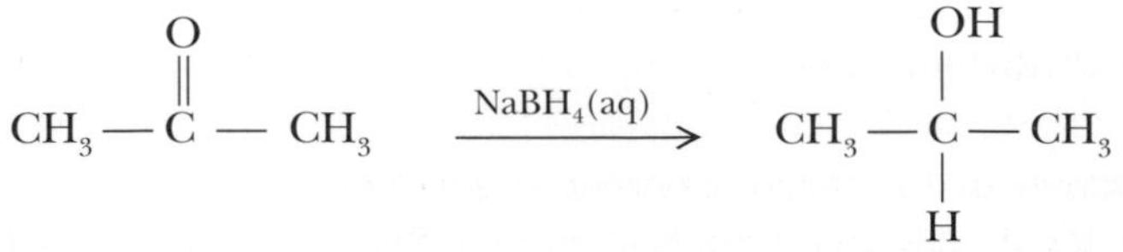

$$CH_3 - \overset{\overset{\displaystyle O}{\|}}{C} - CH_3 \quad \xrightarrow{NaBH_4(aq)} \quad CH_3 - \overset{\overset{\displaystyle OH}{|}}{\underset{\underset{\displaystyle H}{|}}{C}} - CH_3$$

Reduction of propanone to propan-2-ol   propan-2-ol

Ketones, like aldehydes, undergo addition. These are *nucleophilic addition reactions.*

Sometimes addition is immediately followed by elimination of water in *addition–elimination reactions.*

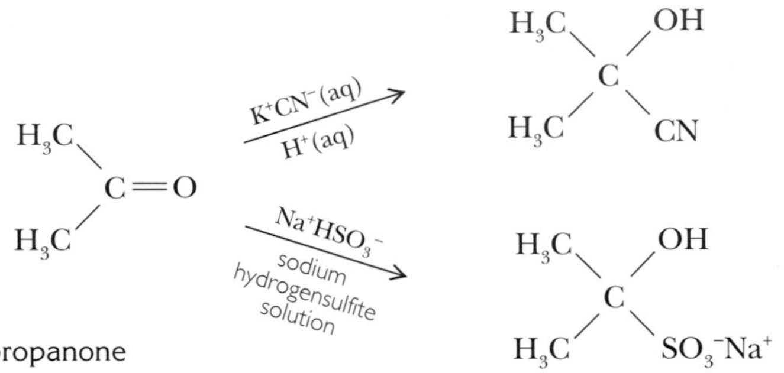

Addition reactions of propanone

Kevlar is an extremely strong, flexible, fire resistant *polymer* with a low *density.* Bullet proof vests are made of Kevlar and punctures are less likely with bicycle tyres which include the polymer. Kevlar ropes are much stronger than the same weight of steel rope. The polymer can also replace steel in motor vehicle tyres.

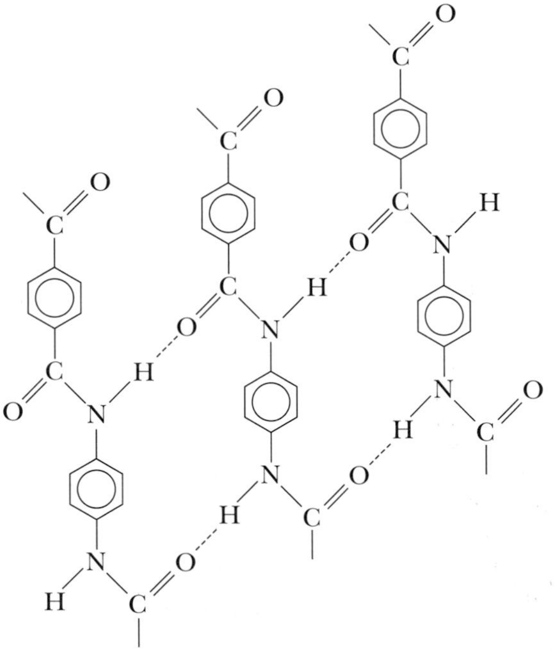

The structure of Kevlar

Kevlar is a *polyamide*. It is a polymer of an aryl amide. Polymers of this kind are called aramids. The rigid, linear polymer chains in Kevlar line up parallel to each other held together by *hydrogen bonding.*

kilogram (symbol kg) is the *SI unit* of mass. 1 kg = 1000 g.

kinetic control: a term used when the product of a reaction is the one which forms faster rather than an alternative product which is more *stable* but forms more slowly.

The production of methanal from methanol and oxygen is an example of kinetic control.

$$2CH_3OH(g) + O_2(g) \longrightarrow 2HCHO(g) + 2H_2O(l)$$

This reaction is fast in the presence of a suitable catalyst at a temperature at which the alternative reaction cannot go. The alternative is burning to carbon dioxide and water. Combustion is more exothermic but has a higher *activation energy.*

$$2CH_3OH(g) + O_2(g) \longrightarrow 2CO_2(g) + 4H_2O(l)$$

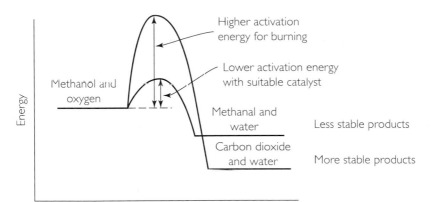

Reaction profiles for alternative reactions illustrating kinetic and thermodynamic control

In the absence of a catalyst, and at a higher temperature, methanol burns to produce the products which are thermodynamically more stable. In these circumstances the system is under *thermodynamic control.*

kinetic inertness (or stability): a term used when a reaction does not go, even though the reaction appears to be *feasible*. There is no change because the rate of reaction is too slow to be noticeable. There is a barrier preventing change – usually a high *activation energy*. The compound or mixture is inert.

Examples of kinetic inertness are:

- a mixture of methane (natural gas) and oxygen at room temperature
- a mixture of CCl_4 and water at room temperature (see *hydrolysis of non-metal chlorides*)
- a solution of *hydrogen peroxide* in the absence of a catalyst
- *aluminium* metal in dilute hydrochloric acid.

$$HO:^- \quad CH_3 \underset{H}{\overset{H}{-C}} Br \longrightarrow HO-\underset{C_3H_7}{\overset{H}{\underset{|}{C}}}-H \quad + \quad Br^-$$

Nucleophilic substitution of a hydroxide ion for a bromide ion from 1-bromobutane. The bromide ion is the leaving group (see page 225).

$$CH_3-\underset{CH_3}{\overset{H}{\underset{|}{C}}}-OH \xrightarrow{H^+} CH_3-\underset{CH_3}{\overset{H}{\underset{|}{C}}}OH_2^+ \longrightarrow Br-\underset{CH_3}{\overset{H}{\underset{|}{C}}}-CH_3 + H_2O$$

Nucleophilic substitution of a bromide ion for a hydroxide ion in propan-2-ol. The reaction goes faster under acid conditions. The — OH group in the alcohol is protonated, so the leaving group is a water molecule (see page 225).

Lewis acid/base theory: a theory which gives a very broad definition of acids and bases in terms of electron pairs. A Lewis acid is a molecule or ion which can form a bond by accepting a pair of electrons. A Lewis base is a molecule or ion which can form a bond by donating a pair of electrons.

The formation of an *oxonium ion* is a Lewis acid–base reaction between the proton (a Lewis acid) and a water molecule (a Lewis base). So, in this theory, it is the proton rather than the proton donor, which is an acid.

Formation of an oxonium ion as the Lewis base, water, forms a *dative covalent bond* with a proton

$$H^+ \quad :\underset{}{\overset{H}{\underset{|}{O}}}-H \longrightarrow H-\underset{\cdot\cdot}{\overset{H}{\underset{|}{O}}}-H^+$$

Lewis acid   Lewis base

This much wider definition of acids and bases describes the formation of a *complex ion* as a reaction between a metal ion (a Lewis acid) and *ligands* (Lewis bases).

$$Ni^{2+} \quad + \quad 6:NH_3 \longrightarrow [Ni(NH_3)_6]^{2+}$$

Complex formation as a   Lewis acid   Lewis base   Complex ion
Lewis acid–base reaction

Aluminium chloride ($AlCl_3$) is a Lewis acid. This accounts for its use as a catalyst in a *Friedel–Crafts reaction*.

Chemists normally use the *Brønsted–Lowry theory*. When they use the terms acid and base they mean 'proton donor' and 'proton acceptor'. To signal that they are using the wider Lewis definition they refer to 'Lewis acids' and 'Lewis bases'.

Lewis acid   Lewis base

Electrophile
for substitution
in benzene

Aluminium chloride acting as a Lewis acid

ligand substitution reactions involve swapping one *ligand* for another in a *complex ion*. These ligand exchange reactions are often reversible.

The ligands NH_3 and H_2O are similar in size and are unchanged. With these ligands exchange reactions take place without change in the *co-ordination number* of the metal ion.

$$[Cu(H_2O)_6]^{2+}(aq) + 4NH_3(aq) \rightleftharpoons [Cu(NH_3)_4(H_2O)_2]^{2+}(aq) + 4H_2O(l)$$

Reversible ligand exchange reaction of copper(II) ions. Note that in this example substitution is incomplete. The aquo complex is pale blue but the ammine complex is a very deep blue.

The chloride ion is larger than uncharged ligands such as water, so that fewer chloride ions can fit round a central metal ion. Ligand exchange involves a change in co-ordination number.

$$[Cu(H_2O)_6]^{2+}(aq) + 4Cl^-(aq) \rightleftharpoons [CuCl_4]^{2-}(aq) + 6H_2O(l)$$

Ligand exchange with change of co-ordination number. The aquo complex is pale blue but the chloro complex is yellow.

ligands are the molecules or ions bound to the central metal ion in a complex ion. Examples of molecules which can act as ligands are water and ammonia. Examples of ions which act as ligands are hydroxide ions, chloride ions and cyanide ions. Ligands have one or more *lone pairs of electrons* which can form *dative bonds* to the metal ion (see *monodentate ligands, bidentate ligands, hexadentate ligands* and *edta*).

limewater is a test reagent for detecting carbon dioxide. The reagent is a saturated solution of calcium hydroxide in water. The reagent is a colourless solution. Bubbling carbon dioxide through the solution produces a white precipitate of calcium carbonate (see *gas tests*).

$$Ca(OH)_2(aq) + CO_2(g) \longrightarrow CaCO_3(s) + H_2O(l)$$

limiting reactant: the chemical in a reaction mixture which is present in an amount which limits the theoretical yield (see *yield calculations*). Often in a chemical synthesis some of the reactants are added in excess to make sure that the most valuable chemical is converted as far as possible to the required product. The limiting reactant is the one which is not in excess and so is used up (if the reaction goes to completion).

lipids: a broad class of biological compounds which are soluble in organic solvents such as ethanol, but insoluble in water. Lipids are very varied and include *fatty acids, fats, vegetable oils,* phospholipids and *steroids.* Lipids release more energy per gram when oxidised than carbohydrates. This makes lipids important as a concentrated energy store in living organisms.

liquids flow to take the shape of their container. The atoms, molecules or ions in a liquid are free to move about but with nothing like the freedom of the molecules of a *gas.* The particles of a liquid are generally slightly more widely spaced than in a solid but the density of packing is only slightly less. Liquids lack the order of crystalline solids. Liquids, like solids, are hard to compress. Unlike gases, they have a definite volume.

liquid chromatography was the first type of *chromatography* developed. The Russian botanist Michel Tswett developed this type of chromatography to separate plant pigments. In 1903 he separated leaf colours (chlorophylls, carotenes and xanthophylls) by dissolving a leaf extract in a hydrocarbon solvent. He added the extract to the top of a column of powdered chalk and then passed more pure solvent through the column to complete the separation. Running solvent through the column to separate the mixture is called 'elution' and the liquid which flows out of the bottom it the eluate.

This is an example of *adsorption* chromatography. Each compound in the mixture has its own equilibrium between adsorption on the solid and solution in the solvent. Compounds which are strongly adsorbed by the stationary phase move slower; compounds which are more soluble in the solvent move faster.

The liquid leaving the bottom of the column is collected in a series of tubes. Tubes containing parts of the mixture can be detected by their colour or some other method such as their appearance in UV light. Components of the mixture can be recovered by evaporating the solvent from these tubes.

liquid crystal: a state of matter which is more ordered than a liquid but less ordered than a solid. Although the liquid crystal state was first noticed as long ago as 1888, it is only since the early 1970s that such crystals have been developed for use in digital watches, calculators and portable computer screens. Much of the pioneering work was done in Hull by a team lead by George Gray.

lithium (Li) is a soft, shiny metal which turns dark grey in air. It is the first member of *group 1* with the electron configuration $[He]2s^1$.

Like other group 1 metals, lithium:

- is stored in oil
- floats on water and reacts, but quite slowly, forming hydrogen and LiOH which is soluble and strongly alkaline
- burns in air with a coloured flame (bright red) forming an oxide (Li_2O)
- forms an ionic, crystalline chloride.

The small size of the lithium ion means that some features of lithium chemistry are not typical of the group as a whole:

- hydration – the Li^+ ion is heavily hydrated in water and lithium has the most negative hydration energy in group 1; as a result the standard electrode potential for $Li^+(aq)|Li(s)$ is most negative for the group

- solubilities – lithium carbonate and lithium fluoride are only slightly soluble while most other simple compounds of group 1 metals are soluble
- thermal stability – lithium carbonate decomposes on heating while all the other group 1 carbonates are stable; lithium nitrate decomposes to the oxide on heating unlike the nitrates of sodium and potassium
- covalent bonding – lithium forms a wide range of covalent molecular compounds such as ethyl lithium, C_2H_5Li.

lithium aluminium hydride: see *lithium tetrahydridoaluminate(III)*.

lithium tetrahydridoaluminate(III) (LiAlH$_4$) is a powerful reducing agent used to reduce *aldehydes, ketones, esters* and *carboxylic acids* to alcohols. Many chemists still use the older name, lithium aluminium hydride.

In its reactions the tetrahydridoaluminate(III) ion can be regarded as a source of hydride ions (H^-). The reagent is rapidly hydrolysed by water so it has to be used in an anhydrous solvent such as dry ether. Where possible *sodium tetrahydridoborate(III)* is preferred because it is easier and safer to use.

lithosphere: the rocks, weathered rocks and soils of the Earth's crust which make up less than 0.0001% of the volume of the planet. The crust consists largely of oxygen (46.6% by mass) mainly combined with silicon (27.7%) in *silicate* minerals. Only a few other elements are abundant: aluminium (8.1%), iron (5%), calcium (3.6%), sodium (2.8%), potassium (2.6%) and magnesium (2.1%). All other elements together make up the remaining 1.4%.

litre (symbol l) is a unit of *volume* equal to $1000 \, cm^3$. A litre is the same as a decimetre cubed (dm^3). $1 \, dm^3 = 10 \, cm \times 10 \, cm \times 10 \, cm = 1000 \, cm^3$. Chemists use the dm^3 when calculating because the units can then be worked through consistently. While the m^3 is the *SI unit* of volume, dm^3 is accepted as convenient, and is preferred to the litre.

localised electrons: electron pairs forming covalent bonds between two atoms. Localised electrons are not free to move through a structure, so giant structures with normal covalent bonds, such as diamond and silicon dioxide, do not conduct electricity, unlike giant structures with *delocalised electrons* such as metals and graphite.

logarithms in chemistry are of two kinds, logarithms to base 10 (lg) and natural logarithms to base e (ln).

Chemists use logarithms to base 10 to handle values which range over several orders of magnitude. Logs to base 10 are defined such that:

- $\lg 1000 = 3$
- $\lg 100 = 2$
- $\lg 10 = 1$
- $\lg 1 = 0$
- $\lg 0.1 = -1$
- $\lg 0.01 = -2$

In general $\lg 10^x = x$

The hydrogen ion concentration in aqueous solutions typically ranges from 1 to $10^{-14} \, mol \, dm^{-3}$. The definition of *pH* ($-\lg [H^+(aq)]$) covers this range in a scale running from 0 to 14.

Epsom salts consist of hydrated magnesium sulfate, $MgSO_4.7H_2O$, which is a laxative.

main group elements: a term used by some chemists to describe the elements in the *s-block* and *p-block* of the periodic table but excluding the transition elements in the *d-block* and *f-block*.

malleability: metals are malleable if they can be hammered into shape without breaking. It is possible to make very thin sheets of a malleable metal such as gold by hammering.

manganese (Mn) is a hard, grey brittle *d-block* metal with the electron configuration $[Ar]3d^54s^2$.

The main oxidation states of manganese are:

- **+7, MnO_4^-** – the purple manganate(VII) ion, which is a strong oxidising agent especially in acid solution
- **+4, MnO_2** – an insoluble, black compound which is an oxidising agent in acid solution
- **+2, Mn^{2+}** – the pink manganese(II) ion in salts such as manganese(II) sulfate.

Under special conditions it is also possible to produce solutions with red manganese(III) ions or green manganate(VI) ions. Manganate(VI) is stable in alkaline solution but *disproportionates* to manganate(VI) and manganese(IV) oxide on adding acid. (See also *potassium manganate(VII)*.)

manometer: an instrument for measuring *pressure*. A manometer consists of a U-tube containing a liquid such as water or mercury.

Typically one arm of a manometer is open to the atmosphere, so the instrument measures the extent to which the pressure inside the apparatus is higher or lower than atmospheric pressure.

margarine, and other non-dairy spreads, are solid emulsions consisting of water finely dispersed in *vegetable oils*, with a high enough proportion of hardened oils to make the product a spreadable solid. *Hydrogenation* is used industrially to add hydrogen to double bonds in oils. This produces saturated fats which are solid at room temperature. So the process is sometimes called 'hardening'. Manufacturers bubble hydrogen though the oil in the presence of finely divided nickel, which is the *catalyst*.

Markovnikov rule: a rule which predicts the main product when a compound HX (such as H — Br, H — OSO_3H or H — OH) adds to an unsymmetrical *alkene* (such as propene). The rule is that the hydrogen atom adds to the carbon atom that already has more hydrogen atoms attached to it. This pattern was first reported by the Russian chemist Vladimir Markovnikov who studied a great many alkene *addition reactions* during the 1860s. (See top diagram on page 233.)

The mechanism for electrophilic addition helps to account for this rule. When HBr is added to propene there are two possible intermediate *carbocations*. The secondary carbocation is preferred because it is slightly more stable than the primary carbocation. The secondary ion has two alkyl groups pushing electrons towards the positively charge carbon atom (see the *inductive effect*). This helps to stabilise the ion by 'spreading' the charge over the ion (see lower diagram on page 233).

Markovnikov's rule

Markovnikov's rule: an explanation

mass (symbol m) is a fundamental physical quantity. The *SI unit* of mass is the kilogram (kg). 1 kg = 1000 g.

Mass can be determined by measuring the pull of gravity on a specimen. This is how chemists usually measure the masses of chemicals with a chemical balance.

Alternatively masses can be found by seeing how much an object accelerates when a force acts on it. This is how chemists measure the masses of ionised atoms and molecules in *mass spectrometry*.

mass number, A: the total number of protons and neutrons in the nucleus of an atom. *Isotopes* have the same atomic number but different mass numbers. The alternative term is nucleon number.

mass spectrometry: an accurate instrumental technique for determining *relative atomic masses* and *relative molecular masses*. Mass spectrometry can also help to determine molecular structures and to identify unknown compounds.

The instrument – inside a mass spectrometer there is a high vacuum so that it is possible to produce and study ionised atoms and molecules including fragments of molecules which do not otherwise exist.

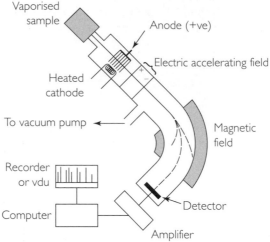

Diagram of a mass spectrometer

The stages in producing a mass spectrum:

- inject a small sample into the instrument where it vaporises
- bombard the sample with a beam of high energy electrons which turns the atoms or molecules into positive ions by knocking out electrons
- accelerate the positive ions in an electric field
- deflect the moving stream of ions with a magnetic field to focus ions with a particular mass onto the detector
- feed the signal from the detector to a computer which prints out a mass spectrum as the magnetic field steadily changes over a range of values to focus the series of ions with different masses one by one onto the detector.

The instrument is calibrated using a reference compound of known structure and molecular mass so that the computer can print a scale on the mass spectrum.

Isotopes of elements – the mass spectrum for an element shows the relative abundance of different *isotopes* of the element. This makes it possible to calculate the *relative atomic mass* for the element.

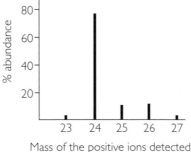

Mass spectrum of magnesium

Molecular masses and structures – when molecular compounds are being analysed, the peak of the ion with the highest mass is usually the whole molecule, ionised. So the mass of this 'parent ion' is the *relative molecular mass* of the compound.

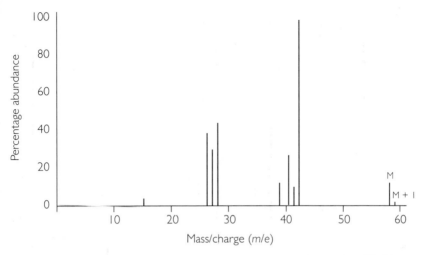

Mass spectrum of butane, C_4H_{10} showing the parent ion with a value 58. This corresponds to the relative molecular mass of 58. The pattern of fragments is characteristic of this compound. The very small peak at 59 is the 'M + 1' peak, which is present because some molecules of butane include carbon-13 atoms. Carbon-13 makes up 1.1% of natural carbon so in a molecule with four carbon atoms the 'M + 1' peak is 4.4% of the molecular ion peak.

Bombarding molecules with high energy electrons splits them into fragments so the mass spectrum is a 'fragmentation pattern'. The computer has a database of mass spectra so it can identify an unknown by matching its spectrum with one in its database. The fragments which show up in the spectrum are positive ions. The highest peaks correspond to positive ions which are relatively more stable, such as tertiary *carbocations* or ions such as RCO+ (the acylium ion).

Chemists who synthesise new compounds can study their fragmentation patterns. They identify the fragments from their masses and then piece together likely structures with the help of evidence from other methods of analysis such as *infra-red spectroscopy* and *nmr spectroscopy*.

The combination of *gas–liquid chromatography* (glc) with mass spectrometry is of great importance in modern chemical analysis. First, glc separates the chemicals in an unknown mixture, such as a sample of polluted water; then mass spectrometry detects and identifies the components.

matches use phosphorus compounds to make a flame. In a match that will strike anywhere, the head contains phosphorus sulfide (P_4S_3) and potassium chlorate(V) as an oxidising agent. The head contains the fuel and the oxidant and needs only friction to heat the match and start a fire.

Safety matches have sulfur and potassium chlorate(V) in the head and red phosphorus in the striking strip on the side of the box.

Maxwell–Boltzmann distribution: the distribution of molecular kinetic energies for a gas at a particular temperature.

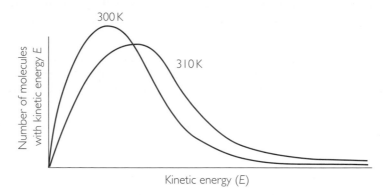

Boltzmann distribution of molecular kinetic energies in a gas at two temperatures. The modal speed gets higher as the temperature rises. The area under the curve gives the total number of molecules. This does not change as the temperature rises so the peak height falls as the curve widens.

The Boltzmann distribution is important in the *collision theory* of reaction rates and helps to account for the effects of temperature changes and *catalysts* on *rates of reaction*.

medicine: a substance or mixture of substances used to treat a disease or to give relief from the symptoms of disease. A medicine normally contains one or more active *drugs* dissolved in water or mixed with an inert solid. Mixing the active ingredient with an inert material makes it easier to give an accurate dose.

Doctors must prescribe some medicines – these are 'prescription only medicines' (POMs). New medicines may start in this category but later be made available 'over the counter' once they have been proved safe.

Medicines which people can buy from a pharmacist without a prescription are called over-the-counter (OTC) drugs. The sale of some of these medicines has to be supervised by a pharmacist – these are P medicines. People can buy other medicines such as mild painkillers with no supervision – these are GSL medicines, as they are on the 'general sale list'.

For a medicine taken by mouth, added flavouring may make the medicine more palatable and colouring can help to identify a drug. Other drugs are taken by injection or drawn into the lungs by a deep breath.

The *pharmaceutical industry* employs many chemists in *drug development* for medicines.

mechanism of a reaction: a description of how a reaction takes place showing step by step the bonds which break and the new bonds which form. Some mechanisms involve *homolytic bond breaking* with *free-radical* intermediates. Other mechanisms involve *heterolytic bond breaking* and ionic intermediates.

Mechanisms with ionic intermediates can be classified according to the type of attacking molecules or ions involved. *Nucleophiles* attack atoms at the δ+ end of *polar covalent bonds*. *Electrophiles* attack regions which are electron rich, especially double bonds and the delocalised electrons in arenes.

Evidence used to support a proposed mechanism include the:

- *rate equation* for the reaction
- identification of *intermediates* using *spectroscopy* or other methods
- use of *isotopic labelling* to track what happens to particular atoms during a reaction.

melting is a change of state from a solid to a liquid. Another word for melting is fusion. The melting point for a pure substance is the temperature at which the solid and liquid are in equilibrium. It is the same temperature as the *freezing point*. Melting points vary with pressure but only very slightly.

Pure compounds have sharp melting points. Measuring melting points is a way of checking the purity and identity of compounds. This technique is especially important in *organic chemistry*.

Molecular substances melt at relatively low temperatures because the intermolecular forces between molecules are weak. Giant structures, with strong bonding between the atoms, have high melting points. Energy is needed to overcome the bonds between particles as a substance melts (See *enthalpy change of melting*.)

meniscus: the curved surface of a liquid in a tube. Water forms a concave meniscus in a glass tube because water molecules can form *hydrogen bonds* with oxygen atoms at a clean glass surface. This allows water to *wet* the glass. Mercury atoms bond strongly with themselves but not with atoms in glass, so mercury has no tendency to wet glass and forms a convex meniscus.

Chemists have to allow for the shape of the meniscus when calibrating and reading graduated glassware such as pipettes and graduated flasks. The correct procedure is to adjust the level until the bottom of the meniscus just touches the mark as seen from eyes on a level with the mark.

mercury (Hg) is the only liquid metal at room temperature. As a dense liquid, mercury (density 13.6 g cm^{-3}) has long been used to measure pressure in *manometers* and barometers. A *pressure* of 1 atmosphere is the pressure at the bottom of a column of mercury 760 mm high.

Mercury forms *amalgams* with metals. In the chlor-alkali industry, a mercury cell for the *electrolysis of brine* has a flowing mercury cathode. The product of electrolysis at the cathode is an amalgam of sodium in mercury. The amalgam flows to a separate vessel where the sodium in the amalgam reacts with water to make sodium hydroxide and hydrogen. This process produces very pure sodium hydroxide but it is being phased out because of the pollution problems caused by the traces of mercury which escape into the environment. Mercury vapour and mercury compounds are very *toxic*.

metal extraction involves two main types of process to reduce metal compounds to metals:

- **pyrometallurgy** – reactions at high temperature often above 1000°C;
  - electrolysis of a molten compound (for example *aluminium extraction*)
  - chemical reduction by coke in a blast furnace (for example *iron extraction*)
  - chemical reduction by a more reactive metal (for example *titanium extraction*)

- **hydrometallurgy** – reactions at low temperature in solution in water;
  - electrowinning using *electrolysis* of an aqueous solution (for example *zinc extraction*)
  - cementation using a *displacement reaction* (for example using iron to displace copper from a solution of copper(II) sulfate).

metallic bonds: the strong bonding between atoms in metal crystals. Each metal atom in a metal crystal contributes electrons from its outer shell to a 'sea' of *delocalised* electrons. Only elements with relatively low first *ionisation energies* form metallic crystals.

By losing electrons the metal atoms become positively charged. So a metal crystal consists of positively charged metal atoms held together by a 'sea' of shared electrons.

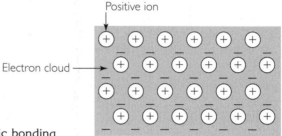

Simplified picture of metallic bonding

Metals conduct because the shared bonding electrons can drift through the crystal structure from atom to atom when there is an electric potential difference.

Metals can bend without breaking because metallic bonding is not highly directional. Lines or layers of metal atoms can shift their position in a crystal without the bonds breaking.

metalloid: an element with properties which are in between those of metals and non-metals. One example of such intermediate behaviour is that metalloids are *semiconductors*. Their electrical conductivities are between those of metals and non-metal insulators.

In *group 4*, the metalloid germanium comes between two non-metals at the top of the group and two metals at the bottom. Other examples of metalloids are *arsenic*, antimony and tellurium. Some chemists also classify silicon and boron as metalloids.

metals are elements on the left-hand side of the *periodic table* with one, two or three electrons in the outer shell which take part in bonding and chemical reactions.

Physically, metals are :

- shiny when freshly polished and free of corrosion
- good conductors of electricity and thermal energy
- *malleable* and *ductile*
- (usually) solids with high melting and boiling points (only six metals melt below 100°C – mercury which is a liquid, gallium and four group 1 metals sodium, potassium, rubidium and caesium).

Chemically, metals tend to:

- lose electrons forming positive ions (they are *reducing agents*)

- form *basic* or *amphoteric oxides* and hydroxides (which are *alkalis* if soluble in water)
- form solid, ionic chlorides.

The more an element shows these properties the greater its 'metallic character'.

methanol, CH₃OH, is a liquid alcohol which is becoming more important as a fuel or *petrol* additive and as a raw material for the chemical industry. Industry can manufacture methanol from the mixture of *carbon monoxide* and hydrogen produced by *steam reforming*. Oxidation of methanol by air, in the presence of an iron or silver catalyst, produces the methanal required to make thermosetting resins.

Methanol reacts with carbon monoxide to make ethanoic acid in the presence of a catalyst made from rhodium or iridium with iodide ions. This process happens in the liquid phase and is an example of *homogeneous catalysis*.

$$CH_3OH(l) + CO(g) \rightleftharpoons CH_3CO_2H(l)$$

micelle: a cluster of *surfactant* molecules in a solution. A micelle has the 'water-hating' (hydrophobic) hydrocarbon chains tangled together inside with the 'water-loving' (hydrophilic) ends of the molecules on the outside.

The inside of a micelle is non-polar. Unlike water, a solution of a surfactant in water can dissolve oil and grease. The molecules of oily dirt diffuse to the inside of the micelles.

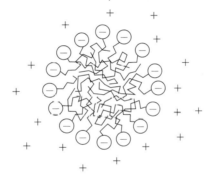

Cross section of a spherical micelle formed when a surfactant dissolves in water. A micelle is likely to consist of about 50 to a 100 molecules. The diameter of a micelle is in the range 5–50 nm.

microwaves: *electromagnetic radiation* with wavelengths in the range 1 mm to 30 cm and frequencies in the range 10^8 to 10^9 Hz. A polar molecule can absorbs microwave radiation. The quanta of microwave radiation correspond to the energy jumps when molecules gain energy and rotate faster (see *quantum theory*).

A microwave oven is 'tuned' so that the radiation is absorbed by water molecules to make them spin. The microwaves can pass through food where they are absorbed by liquid water molecules. As the water molecules gain energy they bump into neighbouring molecules and the energy spreads through the food as the molecules spin, vibrate and move around more, so the food gets hotter.

millilitre (symbol ml) is a unit of *volume* often used on chemical glassware, in medicine and domestically. A millilitre is one thousandth of a *litre*.

$$1000 \text{ ml} = 1000 \text{ cm}^3 = 1 \text{ dm}^3 = 1 \text{ litre}$$

monoprotic acid: an *acid* which can give away (donate) only one *proton* per molecule. Examples of monoprotic acids are: hydrogen chloride, HCl; ethanoic acid, CH_3CO_2H; and nitric acid, HNO_3.

monosaccharide: see *carbohydrates*.

mordant: a chemical which fixes a dye to the fibres of a textile. Alums and other aluminium salts are used as mordants. *Dative bonds* form both between — OH groups of cotton and aluminium ions and between oxygen atoms on the dye and aluminium ions.

Aluminium ions acting as a mordant for the dye alizarin

Alizarin is a natural red dye once extracted from the roots of the madder plant but now made synthetically. Alizarin only sticks fast to cloth in the presence of a mordant and the colour of the dye varies depending on the metal ions in the mordant.

multiple bonds are *double bonds* or *triple bonds* between atoms in molecules. In a double bond there are two shared pairs of electrons. In a triple bond there are three shared pairs.

mutagens are agents that cause mutations. A mutation is a change in the genetic material of living cells. Some chemicals are mutagenic and so are the forms of *ionising radiation*.

names of carbon compounds: modern *IUPAC* names make it possible to work out the name from the formula and the formula from the name.

The name of an organic compound is based on the longest straight chain or main ring of carbon atoms in the skeleton of carbon atoms. If the main part is a straight chain the name is based on the corresponding *alkane*. For ring compounds the name is based on the corresponding *alicyclic* compound or *arene*.

Numbering the carbon atoms identifies the positions of side chains and *functional groups*, with the number repeated if there are two side groups on the same carbon atom. At all times chains are numbered from the end that gives the lowest possible numbers in the name. Numbers are omitted when there is no doubt about the position of the side chains or functional groups as in ethanol.

Prefixes (in front) and suffixes (following) the hydrocarbon name identify the side chains and functional groups.

Where there are two or more of the same side chain or functional group the number is stated as di, tri, tetra and so on.

Prefixes are used for:

- *alkyl groups*, such as 2,3-dimethylbutane or 1,3,5-trimethylbenzene
- *halogenoalkanes*, such as 2-iodopropane or tetrachloromethane.

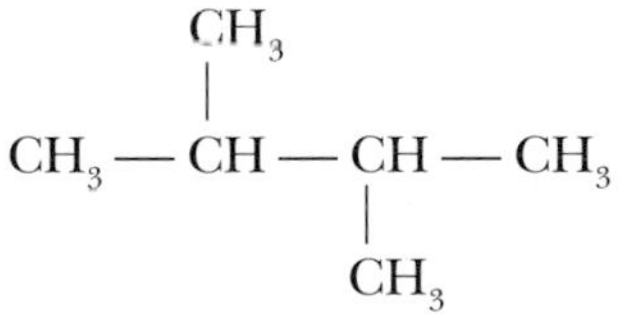

2,3-dimethylbutane

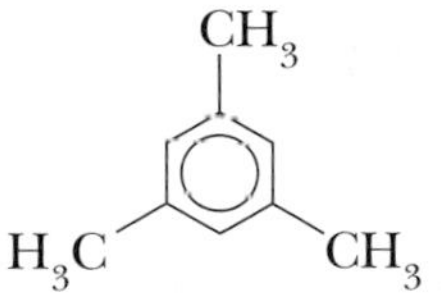

1,3,5-trimethylbenzene

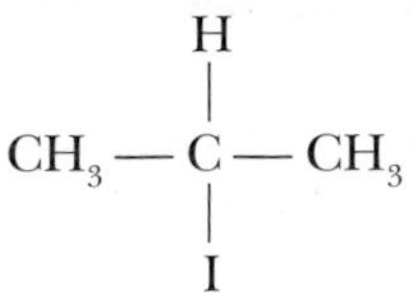

2-iodopropane

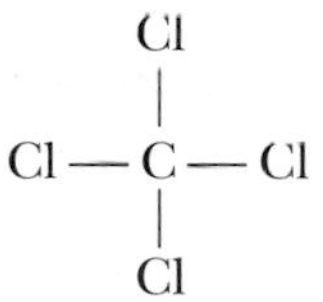

tetrachloromethane

Structures of the examples

Suffixes are used for:

- double bonds in *alkenes*, such as but-2-ene (with a double bond between the second and third carbon atoms)
- *alcohols*, such as propan-1,2,3-triol
- *aldehydes*, such as propanal

- *ketones*, such as pentan-3-one
- *carboxylic acids*, such as hexanoic acid and ethanedioic acid
- *amines*, such as ethylamine and *phenylamine*.

$$CH_3 — CH_2 = CH_2 — CH_3$$

but-2-ene

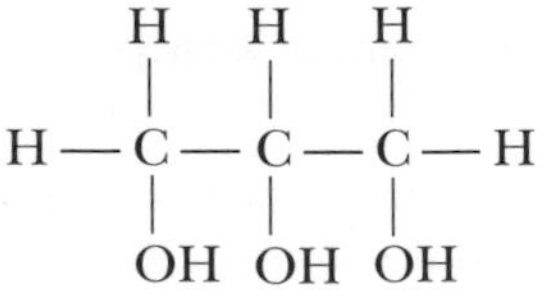

$$H—C—C—C—H$$
with H, H, H on top and OH, OH, OH on bottom

propan-1,2,3-triol

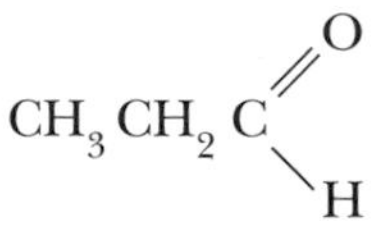

$$CH_3 \, CH_2 \, C \overset{O}{\underset{H}{<}}$$

propanal

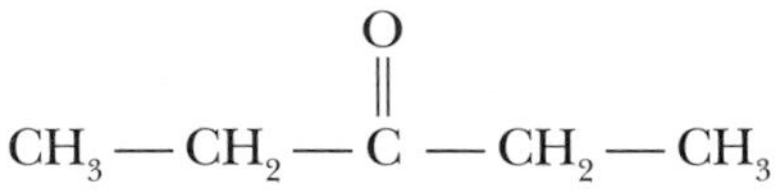

$$CH_3 — CH_2 — \overset{O}{\overset{\|}{C}} — CH_2 — CH_3$$

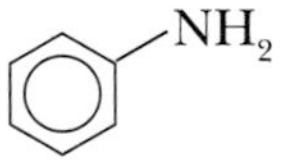

pentan-3-one

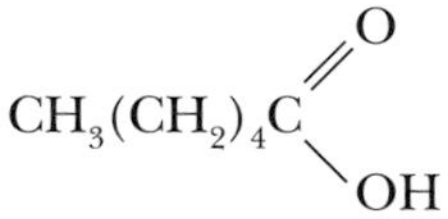

$$CH_3(CH_2)_4 C \overset{O}{\underset{OH}{<}}$$

hexanoic acid

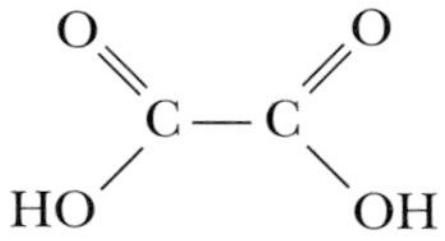

$$\underset{HO}{\overset{O}{\|}} C — C \underset{OH}{\overset{O}{\|}}$$

ethanedioic acid

$$CH_3 — CH_2 — NH_2$$

ethylamine

phenylamine (benzene ring with NH_2)

Structures of the examples

For other examples see entries for the various series of organic compounds.

The systematic names of complex molecules can be hard to pronounce and cumbersome. Chemists still often use older, traditional names which were often based on the Latin name for the source of the compound or where it was discovered. The organic acid 2,3-dihydroxybutanedioic acid, for example, occurs in grape juice. The traditional name is tartaric acid from 'tartar' the name of the recrystallised deposits of tartaric acid salts collected from inside wine containers.

names of complex ions: the systematic names of complex ions show:

- first the number of *ligands*, 'di', 'tri', 'tetra', 'penta', 'hexa'
- then the type of ligands (in alphabetical order if there is more than one type of ligand), such as 'aqua' for water molecules, 'ammine' for ammonia, 'chloro' for chloride ions and 'cyano' for cyanide ions
- next the identity of the central metal atom in a form which shows whether or not the ion is a cation or an anion:
  - for cations (and uncharged complexes) the metal name is normal, such as silver, iron or copper

- for anions the metal name ends in 'ate' and often has an old-fashioned style such as argentate for silver, ferrate for iron and cuprate for copper
- finally the oxidation number of the metal.

Examples:

- $[Ag(NH_3)_2]^+$ is the diamminesilver(I) ion,
- $[Cu(H_2O)_6]^{2+}$ is the hexaaquacopper(II) ion,
- $[CuCl_4]^{2-}$ is the tetrachlorocuprate(II) ion,
- $[Fe(CN)_6]^{4-}$ is the hexacyanoferrate(II) ion,
- $[Fe(CN)_6]^{3-}$ is the hexacyanoferrate(III) ion.

Note that the overall charge on the complex ion is the sum of the charges on the metal ion and the ligands.

names of inorganic compounds are becoming increasingly systematic but chemists still use a mixture of names. Most chemists prefer to call $CuSO_4.5H_2O$ hydrated copper(II) sulfate, or perhaps copper(II) sulfate-5-water but not the fully systematic name tetraaquocopper(II) tetraoxosulfate(VI)-1-water. The systematic names has much more to say about the arrangement of atoms, molecules and ions in the blue crystals but it is too cumbersome for normal use.

These are some of the basic rules for common inorganic names:

- the ending 'ide' shows that a compound contains just two elements mentioned in the name. The more *electronegative* element comes second, for example, sodium sulfide (Na_2S) carbon dioxide (CO_2) and phosphorus trichloride (PCl_3)
- the small Roman numerals in names are the oxidation numbers of the elements, for example iron(II) sulfate ($FeSO_4$) and iron(III) sulfate, $Fe_2(SO_4)_3$
- the names of *oxoacids* end 'ic' or 'ous' as in sulfuric (H_2SO_4) and sulfurous (H_2SO_3) acids and nitric (HNO_3) and nitrous (HNO_2) acids, where the 'ic' ending is for the acid in which the central atom has the higher oxidation number
- the corresponding endings for the salts of oxoacids are 'ate' and 'ite' as in sulfate (SO_4^{2-}) and sulfite (SO_3^{2-}) and in nitrate (NO_3^-) and nitrite (NO_2^-).

To avoid misunderstandings chemists give the name and formula. If necessary they may also give two names: the systematic name and the traditional name.

nanometre: the unit of length used for the sizes of atoms, molecules and ions. One nanometre (1 nm) is one thousand millionths of a metre (10^{-9} m). So there are one million nanometres in a millimetre (1000 nm = 1 mm). One nanometre is roughly five times the diameter of a hydrogen atom.

nanotubes are single-walled cylinders of carbon atoms created as graphite-like sheets of atoms which curl up into tubes, similar to rolls of chicken wire. They take their name from the *nanometer* which indicates their scale. Nanotubes form under similar conditions to *buckminsterfullerene* but with a few per cent of nickel or cobalt atoms added to the mixture of carbon vapour and helium. Research into the properties of nanotubes is exploring the possibility that they could be adapted to make minute diodes, transistors and other electronic devices on an atomic scale – much smaller than existing devices on silicon chips. Nanotubes are also being incorporated into plastics to make them conduct electricity. Nanotube fibres are very strong and tough. In time they may become key ingredients of new composite materials.

naphtha is a mixture of hydrocarbons from the *fractional distillation of oil*. The naphtha fraction is an important feedstock for the *petrochemical industry*. Naphtha contains *hydrocarbons* with 6 to 10 carbon atoms in their molecules.

narcotics are powerful painkillers (*analgesics*) which also lessen anxiety and give a feeling of well-being. Regular use of narcotic analgesics can lead to drug dependence. Morphine is the oldest and most well known narcotic which is obtained from the dried juice of opium poppies (so it is an opiate). Morphine is used medically to treat patients in severe pain.

Diamorphine, otherwise known as heroin, is made from morphine. It is an even more powerful analgesic than morphine. It is also highly addictive so its medical use is limited mainly to the treatment of pain in patients who are going to die.

Codeine is another opiate. It is a much less powerful painkiller than morphine but it does not cause dependence. Unlike morphine, codeine can be taken by mouth. Codeine is an ingredient of some cough medicines because it helps to stop people coughing.

natural gas is a *fossil fuel* used for domestic heating, for raising steam in power stations and as a feedstock for the *petrochemical industry*. The composition of natural gas varies from one gas field to the next. It consists mainly of methane mixed with other *hydrocarbons* such as ethane, propane and butane together with variable amounts of carbon dioxide, nitrogen, helium and hydrogen sulfide.

Natural gas is processed before being supplied to homes. Hydrocarbons other than methane are separated for used in the petrochemical industry. Sulfur compounds are removed and a trace of a smelly chemical is added so that people notice gas leaks.

natural product chemistry is the study of chemicals produced by living things. Natural products include *carbohydrates, proteins, nucleic acids* and *lipids*. Before the start of the modern organic chemical industry, natural products from plants were the important ingredients of perfumes, dyes, oils, food flavours and drugs.

Plants may become important again as a large scale source of chemicals now that genetic engineering makes it possible to transfer genes from one species to another.

neon (Ne) is the second member of the family of *noble gases*, coming below helium, with the *electron configuration* $[He]2s^22p^6$. Neon is best known for the red glow of neon lamps and tubes. It is separated from air as the other gases are liquefied. Neon boils at 27 K so it does not condense at the temperatures used to liquefy oxygen, nitrogen and argon.

neutralisation reaction: a reaction in which an acid reacts with a base to form a *salt*.

$$HCl(aq) + NaOH(aq) \longrightarrow NaCl(aq) + H_2O(l)$$

Mixing equal amounts (in moles) of hydrochloric acid with sodium hydroxide produces a *neutral solution* of sodium chloride.

Strong acids, such as hydrochloric acid, and *strong bases*, such as sodium hydroxide, are fully ionised in solution, as is the salt formed, sodium chloride. Writing ionic equations for these examples shows that neutralisation is essentially a reaction between aqueous hydrogen ions and hydroxide ions. This is supported by the values for the *enthalpy changes of neutralisation.*

$$H_3O^+(aq) + OH^-(aq) \longrightarrow 2H_2O(l)$$

The surprise is that neutralisation reactions do not always produce neutral solutions. Neutralising a weak acid such as ethanoic acid with an equal amount in moles of a strong base sodium hydroxide produces a solution of sodium ethanoate which is alkaline.

Neutralising a weak base, such as ammonia, with an equal amount of the strong acid hydrochloric acid produces a solution of ammonium chloride which is acidic.

Where a salt has either a 'parent acid' or a 'parent base' which is weak it dissolves to give a solution which is not neutral (see *hydrolysis of salts*). The 'strong parent' in the partnership 'wins':

- weak acid/strong base – the salt is alkaline in solution
- strong acid/weak base – the salt is acidic in solution.

neutral oxides are non-metal oxides which do not react with water to form acids. The common examples are carbon monoxide (CO), nitrogen monoxide (NO) and dinitrogen oxide (N_2O). All three of these oxides are insoluble in water.

neutral solution: a solution with pH 7. Since pH = $-\lg [H_3O^+]$, this means that the concentration of aqueous hydrogen ions in a neutral solution is 10^{-7} mol dm^{-3} at 298 K.

neutrons are the uncharged particles in the nuclei of *atoms*. The number of neutrons in an atom can vary without changing its chemical properties. This accounts for the existence of *isotopes*.

nickel (Ni) is a hard, greyish but shiny *d-block* metal with the electron configuration $[Ar]3d^84s^2$. Nickel is relatively unreactive so it is used to make spatulas and crucibles. Nickel is a constituent of many alloys including some alloy *steels* and the ferromagnetic alloy Alnico in permanent magnets.

The common oxidation state of nickel is +2. Nickel(II) salts, such as the sulfate $NiSO_4$, are green.

Finely divided nickel metal is a good catalyst for *hydrogenation* reactions. It is a *heterogeneous catalyst*. It is used to 'harden' unsaturated *vegetable oils* by adding hydrogen across the double bonds.

After extraction from its ores, nickel is purified by *electrolysis*. An older process for purifying nickel took advantage of the fact that the metal forms a volatile, neutral complex with carbon monoxide. The oxidation state of the metal is zero, so the complex is called tetracarbonylnickel(0). The impure metal reacts with carbon monoxide at 50°C. The complex evaporates leaving the impurities behind. The vapour of the complex passes to a second vessel where it decomposes at about 200°C.

$$Ni(s) + 4CO(g) \rightleftharpoons [Ni(CO)_4](l)$$

This was an effective but hazardous process because both carbon monoxide and the nickel complex are highly toxic.

nickel–cadmium cells are now widely used as rechargeable 'NiCad' cells. As in a *lead–acid cells,* when a current flows the products formed at the electrodes are insoluble solids which stay put instead of dissolving in the electrolyte. This means that the electrode processes can be reversed as the cell is recharged. The electrode processes as a NiCad cell supplies a current are:

anode (oxidation): $Cd(s) + 2OH^-(aq) \longrightarrow Cd(OH)_2(s) + 2e^-$

cathode (reduction): $NiO(OH)(s) + H_2O(l) + e^- \longrightarrow Ni(OH)_2(s) + OH^-(aq)$

Unlike lead–acid cells, 'NiCad' cells must be regularly discharged fully and then recharged if they are to retain their full capacity.

nitrates are *salts* of nitric acid (HNO_3) which all contain the nitrate ion (NO_3^-). Nitrates are common in laboratory chemistry because they are all soluble in water. Since nitric acid is a *strong acid,* the nitrate ion is a very weak base and does not change the pH when dissolved in water.

The reason that the nitrate ion is a weak base is that the negative charge is *delocalised* over the planar ion. Delocalisation stabilises the ion (see *oxoacids*).

Heating decomposes nitrates. Hydrated nitrates first give off steam, then the usual products are the metal oxide, oxygen and the brown gas nitrogen dioxide.

$$2Mg(NO_3)_2(s) \longrightarrow 2MgO(s) + O_2(g) + NO_2(g)$$

The exceptions are the nitrates of sodium and potassium which are hard to decompose and give only the nitrite and oxygen.

$$2NaNO_3(s) \longrightarrow 2NaNO_2(s) + O_2(g)$$

Nitrates are *oxidising agents.* In alkaline conditions they are reduced to ammonia by aluminium metal on heating. This reaction is used as a test for nitrates since the basic ammonia gas can be detected by turning litmus paper blue (see *anion tests*). More effective than aluminium is an alloy of aluminium, zinc and copper (Devarda alloy).

$$NO_3^-(aq) + 6H_2O(l) + 8e^- \longrightarrow NH_3(g) + 9OH^-(aq)$$

Plants take up nitrogen from the soil in the form of nitrate ions. So a diet rich in vegetables is rich in nitrate too. Nitrates have had a bad press because they are associated with the environmental problem *eutrophication.*

nitration of benzene is an *electrophilic substitution* reaction which takes place in the presence of a nitrating mixture of concentrated nitric and sulfuric acids. The product is nitrobenzene.

The purpose of the concentrated sulfuric acid is to produce an electrophile which is the nitronium ion, NO_2^+. The concentrated sulfuric acid gives a proton to the $—OH$ group in nitric acid which then loses a molecule of water, forming the nitronium ion.

$$HO—NO_2 + H_2SO_4 \longrightarrow \underset{H}{HO^+}—NO_2 + HSO_4^-$$

$$\downarrow$$

$$H_2O + NO_2^+$$

Mechanism for the
nitration of benzene

Nitration of arenes is important because it produces a range of useful products including the explosive tnt (trinitrotoluene, or 1-methyl-2,3,5-trinitrobenzene). Nitro compounds are also intermediates in the synthesis of chemicals used to make *polyurethanes*. Nitro groups are easily reduced to amine groups. Aryl amines, such as *phenylamine*, are important intermediates in the production of dyes, such as *azo dyes*.

nitric acid (HNO$_3$): pure nitric acid is a colourless, fuming liquid but it gradually turns yellow as it decomposes forming nitrogen dioxide. It is a highly corrosive but important chemical reagent because it can act as a:

- **strong acid** – nitric acid ionises fully as it dissolves in water:
  $HNO_3(l) + H_2O(l) \longrightarrow H_3O^+(aq) + NO_3^-(aq)$.
  Dilute nitric acid neutralises bases producing soluble salts called *nitrates*.
- **powerful oxidising agent** – in this role nitric acid oxidises:
  - most metals including metals such as copper which do not react with non-oxidising acids
  - non-metals, forming oxides in the highest oxidation state of the element, giving I$_2$O$_5$ with iodine, for example
  - ions in solution such as iodide ions to iodine and iron(II) ions to iron(III)
- **nitrating agent** – in this role it is used for the *nitration of benzene* and other arenes.

nitric acid manufacture: a process for converting ammonia to nitric acid in two stages:

- oxidation of ammonia by oxygen on the surface of a catalyst gauze made of an alloy of platinum and rhodium:
  $4NH_3(g) + 5O_2 \rightleftharpoons 4NO(g) + 6H_2O(g) \qquad \Delta H = -909 \text{ kJ mol}^{-1}$
- absorption in water in the presence of oxygen to make nitric acid:
  $2NO(g) + O_2(g) \rightleftharpoons 2NO_2(g)$
  $4NO_2(g) + O_2(g) + 2H_2O(l) \longrightarrow 4HNO_3(g)$

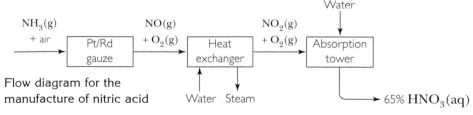

Flow diagram for the
manufacture of nitric acid

About 80% of the three to four million tonnes of nitric acid produced each year in the UK is converted to ammonium nitrate. Ammonium nitrate is largely used as a nitrogen *fertiliser*. Ammonium nitrate is also widely used as part of most *explosives* for mining and quarrying. Nitric acid is used to make other explosives such as nitrocellulose and nitroglycerine.

- during thunderstorms the energy in lightning allows nitrogen and oxygen to combine to form oxides of nitrogen which are washed into the soil by rain and taken up by plants as nitrates
- some soil bacteria have an enzyme nitrogenase which can harness the energy of *ATP* to convert nitrogen from the air into ammonium compounds
- bacteria in the nodules of leguminous plants such as peas, beans and clover can also make ammonium salts from nitrogen.

As agriculture became more intensive the demand for nitrogen by crops outstripped the rate it could be produced by natural processes. Manures could not meet the demand either, so the challenge for chemists was to find a practicable and economic way to fix nitrogen industrially. This is the problem which was solved by Fritz Haber, leading to *ammonia manufacture* by the Haber process. Some of the ammonia is used for *nitric acid manufacture* and the two chemicals combine to make nitrogen *fertilisers*.

nitrogen oxides: there are six compounds of nitrogen with oxygen. All the oxides have positive enthalpies of formation. They are unstable relative to the elements and are easily decomposed by heating. Two of the oxides, N_2O_3 and N_2O_5, are particularly unstable and of little practical importance. Oxides of nitrogen contribute to air pollution. NO_x is a general formula used in accounts of air pollution to stand for any of the various nitrogen oxides when fuels burn at high temperatures in engines and furnaces. NO_x contributes to *acid rain* and the formation of *photochemical smog*.

Oxidation state	Formula	Name	Type	$\Delta H_f^{\ominus}$ / kJ mol^{-1}	Notes
+4	$2NO_2 \rightleftharpoons N_2O_4$	nitrogen dioxide and dinitrogen tetroxide	acidic	+33.2 (for NO_2)	NO_2 is the brown gas formed on heating nitrates. It is used as an oxidant in rocket fuels. Lowering the temperature makes the gas paler as the equilibrium favours the colourless N_2O_4.
+2	NO	nitrogen monoxide	neutral	+90.2	Colourless but reacts with oxygen in the air to make NO_2; an intermediate in the manufacture of nitric acid from ammonia.
+1	N_2O	dinitrogen monoxide	neutral	+82.0	Colourless; also called laughing gas.

Laughing gas is the traditional name for dinitrogen oxide (N_2O). The gas can be used as an *anaesthetic* for dentistry and minor surgery. Dinitrogen oxide is soluble in fats. It is tasteless and non-toxic, and is used as the foaming agent and propellant in cans of whipped cream.

In recent years scientists have discovered that NO is a vital messenger molecule in the human body. When tiny quantities of NO are released from cells lining blood vessels the muscles around the vessels relax so that they can carry more blood. This prevents the blood pressure rising during exercise and avoids strain on the heart.

nitrous acid (HNO$_2$) is an unstable, *weak acid* which is an important reagent for producing the *diazonium salts* needed to make *azo dyes*. Nitrous acid is too unstable to be stored so it is made in solution when needed by adding a strong acid, such as hydrochloric acid, to a solution of sodium nitrite:

$$NO_2^-(aq) + H_3O^+(aq) \longrightarrow HNO_2(aq) + H_2O(l)$$

The solution of the acid is blue. It starts to decompose at room temperature, giving off nitrogen monoxide (NO) which turns brown as it meets the air and turns into nitrogen dioxide (NO$_2$).

The systematic name of the acid is dioxonitric(III) acid which is sometimes abbreviated to nitric(III) acid. This shows that nitrogen is in the +3 oxidation state. When it decomposes it *disproportionates* to the +5 and the +2 states.

$$3HNO_2(aq) \longrightarrow HNO_3(aq) + 2NO(g) + H_2O(l)$$

| +3 | +5 | +2 | oxidation states of nitrogen |

nmr: see *nuclear magnetic resonance spectroscopy*.

noble gases: the unreactive gases in *group 8* of the periodic table. They were called the 'inert gases' until the discovery that *krypton* and *xenon* can form compounds. Even so the gases keep apart from most chemical changes and react only with highly reactive chemicals such as fluorine and oxygen.

The term 'noble' has been used for a long time to describe metals such as gold and silver. These metals were *inert* to the reagents used by alchemists and early chemists. Like the nobility in society, these elements were seen to be special by standing apart from everyday changes and ordinary events.

nomenclature: see *names of carbon compounds, names of complex ions* and *names of inorganic compounds*.

non-aqueous solvent: any *solvent* other than water. A non-aqueous solvent has to be used if the *solute* will not dissolve in water. For this reason white spirit (a *hydrocarbon* solvent) is used to make oil paint while other organic solvents are used as nail-varnish removers, stain removers and dry-cleaning fluids. A non-aqueous dry cleaning fluid removes dirt without causing shrinkage or other damage to fabrics harmed by water.

A non-aqueous solvent has to be used for a reaction in solution if one of the reactants or products is rapidly hydrolysed by water. For this reason *lithium tetrahydridoaluminate(III)* reductions take place in ether solution.

The choice of solvent can affect the outcome of a reaction. A solution of potassium hydroxide or sodium hydroxide in water hydrolyses *halogenoalkanes* to alcohols. If the solvent is ethanol it is much more likely that an *elimination* reaction will happen and the product will be an alkene.

non-metals are the elements towards the right-hand side of the *periodic table* which do not show the characteristic properties of metals. In general non-metals:

- consist of small molecules with the atoms linked by *covalent bonding* (H_2, S_8, P_4 and Cl_2); the exceptions include carbon, silicon and red phosphorus in which covalent bonding is continuous throughout the *giant structure* of atoms
- are mostly gases at room temperature because of the weak *intermolecular forces* (for example hydrogen, oxygen, nitrogen, chlorine and all the noble gases)
- if solid, are not shiny like metals; they are brittle rather than bendable and do not normally conduct electricity or thermal energy
- tend to form negative ions by gaining electrons when they react with metals
- form *acidic oxides* which react with water to produce *oxoacids*
- form covalent molecular chlorides which are liquids and usually rapidly hydrolysed by water, such as PCl_3 and $SiCl_4$
- form covalent molecular hydrides which are gases at room temperature, such as CH_4, SiH_4, NH_3, HCl, HBr and HI – the notable exception is water which is a liquid because of *hydrogen bonding*.

non-polar solvent: a solvent in which the molecules are not *polar*. Examples of non-polar solvents are the many liquids *hydrocarbons* and the mixtures of hydrocarbons obtained by refining crude oil. These are also *non-aqueous solvents*.

Following the general rule that 'like dissolves like', non-polar solvents dissolve many compounds which consist of small molecules. Chlorine, bromine and iodine, for example, dissolve freely in hexane. The stain removers and dry cleaning fluids intended to remove oily grease from clothes consist of non-polar solvents.

Non-polar solvents do not dissolve ionic crystals. The interaction between non-polar molecules and ions is much too weak to surround ions and pull them away from the crystal against the *electrostatic forces* between ions with the opposite charge.

non-stoichiometric compounds are compounds with formulae which do not have simple whole number ratios of atoms. The formulae of non-stoichiometric compounds vary. Many oxides and sulfides of *d-block* elements have variable formulae for instance iron(II) sulfide ($Fe_{1.1}S$ to $FeS_{1.1}$) and iron(II) oxide ($Fe_{1.06-1.19}O$). This variation is a result of variable oxidation states and irregularities in the crystal lattice. Other examples are the *interstitial hydrides* of *d*-block elements such as vanadium hydride, $VH_{0.6}$.

NO$_x$: see *nitrogen oxides*.

nuclear fusion is the process which produces the energy of the Sun and other stars. It is also the process which accounts for the origin of all the elements.

In the Sun, at a temperature of ten million degrees or so, hydrogen atoms fuse to make helium atoms releasing about 10^9 kJ per mole of helium formed.

$$4\,{}^1_1H \longrightarrow {}^4_2He + 2 \text{ positrons}$$

The pressure and temperature at the centre of very large stars is high enough for helium nuclei to fuse to make heavier elements such as carbon, silicon and iron. Since helium has an even number of *nucleons*, it turns out that elements with even *atomic numbers* are more abundant than other elements.

Iron is the final product of the series of *exothermic* fusion reactions. Energy is needed to make heavier elements. Heavier elements form during the highly energetic explosion when massive stars run out of nuclear fuel and start to collapse and then burst apart. The massive explosion is a supernova which scatters dust and gas through the universe where it mixes with hydrogen and helium.

In time the remains of 'dead' stars start to coalesce into new stars and the process starts all over again. The Sun is an example of a second generation star; it and the planets of the solar system formed from the remains of supernovae. This explains the variety of elements on Earth in a Universe where over 90% of all atoms are hydrogen.

nuclear magnetic resonance spectroscopy (nmr) is a powerful analytical technique for finding the structures of carbon compounds. The technique is used to identify unknown compounds, to check for impurities and to study the shapes of molecules.

In medicine, magnetic resonance imaging uses nmr to detect the hydrogen nuclei in the human body, especially in water and lipids. A computer translates the information from a body scan into 3-D images of the soft tissue and internal organs which are normally transparent to X-rays.

The name of the technique summarises its key features:

- **nuclear** – the technique detects nuclei of atoms such as hydrogen-1 (protons)
- **magnetic** – the nuclei detected by the technique are the ones which act like tiny magnets which can line up either in the same direction or in the opposite direction to an external magnetic field
- **resonance** – the absorption of energy is from radiowaves with the frequency corresponding to the size of the energy jump as the nuclei flip from one alignment in a magnetic field to the other.

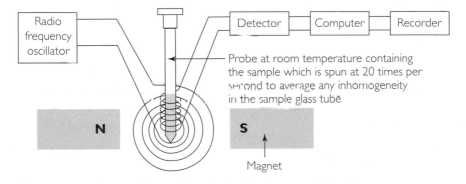

Diagram of an nmr spectrometer

Samples for analysis are dissolved in a solvent with no hydrogen-1 atoms. Also in the solution is some tetramethyl silane (TMS) which is a standard reference compound which produces an absorption peak well away from the sample peaks.

The tube with the sample is supported in a strong magnetic field. The operator turns on an oscillator which produces radiation at radio frequencies (rf). The rf detector records the intensity of the signal as the oscillator scans across a range of wavelengths.

The recorder prints out a spectrum with peaks wherever the sample absorbs radiation strongly. The zero on the scale is fixed by the absorption of hydrogen atoms in the reference chemical TMS. The distances of the sample peaks from zero are called their 'chemical shifts' (δ).

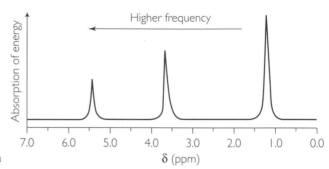

nmr spectrum for
ethanol at low resolution

Each peak corresponds to a hydrogen nucleus in a different chemical situation. The area under a peak is proportional to the number of nuclei in each situation. In ethanol (CH_3CH_2OH) there are hydrogen nuclei in three environments which have different chemical shifts.

At high resolution it is possible to produce nmr spectra with more detail, which provide even more information about molecular structures.

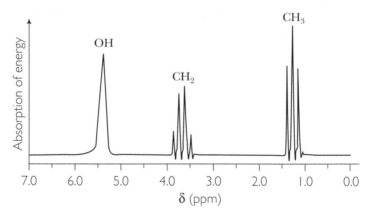

High resolution nmr spectrum for ethanol. Note the extra peaks compared with the low resolution spectrum. This illustrates the '$n + 1$' rule. A peak from protons bonded to an atom which is next to an atom with two protons splits into three lines. A peak from protons bonded to an atom which is next to an atom with three protons splits into four lines.

Protons connected to neighbouring atoms interact with each other. Chemists call this interaction 'coupling' and they find that the effect is to split the peaks into a number of lines.

nuclear reactions: changes affecting the nuclei of atoms. Nuclear reactions are brought about by bombarding materials with high energy particles in particle accelerators or with beams of neutrons from nuclear reactors.

Nuclear reactions can be used to make elements which do not occur naturally, with higher proton numbers than uranium. Bombarding uranium-238 with neutrons produces uranium-239 which decays by *beta decay* to neptunium. Neptunium in turn decays to plutonium.

$$^{238}_{92}U + ^{1}_{0}n \longrightarrow ^{239}_{92}U \xrightarrow{\text{decay}} ^{239}_{93}Np + ^{0}_{-1}e$$

neutron

β–particle

decay

$$^{239}_{94}Pu + ^{0}_{-1}e$$

β–particle

Nuclear reactions producing the transuranium elements neptunium and plutonium. Note that the *mass numbers* and *atomic numbers* balance.

Nuclear reactions are used to make radioactive isotopes for use as tracers.

nucleic acids are the molecules which carry genetic information and control protein synthesis in cells. There are two main types of nucleic acid, *DNA* and *RNA*. Nucleic acids are polynucleotides formed as *nucleotides* link together in long chains.

nucleon: a particle in the nucleus of an atom – either a *proton* or a *neutron*.

nucleon number: the nucleon number for an atom is an alternative term to *mass number*.

nucleophiles are molecules or ions with a lone pair of electrons which can form a new covalent bond. Nucleophiles are reagents which attack molecules where there is a partial positive charge, δ+. So they seek out positive charges – they are 'nucleus-loving'. Some examples of nucleophiles are:

H—O:⁻ hydroxide ion

H—O: (with H above) water molecule

:C≡N cyanide ion

H—N—H (with H above) ammonia molecule

:Br:⁻ bromide ion

CH_3CH_2—O:⁻ ethoxide ion

Examples of nucleophiles

Nucleophiles take part in *nucleophilic addition* and *nucleophilic substitution reactions*.

nucleophilic addition reaction: the attack of a nucleophile on a *carbonyl compound* leading to an *addition reaction*. The *electronegative* oxygen draws electrons away from carbon so that the $C = O$ bond in the aldehyde or ketone is *polar*. The incoming nucleophile uses its lone pair to form a new bond with the carbon atom. This displaces one pair of electrons in the double bond onto oxygen. Oxygen has thus gained one electron from carbon and now has a negative charge.

$$CN^- \overset{H}{\underset{H_3C}{\diagdown}} C = O \longrightarrow NC - \overset{H}{\underset{CH_3}{\overset{|}{C}}} - \overset{..}{\underset{}{O}}{}^-$$

intermediate

First step of nucleophilic addition of hydrogen cyanide to ethanal

To complete the reaction, the negatively charged oxygen acts as a *base* and gains a proton.

$$NC - \overset{H}{\underset{CH_3}{\overset{|}{C}}} - \overset{..}{O}{}^- \quad H - CN \longrightarrow NC - \overset{H}{\underset{CH_3}{\overset{|}{C}}} - OH + {}^-CN$$

intermediate

Second step of nucleophilic addition of hydrogen cyanide to ethanal

nucleophilic substitution in derivatives of carboxylic acids: nucleophilic attack on the carbon atom of the $C = O$ bond in the functional group which leads to a *substitution reaction*. The first step is similar to *nucleophilic addition*.

The *electronegative* oxygen draws electrons away from carbon so that the $C = O$ bond in the aldehyde or ketone is *polar*. The incoming nucleophile uses its lone pair to form a new bond with the carbon atom. This displaces one pair of electrons in the double bond onto oxygen. Oxygen has thus gained one electron from carbon and now has a negative charge.

$$CH_3 - \overset{O}{\underset{HO:\diagdown X}{\overset{\diagup}{C}}} \longrightarrow H_3C - \overset{:O^-}{\underset{OH}{\overset{|}{C}}} - X$$

The first step in the attack of a nucleophile on a derivative of a carboxylic acid. If X = OC_2H_5 the compound is an ethyl *ester*, if X = Cl it is an *acyl chloride*, if X = NH_2 it is an *amide*.

At this point the negative oxygen does not gain a proton; instead it uses a pair of electrons to reform the double bond and displace X.

The second step which eliminates X as an X⁻ ion

$$H_3C - \overset{:O^-}{\underset{OH}{\overset{|}{C}}} - X \longrightarrow CH_3 - \overset{O}{\overset{\diagup}{C}}{\diagdown}_{OH} + X^-$$

Nucleophiles which react with esters, acyl chlorides and amides in this way are hydroxide ions and hydride ions from the tetrahydridolithium(I) ion.

nucleophilic substitution in halogenoalkanes: *nucleophilic attack* on the carbon atom of the C—X bond in a *halogenoalkane*, RX, which leads to a *substitution reaction*, where X = Cl, Br or I. Study of the *rate equations* suggests that there are two different mechanisms.

Hydrolysis of *primary* **halogenoalkanes** such as bromobutane, is overall *second order*. The rate equation has the form: rate = $k[C_4H_9Br][OH^-]$.

The suggested *mechanism* shows the C—Br bond breaking as the nucleophile, OH^-, forms a new bond with carbon.

The S_N2 mechanism. Substitution–Nucleophilic–2 (for bimolecular – that is two molecules or ions involved in the rate determining step).

Hydrolysis of *tertiary* **halogenoalkanes** such as 2-bromo-2-methylpropane, is *first order*. The rate equation has the form: rate = $k[C_4H_9Br]$.

The suggested mechanism shows the C—Br bond breaking first to form a *carbocation intermediate*. Then the nucleophile OH^- forms a new bond with carbon.

The S_N1 mechanism. Substitution–nucleophilic–1 (for unimolecular – that is one molecule or ion involved in the rate determining step).

nucleotide: the monomer which polymerises to make *nucleic acids, DNA* and *RNA*. A nucleotide consists of three parts:

- **a five-carbon sugar** – ribose in RNA and deoxyribose in DNA
- **a phosphate group**
- **a nucleotide base** – there are five different bases, adenine, cytosine, guanine, thymine and uracil often abbreviated as A, C, G, T and U.

In the nucleotides which make up DNA the base is one of A, C, G or T. In RNA, thymine, T, is replaced by uracil, U.

octet rule: this was first suggested as a guide by the US chemist Gilbert Lewis in 1916. The rule says that atoms tend to gain, lose or share electrons when they combine with other atoms to acquire a stable octet of electrons. The 'stable octet' is the eight $s^2 p^6$ electrons, corresponding to the outer *electron configuration* of the nearest noble gas in the periodic table.

Dot and cross diagrams for molecules and ions showing outer shells with eight electrons as in the nearest noble gas

There are many exceptions to the octet rule, so it is not a safe guide. The rule works pretty well for the elements Li to F in period 2 because there are only four orbitals in the second shell (one *s* and three *p* orbitals). The octet rule also works for ionic compounds of s-*block* elements with the halogens, oxygen and sulfur. Exceptions arise in period 3 and beyond, because *d-orbitals* in the third shell can become involved in bonding.

Born–Haber cycles can help to account for the octet rule in ionic compounds. When magnesium, for example, forms ionic compounds with the Mg^{2+} ion the extra *ionisation energy* needed to remove two rather than one electron from the outer shell is more than compensated for by the big increase in *lattice energy*. An Mg^{2+} ion is much smaller than an Mg^+ ion because of the loss of the outer shell and the greater charge.

Removing a third electron from a magnesium atom needs much more energy because it involves taking an electron from the next shell where the electrons are more strongly held by the nucleus. If it could form, an Mg^{3+} ion would be little smaller than an Mg^{2+} ion so the increased lattice energy is not enough to compensate for the very large third ionisation energy.

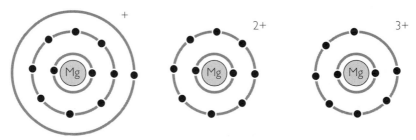

Electron configurations of Mg^+, Mg^{2+} and Mg^{3+} ions

odd-electron compounds: stable compounds with an odd number of electrons in an outer shell. Examples are two *nitrogen oxides*: nitrogen oxide and nitrogen dioxide. Both are exceptions to the *octet rule.*

Dot and cross diagrams showing the odd-electrons in molecules

optical isomers: *isomers* which have opposite effects on *polarised light.* One isomer rotates the plane of polarised light clockwise (the + isomer). The other isomer rotates the plane of polarised light in the opposite direction (the − isomer). Rotations are measured with *monochromatic light* in an instrument called a polarimeter. Optical activity is characteristic of *chiral compounds.*

Optical isomerism is shown by molecules with the same structure but different three-dimensional shapes, so it is a type of *stereoisomerism.*

orbitals: see *atomic orbitals* and *molecular orbitals.*

order of reaction: see *rate equation.*

ore: a *mineral* mass which can be profitably mined and processed to produce a metal. Prospectors seek ore bodies which are local concentrations of ores. After mining or quarrying the first stages of processing separate the valuable minerals from waste rock.

The work of the mineral industry would be impossible if minerals were evenly distributed throughout the *lithosphere.* Fortunately, natural processes have produced concentrations of valuable minerals with much higher percentages of rare metals than in the Earth's crust.

organic acids: see *carboxylic acids.*

organic analysis consists of chemical methods for *qualitative* and *quantitative analysis* to identify organic compounds and work out their composition and structure.

In a modern laboratory organic analysis is based on a range of automated and instrumental techniques including *combustion analysis, mass spectrometry, infra-red spectroscopy, ultraviolet spectroscopy* and *nuclear magnetic resonance spectroscopy.*

Melting points and boiling points also provide a check on the identity and purity of compounds.

Traditionally, chemical tests helped to identify *functional groups* in organic molecules. Preliminary tests typically involve observing:

- the state and appearance of the compound
- its solubility in water and the pH of any solution
- the type of flame when a small sample burns.

Functional group	Test	Observations
$\diagup$C$=$C$\diagdown$ in an *alkene*	• Shake with a dilute solution of bromine • Shake with a very dilute, acidic solution of potassium manganate(VII)	• Orange colour of the solution decolorised • Purple colour fades and the solution turns colourless

(cont'd)

All the reactions in organic chemistry convert one compound to another, but there are some reactions which are particularly useful for developing a synthetic route, such as:

- oxidation of primary *alcohols* to aldehydes and then carboxylic acids
- hydrolysis of *halogenoalkanes*
- addition of hydrogen halides to *alkenes*
- elimination of a hydrogen halide from a *halogenoalkane*
- lengthening of a carbon chain by forming a *nitrile* or using a *Grignard reagent.*

organometallic compounds: organic compounds which contain a metal atom. *Grignard reagents* are organometallic compounds formed with magnesium.

osmosis is the movement of a solvent by diffusion through a *selectively permeable membrane.* The membrane allows small solvent molecules through but not dissolved molecules or ions.

The concentration of solvent molecules is higher in the pure solvent than in the solution. Solvent molecules diffuse in both direction but overall they move from the solvent into the solution

The membranes which surround living cells are selectively permeable. This makes osmosis important in biology. Solutions for medical treatment of the eyes and other delicate parts of the body are formulated to have the same osmotic pressure as body fluids, as are isotonic drinks.

osmotic pressure: the pressure needed to stop *osmosis* when the solution is separated from pure solvent by a selectively permeable membrane. Osmotic pressure depends on the concentration of solute particles and not on their chemical nature. There is a formula relating osmotic pressure to concentration at a given temperature. With the help of this formula osmotic pressure measurements can be used to determine the *molar masses* of large molecules such as *proteins* and *polymers.*

If the pressure applied to the solution is greater than the osmotic pressure, the direction of net flow reverses and pure solvent moves out of the solution. This is reverse osmosis which can make drinking water from salty water.

oxidant: an alternative term to *oxidising agent* which is convenient when describing half-equations for *redox reactions* and *standard electrode potentials.* By convention, when electrode potential values are being assigned the half-equation takes the form:

oxidant + ne^- ⇌ reductant

Every half-equation involves an oxidant and a reductant.

oxidation: originally this meant combination with oxygen but the term now covers all reactions in which atoms, molecules or ions lose electrons. The definition is further extended to cover molecules as well as ions by defining oxidation as a change which makes the *oxidation number* of an element more positive, or less negative.

Oxidation and reduction always go together in *redox reactions.*

Oxidation number rules apply in principle in organic chemistry but it is often easier to use the older definitions. Oxidation is either addition of oxygen to a molecule or removal of hydrogen.

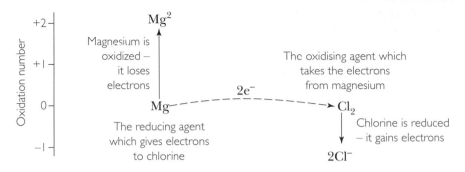

Chlorine oxidises magnesium by taking two electrons from each magnesium atom

Two stage oxidation of an organic compound

oxidation numbers are used by chemists to define *oxidation* and *reduction* and to identify *redox reactions*. The advantage of oxidation numbers is that they apply to molecules and complex ions as well as to simple atoms and ions. The sum of the oxidation numbers in an uncharged compound is zero.

Chemists have agreed a set of rules for deciding on the oxidation numbers of elements:

- the oxidation numbers of uncombined elements are zero
- in a simple ion the oxidation number is the charge on the ion
- the sum of oxidation numbers in an uncharged compound is zero
- the sum of the oxidation numbers in an ion made of several atoms is equal to the charge on the ion
- some elements have fixed oxidation numbers (see Table).

Metals		Non-metals	
group 1 (Li, Na, K)	+1	hydrogen	+1
group 2 (Mg, Ca, Ba)	+2	(except in metal hydrides)	
group 3 (Al)	+3	fluorine	−1
		oxygen	−2
		(except in peroxides and compounds with fluorine)	
		chlorine	−1
		(except in compounds with fluorine and oxygen)	

Oxidation numbers appear in the *names of inorganic compounds* and the *names of complex ions*. They also help with writing balanced redox equations.

MgO	NH_4^+	CCl_4	MnO_4^-
+2 −2	−3 +1	+4 −1	+7 −2

Al_2O_3	SO_4^{2-}	H_2SO_4	$Cr_2O_7^{2-}$
+3 −2	+6 −2	+1 +6 −2	+6 −2

Examples of the application of oxidation number rules. Note that each element in a compound is assigned an oxidation number

Oxidation number rules are a formal way of keeping track of electrons in redox reactions. For the *s*-block elements the oxidation numbers are the same as the charges on the ions. Magnesium has an oxidation state of +2 and the metal atoms turn into Mg^{2+} ions in its compounds. The higher oxidation states of *d-block elements*, however, are certainly not ionic. Manganese in the +7 state cannot form an Mn^{7+} ion. The bonding between manganese and oxygen in the MnO_4^{2-} ion is covalent.

oxidation states: the states of oxidation or reduction shown by an element in its chemistry. The states are labelled with the oxidation numbers of the element in each state.

Describing the chemistry of an element according to oxidation state is a useful way of making sense of a large number of compounds and reactions. See for example the entries for *chlorine, manganese, hydrogen peroxide* and *nitrogen oxides*.

oxides are compounds of elements with oxygen. The compounds of *metals* with oxygen are *basic oxides* or *amphoteric oxides*. The compounds of *non-metals* with oxygen are either *acidic oxides* or *neutral oxides*.

oxidising agents are chemical reagents which can oxidise other atoms, molecules or ions by taking away electrons from them. Common oxidising agents are *oxygen, chlorine, bromine, hydrogen peroxide*, the manganate(VII) ion in *potassium manganate(VII)*, and the dichromate(VI) ion.

Some reagents change colour when they are oxidised which makes them useful for detecting oxidising agents. In particular, a colourless solution of iodide ions is oxidised to iodine which turns the solution to a yellow–brown colour.

$$2I^-(aq) \longrightarrow I_2(aq) + 2e^-$$

electrons taken by the oxidising agent

This can be a very sensitive test if *starch* is also present because starch forms an intense blue–black colour with iodine. Moistened starch–iodide paper is a version of this test which can detect oxidising gases such as chlorine and bromine.

Chlorine acting as an oxidising agent by taking electrons from iron(II) ions. An oxidising agent is itself reduced when it reacts. Oxidation and reduction always go together; hence the term *redox reactions*.

oxidised

$$2Fe^{2+}(aq) \longrightarrow 2Fe^{3+}(aq) + 2e^-$$

$$Cl_2(aq) + 2e^- \longrightarrow 2Cl^-(aq)$$

reduced

oxidising substances: substances which are hazardous if they come into contact with materials that burn – especially if they are flammable liquids. The danger is that oxidising substances will start a fire in contact with materials that can burn. (See *hazard symbols*.)

oxoacids are acids which form when acidic non-metal *oxides* react with water.

Oxoacid	Structure	Oxoanion	Formula
sulfuric acid, H_2SO_4		sulfate ion hydrogensulfate ion	SO_4^{2-} HSO_4^{-}
sulfurous acid, H_2SO_3		sulfite ion	SO_3^{2-}
nitric acid HNO_3		nitrate ion	NO_3^{-}
nitrous acid, HNO_2		nitrite ion	NO_2^{-}

The systematic names for oxoacids are not widely used but they make it possible to work out the formulae from the names. The systematic name for sulfuric acid is tetraoxosulfuric(VI) acid.

Oxoacids ionise in solution by giving hydrogen ions to water molecules.

Ionisation of nitric acid in water showing delocalisation of the charge on the nitrate ion

Delocalisation spreads the charge on the oxoanion, thus stabilising the ion and favouring ionisation. The more oxygen atoms doubly-bonded to the central atom, the stronger the acid. As a result, sulfuric acid is a stronger acid than sulfurous acid and nitric acid is a stronger acid than nitrous acid. So where an element forms two oxoacids, the one with the element in the higher oxidation state is the stronger.

oxonium ions are aqueous hydrogen ions or hydrated *protons* formed when an *acid* dissolves in water. A lone pair of electrons on a water molecule forms a *dative covalent* bond with a hydrogen ion from an acid.

The formula of the oxonium ion is H_3O^+. It is often convenient to write $H^+(aq)$ instead, but remember that the hydrogen ion is *hydrated*.

(Alternative names for the oxonium ion used by some authors are hydroxonium ion and hydronium ion.)

oxygen (O) is a colourless, odourless and highly reactive gas which makes up about a fifth of the *air* by volume and is essential to life. It is the first element in group 6 with the electron configuration $[He]2s^22p^4$. Most oxygen occurs in the form of O_2 molecules but *ozone*, O_3, is a second *allotrope*.

All living organisms need oxygen for respiration. Green plants produce oxygen by *photosynthesis* in sunlight. Many biochemical molecules include oxygen atoms, for example *carbohydrates*, *fats*, *proteins* and *nucleic acids*.

Oxygen is the most abundant element in the *lithosphere*, mainly in the silicon–oxygen giant structures of *silicates*.

Oxygen combines with most other elements to form *oxides*. Particularly important is *water*, the oxide of hydrogen.

Air separation plants use *fractional distillation* to separate oxygen and other gases. Alternatively, where only oxygen is required, the nitrogen from the air can be absorbed in a bed of a molecular sieve such as a *zeolite*.

Two millions tonnes of oxygen are separated from the air each year in the UK. Over half is used in making *steel*. The manufacture of chemicals is another major use. Smaller quantities are used with a fuel for high temperature metal cutting and welding. Oxygen is required medically by people with lung disease and is also used for *water treatment* to improve the quality of water contaminated with organic pollutants.

ozone (O₃): a colourless, reactive and unstable gas which is an *allotrope* of oxygen.

Ozone is an *oxidising agent* and, like oxidising bleaches, it will destroy micro-organisms. Ozone is increasingly used to treat drinking water and swimming pools as an alternative to chlorine.

Ozone often features in stories about *environmental issues*. The ozone layer in the upper *atmosphere* is a 'good thing'. The ozone high in the stratosphere protects living things by absorbing harmful UV radiation from the Sun. *CFCs* and other pollutants, such as the pesticide methyl bromide tend to destroy the protective ozone layer.

In the lower atmosphere ozone is a 'bad thing' because it is harmful to living things and helps to cause *photochemical smog*. Ozone attacks and splits $C = C$ bonds. This is ozonolysis which can damage and destroy materials such as natural rubber.

paint is a product formulated to protect and decorate surfaces. A paint has three ingredients:

- pigments which scatter and absorb light so that the paint covers up the surface underneath and decorates it with colour
- *polymers* which hold the pigment to the surface by forming a smooth plastic film as the paint dries and sets – in gloss paints the film-forming polymers are alkyd resins (see *polyesters*); in emulsion paints they are a latex polymer
- a vehicle which is a liquid in which the other ingredients are dissolved or dispersed – in gloss paint the vehicle is traditionally a *hydrocarbon* solvent such as white spirit; in emulsion paints it is water.

paper chromatography: a type of *chromatography* in which the stationary phase is water in the fibres of paper and the moving phase is another solvent such as ethanol or propanone. The technique is used to analyse mixtures such as inks, food colours, dyes and amino acids.

During paper chromatography the chemicals in a mixture *partition* themselves between two solvents: water and the moving solvent. Each component has a different equilibrium constant (partition constant) which decides whether it has a greater tendency to dissolve in the stationary phase or in the mobile phase. As a result the mixture separates as the chemicals move at different speeds.

If the conditions are kept the same, each chemical in a mixture will move a fixed fraction of the distance moved by the solvent. The R_f *value* for the substance is a measure of this fraction.

If the chemicals in a mixture are colourless they are invisible on the paper. If so, the analyst has to 'develop' the plates with a suitable 'locating agent'. After chromatography of amino acids, for example, the paper can be sprayed with ninhydrin solution and warmed. The amino acids then show up as purple spots.

paracetamol is a mild painkiller (*analgesic*) which can also help to lower the body temperature during fever. It is readily available as an over-the-counter drug. Unlike aspirin it does not reduce inflammation.

Paracetamol is made by *acylating* 4-aminophenol with ethanoic anhydride.

The advantage of paracetamol over aspirin is that it does not irritate the lining of the stomach and it has few side effects.

The great danger is that an overdose with *medicines* containing paracetamol causes severe liver damage which can be fatal. The symptoms do not appear until after about two days. Unfortunately with paracetamol, the dose that causes poisoning is not many times the dose required to relieve pain or lower fever.

paramagnetism: a type of magnetism shown by substances with unpaired electrons. Paramagnetic substances are weakly attracted into a magnetic field. The effect is very much weaker than the *ferromagnetism* of iron. Most compounds have orbitals filled by

have been widely used in transformers, in hydraulic equipment, as *plasticisers* and lubricants. PCBs are suited to these applications because they are very stable, non-flammable, involatile and particularly suited for use in some electrical equipment (because they have high dielectric constants). Some PCBs are also toxic pesticides.

Structure of a PCB

The problem with PCBs is that they escape into the environment where they are very persistent. PCBs can be a hazard too if they are mixed with waste burnt in an incinerator at too low a temperature. The danger is that incomplete combustion produces highly toxic polychlorobenzodioxins and polychlorobenzofurans. Modern incinerators are designed to operate at a high enough temperature to break $C—Cl$ bonds and destroy PCBs.

peptides are compounds made up of chains of *amino acids*. The simplest example is a dipeptide with just two amino acids linked. For chemists this is an example of an *amide* bond but the tradition in biochemistry is to call it a 'peptide bond'. The formation of a peptide bond is an example of a *condensation reaction*.

Formation of a peptide bond between two amino acids

Polypeptides are long-chain peptides. There is no precise dividing line between a peptide and a polypeptide. A *protein* molecules consists of one or more polypeptide chains.

Note that some writers make a distinction between polypeptides and the longer amino acid chains in proteins. They restrict the definition of polypeptides to chains with 10 to 50 or so amino acids.

Heating a peptide with dilute sulfuric acid leads to hydrolysis of the peptide bonds. In time this process splits the peptide into separate amino acids.

percentage composition: the percentage by mass of each of the elements in a compound. Percentage composition is one way of expressing the results of chemical analysis, such as *combustion analysis*. The *empirical formula* of the compound can be calculated from these results.

Worked example:

What is the empirical formula of copper pyrites which has the analysis 34.6% copper, 30.5% iron and 34.9% sulfur?

Notes on the method

Follow the procedure in the worked example for finding an *empirical formula*.

The percentages show the combining masses in a 100 g sample.

Answer

	copper	iron	sulfur
Combining masses	34.6 g	30.5 g	34.9 g
Molar masses of elements	64 g mol^{-1}	56 g mol^{-1}	32 g mol^{-1}
Amounts combined	$\dfrac{34.6 \text{ g}}{64 \text{ g mol}^{-1}}$	$\dfrac{30.5 \text{ g}}{56 \text{ g mol}^{-1}}$	$\dfrac{34.9 \text{ g}}{32 \text{ g mol}^{-1}}$
	= 0.54 mol	= 0.54 mol	= 1.09 mol
Simplest ratio of amounts	1	1	2

The formula is $CuFeS_2$.

The percentage composition of a compound can be worked out from its chemical formula. This is a guide to people who formulate products such as fertilisers, medicines and cleaning agents.

Worked example:

Two common nitrogen fertilisers are urea $(H_2N)_2CO$, and ammonium nitrate, NH_4NO_3. Compare the percentage of nitrogen in the two compounds.

Notes on the masses

Look up the relative atomic masses of the elements and use them to determine the two *relative formula masses*. (Note that when part of a formula is in brackets the number outside refers to all the atoms in the bracket.)

Answer

Relative formula mass of urea $= 2 \times (2 + 14) + 12 + 16 = 60$

of which $(2 \times 14) = 28$ is nitrogen

Percentage of nitrogen in urea $= \dfrac{28}{60} \times 100\% = 46.7\%$

Relative formula mass of ammonium nitrate $= 14 + 4 + 14 + (3 \times 16) = 80$
of which $(2 \times 14) = 28$ is nitrogen

Percentage of nitrogen in ammonium nitrate $= \dfrac{28}{80} \times 100\% = 35\%$

Urea contains the higher percentage of nitrogen.

percentage yield: see *yield calculations.*

perfumes: fragrant chemicals used not only for expensive cosmetics but also to give a smell to everyday household products. Many natural perfumes are essential oils separated from plants by *steam distillation* or *solvent extraction*. Increasingly, however, perfumes now contain synthetic chemicals, some of which are identical to those extracted from natural sources.

Experts use an analogy with music to describe the odours from complex perfumes. 'Top' notes are the most volatile chemicals, the main effect depends on the 'middle notes' and the most long lasting are the 'end notes'. So a perfume chemist has to understand the volatility of perfume chemicals.

undecanal
top note; boiling point 117°C

geraniol
middle note;
boiling point 146°C

indane musk
bottom note; melting point 53°C

Skeletal structures of some perfume chemicals showing top, middle and end notes

The connection between odour and chemical structure is little understood so there is no easy way to predict smells by examining molecular models.

period 3: the third horizontal row of elements in the *periodic table* which starts with sodium and ends with argon. In each main group of the periodic table the element in period 3 is often regarded as the typical element in the group. Chemists then generalise about trends down the group by comparing the elements in periods 4, 5 and 6 with the period 3 element. The first member of each group (usually an element in period 2) is often exceptional in some way.

The changes in properties from Na to Ar (across period 3) show periodic patterns clearly (see *periodicity of physical properties* and *periodicity of chemical properties*).

periodic table: a table of the elements arranged in order of *atomic number* (proton number). *Periods* are the horizontal rows in the table. Each period ends with a *noble gas*. The vertical columns are *groups* arranged in blocks – the *s-block, p-block, d-block* and *f-block* based on the electron configurations of the elements.

The periodic table was a triumph for nineteenth-century chemistry. In 1869 Dmitri Mendeleev published a version of the table based on his periodic law that when elements are arranged in order of atomic mass, similar properties recur at intervals. He left gaps in his table for undiscovered elements and predicted their properties. His success with these predictions helped to persuade other scientists of the merits of his ideas when the missing elements were discovered.

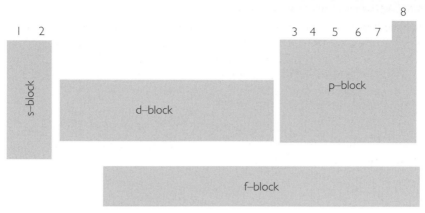

Outline periodic table showing the main blocks

periodicity is a pattern which repeats itself. The most obvious repetition in the periodic table is from metals on the left of each period to *non-metals* on the right. Another repeating pattern from one period to the next is shown by the main *oxidation states* of the elements in their common compounds.

Periodicity of oxidation states of the period 2 and 3 elements in their oxides. Note that the pattern repeats, but with variations.

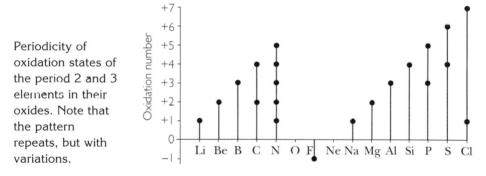

periodicity of chemical properties: the repeating patterns from one period to the next in the chemical properties of the elements and their compounds.

The *oxides* of reactive metals are ionic and basic. The oxides of non-metals are covalently bonded and acidic with *amphoteric oxides* in between.

Ionic crystals		Covalent giant structures	Covalent molecular gases and solids			
$Li_2O(s)$	$BeO(s)$	$B_2O_3(s)$	$CO_2(g)$	$N_2O_5(s)$	$O_2(g)$	$F_2O(g)$
$Na_2O(s)$	$MgO(s)$	$Al_2O_3(s)$	$SiO_2(s)$	$P_4O_{10}(s)$ $P_4O_6(s)$	$SO_3(s)$ $SO_2(g)$	$Cl_2O(g)$
	Basic	Amphoteric		Acidic		

Formulae, structures, bonding and acid–base character of oxides in periods 2 and 3

The chlorides of reactive metals are ionic. They dissolve in water to give neutral solutions. The chlorides of *non-metals* consist of covalent molecules and they are normally *hydrolysed* by water.

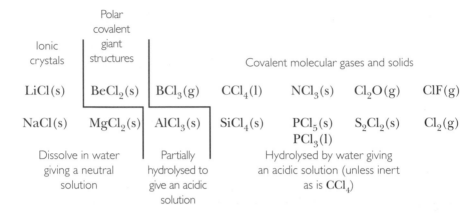

Periodicity: formulae, structures, bonding and behaviour in water of chlorides in periods 2 and 3

periodicity of physical properties: the repeating patterns in the *physical properties* of the elements and their compounds from one period to the next. Metals on the left of a period conduct electricity well. *Non-metals* to the right of a period are non-conductors. The in-between elements are the *metalloids*, many of which are *semiconductors*.

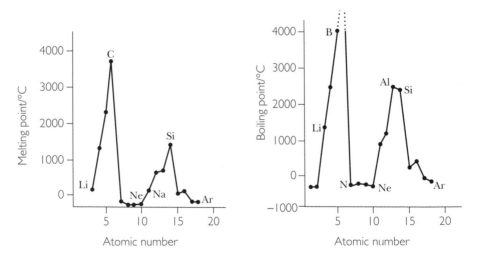

Plots of the melting points and boiling points against atomic number for periods 2 and 3. Note the repeating pattern. This is an example of periodicity. A plot of first *ionisation energy* against proton number is another example of a periodic pattern.

The periodicity in the physical properties of the elements in periods 2 and 3 reflects the underlying patterns of structure and bonding.

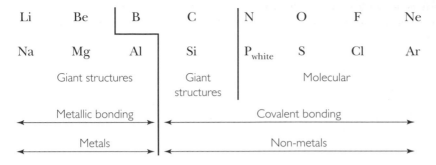

Li	Be	B	C	N	O	F	Ne
Na	Mg	Al	Si	P_{white}	S	Cl	Ar

| Giant structures | | Giant structures | | Molecular | | | |

| Metallic bonding | | ← Covalent bonding → | | | | | |

| Metals | | Non-metals | | | | | |

Periodicity: structure and properties of elements in periods 2 and 3

permanent dipole: see *polar molecules* and *intermolecular forces*.

peroxides are compounds related to *hydrogen peroxide*, H_2O_2. The inorganic peroxides of *s-block elements* are ionic and contain the O_2^{2-} ion. When sodium burns in air it forms sodium peroxide Na_2O_2. Barium similarly produces barium peroxide, BaO_2.

Inorganic peroxides are powerful *oxidising agents*. The peroxide ion is a *strong base* so they also react with water and dilute acids to make hydrogen peroxide.

Organic peroxides can act as a source of *free radicals* to initiate *addition polymerisation* of unsaturated compounds.

A benzoyl peroxide molecule splitting to form two free radicals. The O — O bond is relatively weak.

pesticides are *agrochemicals* which kill the living organisms that cause damage to crops, food, wood, fabrics and other materials. They include:

- herbicides such as paraquat and glyphosate
- insecticides including organochlorines such as DDT, organophosphorus compounds such as malathion, and pyrethroids
- fungicides.

Research chemists make and test many new compounds in search of pesticides which are:

- fast-acting
- specific so that they kill only the pests without harming other organisms
- effective at low doses
- biodegradable so that they quickly break down in the environment.

petrochemical industry: the part of the *chemical industry* based on crude oil and natural gas. The industry has developed and expanded massively since the late 1930s.

petrol consists mainly of a mixture of *hydrocarbons* and is the fuel for motor engines. Petrol has to be carefully blended if modern engines are to start reliably and run smoothly. The proportion of *volatile* hydrocarbons added to petrol is higher in winter to help cold starting but lower in summer to prevent vapour forming before the fuel gets to the carburettor.

The starting point for making petrol is the gasoline fraction from the *fractional distillation of oil*. Hydrocarbons in this fraction are mainly *alkanes* with 5 to 10 carbon atoms. Fractional distillation does not provide enough gasoline and produces more than enough of the heavier fractions, so oil refineries run *catalytic crackers* to make more hydrocarbons of the right size.

For smooth running petrol has to burn smoothly without *knocking*. The octane number of a fuel measures its performance. The higher the compression of fuel and air in the engine cylinders, the higher the octane number has to be to stop knocking.

The octane number scale was devised by Thomas Midgley (1889–1944) and he discovered anti-knock additives based on lead, which were used for many years. Leaded fuel is now being phased out. There are two main objections to adding lead compounds to petrol. The first is that lead compounds are *toxic* and harmful, especially to young children, when released into the air from car exhausts. The second is that lead compounds 'poison' the catalysts in *catalytic converters*.

The oil companies now produce high-octane fuel by increasing the proportions of both branched alkanes and arenes plus blending in some oxygen compounds. The four main approaches are:

- *cracking* which not only makes more small molecules but also forms hydrocarbons with branched chains
- *isomerisation* which turn straight chain alkanes into branched chain compounds by passing them over a platinum catalyst
- *reforming* turns cyclic alkanes into *arenes* such as benzene and methyl benzene
- **adding** *alcohols* **or** *ethers* such as MTBE (initials based on its older name methyl tertiary butyl ether; it is now called 2-methoxy-2-methylpropane).

Structure of the ether MTBE which has an octane number of 120. Adding MTBE and other methods raise the octane number of gasoline from about 70 to 95 as required for unleaded premium petrol.

$$CH_3 - \underset{\underset{CH_3}{|}}{\overset{\overset{CH_3}{|}}{C}} - O - CH_3$$

petroleum: a name for crude oil which is also used for some products from the *fractional distillation* and refining of oil. The *naphtha* fraction which is an important feedstock for the chemical industry is sometimes called the 'light petroleum fraction'.

Petroleum ether is a laboratory solvent which does not contain *ether*. Petroleum ether which distils in the range 40–60°C consists mainly of C_5 *hydrocarbons*. The solvent boiling in the 60–80°C range contains mainly C_6 hydrocarbons.

pH changes during acid-base titrations: the variation in *pH* as a solution of a base flows from a burette into a flask containing a measured volume of acid. Plotting a graph of *pH* against volume of alkali added gives a shape determined by the nature of the acid and the base. (See diagram below.)

The indicator chosen to detect the end-point must change colour completely in the *pH* range of the near vertical part of the curve. Note that at the equivalence point, when exactly equal amounts of acid and base have been added, the *pH* is not always neutral. (See also *neutralisation reactions*.)

It is possible to read off the value of K_a for a weak acid from the corresponding titration curve (see *Henderson–Hasselbalch equation*).

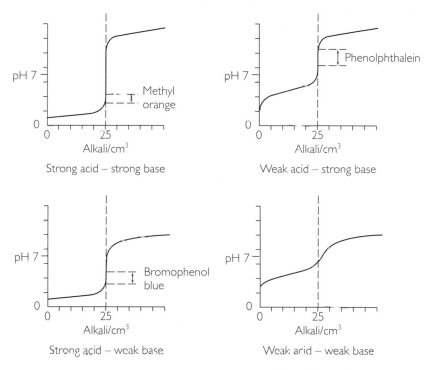

Graphs to show the *pH* change on adding 0.1 mol dm^{-3} of the named alkali to 25 cm^3 of a 0.1 mol dm^{-3} of the named acid. Note the sharp change of *pH* around the equivalence point except when both the acid and the base are weak. The graphs show the *pH* range for the colour change of an *acid–base indicator* for detecting the *end-point* of the titration.

pH scale: a logarithmic scale for measuring the concentration of aqueous hydrogen ions in solutions.

$$pH = -\lg [H_3O^+(aq)]$$

pH	0	1	2	3	4	5	6	7	8	9	10	11	12	13	14
[H$_3$O$^+$(aq)]/mol dm^{-3}	10^0	10^{-1}	10^{-2}	10^{-3}	10^{-4}	10^{-5}	10^{-6}	10^{-7}	10^{-8}	10^{-9}	10^{-10}	10^{-11}	10^{-12}	10^{-13}	10^{-14}

increasingly acidic   neutral   increasingly alkaline

Worked example:

What is the pH of 0.02 mol dm^{-3} hydrochloric acid?

Notes on the method

Hydrochloric acid is a strong acid so it is fully ionised. Note that 1 mol HCl gives 1 mol $H_3O^+(aq)$.

Enter the value of $[H_3O^+(aq)]$ in your calculator, then press the log button, finally press +/– to reverse the sign.

Answer

$[H_3O^+(aq)]$ = 0.02 mol dm^{-3}

pH = – log (0.02) = 1.7

Worked example:

What is the aqueous hydrogen ion concentration in a cola drink with pH 2.3?

Notes on the method

pH = – log $[H_3O^+(aq)]$

From the definition of *logarithms* this rearranges to $[H_3O^+(aq)] = 10^{-pH}$

Enter the pH value in your calculator, press +/– to reverse the sign and then the inverse log button (10^x). (The order of pressing the buttons matters.)

Answer

pH = 2.3

$[H_3O^+(aq)]$ = $10^{-2.3}$ = 5×10^{-3} mol dm^{-3}

pharmaceutical industry makes drugs and *medicines*. In the UK the production of pharmaceuticals accounts for about a third of the wealth created by the chemical industry. (See also *drug development*.)

phase: the three states of matter – solid, liquid or gas. A phase diagram is a map showing the states of a compound over a range of temperatures and pressures.

The 'liquid' area on the phase diagram for water shows all the conditions of temperature and pressure when water is a liquid. Similarly the 'solid' area shows the conditions for water to be solid. The 'vapour' region shows the conditions under which water is a gas.

The lines show the conditions when two phases can be together in equilibrium. At 1 atmosphere pressure, ice and liquid water are in equilibrium at the melting point, 0°C; while liquid water and steam are in equilibrium at the boiling point, 100°C.

Chemical systems often have more than one phase. Each phase is distinct but need not be pure:

- a solid in equilibrium with its saturated solution is a two phase system
- in the reactor for *ammonia manufacture* the mixture of nitrogen, hydrogen and ammonia gases is one phase with the iron catalyst being a separate solid phase

- *chromatography* separates mixtures by letting the components in a mixture come to equilibrium between two phases – the stationary phase and the mobile phase.

The same material broken up into little pieces still counts as one phase. In a *colloid* one phases is finely dispersed as specks, droplets or bubbles in a continuous phase.

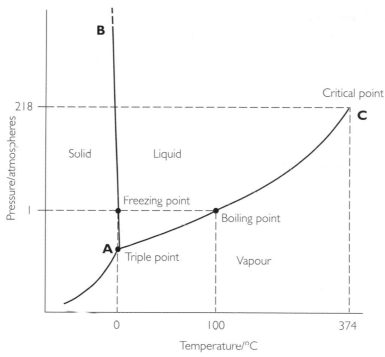

Phase diagram for water. Line AB shows how the melting point of ice varies with pressure. Line AC shows how the boiling point varies with pressure. Point A is the triple point which defines the only conditions of temperature and pressure at which ice, water and water vapour can be in equilibrium.

phenols: a series of compounds with one or more — OH groups directly attached to a benzene ring. The simplest example is phenol, C_6H_5 — OH.

Structure of phenol and other phenols

The UK chemical industry produces about 100 000 tonnes of phenol each year. The main uses are to make:

- *thermosetting polymers*
- *epoxy resins* for adhesives
- *intermediates* for the manufacture of nylon.

The hydroxyl group and the benzene ring in phenol interact. The benzene ring makes the hydroxyl group more acidic in phenol than it is in *alcohols*. Phenol is acidic enough to form salts with alkalis. So phenol dissolves easily in a solution of sodium hydroxide.

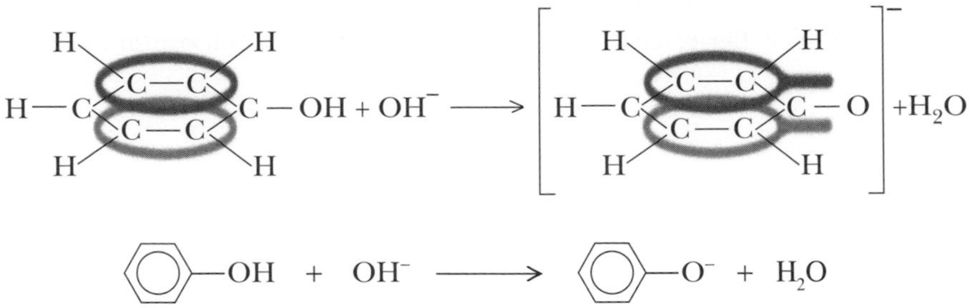

$$\text{C}_6\text{H}_5\text{—OH} + \text{OH}^- \longrightarrow \text{C}_6\text{H}_5\text{—O}^- + \text{H}_2\text{O}$$

Phenol reacting with alkali. Note that the phenoxide ion is stabilised because the negative charge can be spread by delocalisation over the whole molecule. This cannot happen in alcohols.

Phenol is not as acidic as *carboxylic acids*. It does not ionise significantly in water and it does not react with carbonates to produce carbon dioxide. (See also *acid strength of organic hydroxy compounds*.)

Unlike alcohols, phenols do not react directly with carboxylic acids to form esters. It is, however, possible to convert a phenol to an ester with an *acyl chloride*.

The hydroxyl group makes the benzene ring more reactive. *Electrophilic substitution* takes place under much milder conditions with phenol than with benzene. Adding aqueous bromine to a solution of phenol produces an immediate white precipitate of 2,4,6-tribromophenol as the bromine colour fades.

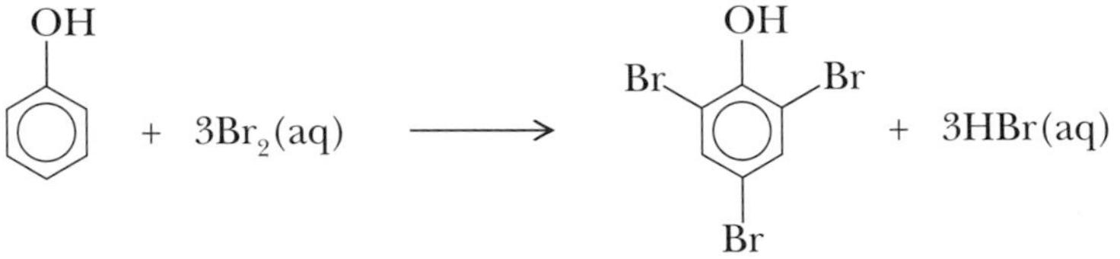

Reaction of phenol with bromine. The reaction is rapid at room temperature and there is no need for a catalyst.

Many phenols form coloured complexes with Fe^{3+}(aq) ions in neutral iron(III) chloride solution. Phenol itself forms a violet complex.

Halogenated phenols are used as *antiseptics* and *disinfectants*.

phenyl group: the group C_6H_5— which is present in many compounds where one of the hydrogen atoms in benzene has been replaced by another atom or group.

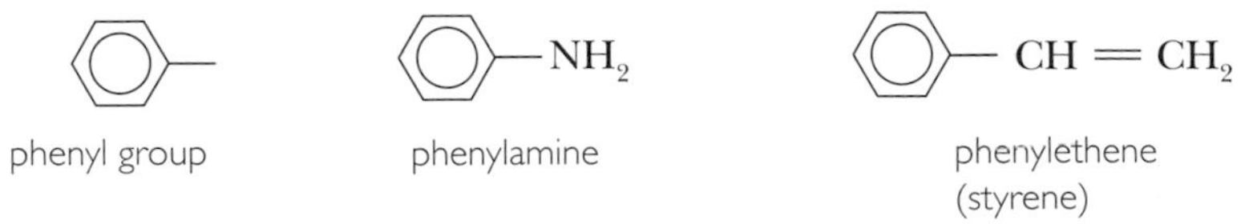

| phenyl group | phenylamine | phenylethene (styrene) |

Structure of the phenyl group and compounds derived from benzene

The use of phenyl in this way dates back to the first studies of benzene when 'phene' was suggested as an alternative name for the compound, based on a Greek word for 'giving light'. Benzene was found in the tar formed on heating coal to produce gas for lighting.

Note that 'benzyl' refers to aryl compounds with a carbon atom attached directly to the benzene ring.

CH₂OH   CHO   CO₂H

benzyl alcohol   benzaldehyde   benzoic acid

Benzyl alcohol and other compounds related to $C_6H_5.CH_2$ —

phenylamine is a primary *amine* with an — NH_2 group attached to a benzene ring. There is a two step laboratory route from benzene to aniline. First *nitration* then reduction.

The chemical industry makes phenylamine from phenol and ammonia with an alumina catalyst. Phenylamine and other arylamines are important intermediates in the manufacture of *azo dyes*. This is because they react with nitrous acid below 10°C to produce *diazonium salts*.

NO₂   NH₂

conc HNO₃ + conc H₂SO₄   Sn or Fe / HCl(aq)

Formation of phenylamine from benzene

Phenylamine is a weaker base than ammonia, unlike alkyl amines which are stronger bases than ammonia. The lone pair on the nitrogen atom interacts with the delocalised electrons of the benzene ring. This extended delocalisation stabilises the free amine and lessens the tendency of the molecule to form a dative covalent bond with a proton.

phosphorescence: see *luminescence*.

phosphoric(v) acid (H_3PO_4) is manufactured on a large scale from phosphate rock to make *fertilisers*, *detergent* phosphates and phosphates for food and drink. The pH of a typical cola drink is 2.3 because it contains 0.05% phosphoric acid. Another use of phosphoric acid is to make iron more resistant to corrosion.

The salts of phosphoric acid have many uses. These are a few examples:

- NaH_2PO_4 – used as a source of phosphate to make drugs and for *pH* control in toothpaste
- Na_2HPO_4 – used to make processed cheese (by a method patented by J. L. Kraft in 1916), also a *corrosion* inhibitor and an ingredient of enamels and glazes

- Na_3PO_4 – the most basic of the sodium salts used in heavy duty cleaners, paint strippers and as a water softener
- $Ca(H_2PO_4)_2$ – the acid in baking powders
- $CaHPO_4$ – the abrasive in toothpaste.

Organic phosphates play a vital part in biochemistry (see *ATP, nucleotides* and *nucleic acids*).

phosphorus (P) is a highly reactive element and important *non-metal*. It is the second element in group 5 of the periodic table, $[Ne]3s^22p^3$.

There are two common *allotropes*:

- **white phosphorus** which is molecular (P_4) and has to be stored under water otherwise it catches fire in air
- **red phosphorus,** which is made up of long chains of P_4 units, is more stable and does not catch fire in air at room temperature.

There is also a black allotrope.

White phosphorus condenses when phosphorus vapour cools. This happens in the manufacture of phosphorus from phosphate rock (fluorapatite). The red form is more stable and, at 540 K, white phosphorus changes rapidly to the red form without a catalyst.

Phosphorus forms two oxides. They are both molecular and both *acidic oxides*. The more important oxide is P_4O_{10} which forms as a white solid when white phosphorus burns in excess air (see *phosphoric acid*).

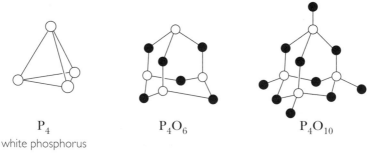

$$P_4$$
white phosphorus

$$P_4O_6$$

$$P_4O_{10}$$

Structures of white phosphorus, phosphorus(III) oxide, P_4O_6, and phosphorus(V) oxide, P_4O_{10}

As a laboratory reagent P_4O_{10} is used as a drying and dehydrating agent. It dehydrates *amides* to *nitriles*.

Phosphorus also forms two chlorides. Phosphorus trichloride (PCl_3) is made commercially from white phosphorus and chlorine. It is an important intermediate in the manufacture of medical drugs, insecticides and flame retardants. PCl_3 is a molecular compound. At room temperature it is a colourless fuming liquid which is rapidly *hydrolysed* by water.

Phosphorus pentachloride (PCl_5) forms when PCl_3 reacts with excess chlorine. It is a volatile solid. It too is rapidly hydrolysed by water and is a laboratory reagent for replacing hydroxyl groups with chlorine atoms in *alcohols* and *carboxylic acids*.

photochemical smog is produced by sunlight and the pollutants from the exhaust gases of motor vehicles. This type of *smog* forms on still, sunny days when there is no wind to blow away the gases. It is severe in cities such as Los Angeles where weather conditions and the local geography tend to trap pollutants in the city.

The primary pollutants are *nitrogen oxides* and unburnt *hydrocarbons* which are emitted in large amounts during the morning rush hour in a city. Bright sunlight during the middle of the day sets off photochemical reactions involving oxygen in the air. The products are the secondary pollutants which create smog.

The level of *ozone* in the air rises and oxidising *free radicals* form. The unburnt hydrocarbons are then oxidised to *aldehydes, ketones* and other chemicals such as organic nitrates which irritate the eyes and lungs.

Catalytic converters are designed to lessen this kind of pollution.

photochemistry: the study of reactions caused by light or other forms of *electromagnetic radiation*. Photons of ultraviolet light have enough energy to break covalent bonds and form *free radicals*. These reactive intermediates can then start *free-radical chain reactions*. Photochemistry is important in the environment as well as in the laboratory (see *ozone* and *photochemical smog*).

photography uses chemical reactions to form images. Photography is based on the chemistry of silver and shows that, like other *d-block* elements, silver can act as a *catalyst*, form *complex ions* and take part in *redox reactions*. Photographic film consists of tiny silver bromide crystals in a film of gelatin on a strip of transparent plastic (see *ionic precipitation*). The stages of photography are as follows.

1   **Exposure** – in both black and white and colour photography the first step is a photochemical reaction. Light produces a few atoms of silver on the surface of exposed silver bromide crystals:

$$2AgBr(s) \longrightarrow 2Ag(s) + Br_2(l).$$

At this stage there is no visible image.

2   **Development** – the atoms of silver are the catalyst for the next stage when the film is taken out of the camera and treated with a solution of the developer. In black and white photography the developer is a *reducing agent* which turns many more of the silver ions in exposed crystals to silver. Exposed parts of the film turn black. In colour photography the developer is oxidised to an intermediate which then reacts with other chemicals in the gelatin layers of the film to produce coloured dyes.

3   **Fixation** – this removes unexposed and undeveloped silver bromide to stop the whole film turning black when it is brought out into the light. The fixer is a solution of sodium thiosulfate. Thiosulfate ions form a complex with silver ions, turning insoluble silver bromide into a soluble complex:

$$AgBr(s) + 2S_2O_3{}^{2-}(aq) \longrightarrow [Ag(S_2O_3)_2]^{3-}(aq) + Br^-(aq)$$

4   **Bleaching** – this is needed only in colour photography. In black and white photography the grains of silver formed by the developer stay to form the black areas of the negatives and prints. In colour photography the silver has done its job once it has helped to make the coloured dyes. So the silver is removed by treatment with a solution of iron(III) ions. The silver ions

Worked example:

What is the pH of a 0.02 mol dm^{-3} solution of sodium hydroxide?

Notes on the method

Sodium hydroxide (NaOH) is a *strong base* so it is fully ionised.

Find the values of logarithms by entering the value in a calculator and then pressing the lg button.

Answer

$[OH^-]$ = 0.02 mol dm^{-3}

pOH = $-$ lg 0.02 = 1.7

pH = 14 $-$ pOH = 14 $-$ 1.7 = 12.3

planar molecules are flat molecules with all the atoms in the same plane (see *shapes of molecules*).

Planck's constant: see *quantum theory*.

plaster of Paris is the main ingredient of building plasters and much is used to make plasterboard. The white powder is made by heating the mineral gypsum in kilns to remove most of the water of crystallisation.

$$CaSO_4.2H_2O(s) \longrightarrow CaSO_4.\frac{1}{2}H_2O(s) + \frac{3}{2}H_2O(g)$$

Stirring plaster of Paris with water produces a paste which soon sets as it turns back into interlocking grains of gypsum. Plaster makes good moulds because it expands slightly as it sets so that it fills every crevice.

plastic materials are materials which can be moulded by gentle pressure. Examples are potter's clay and Plasticine. Once moulded, a plastic material keeps its new shape, unlike an *elastomer* which tends to spring back to its original shape.

plasticisers are additives mixed with *polymer* materials to make them more flexible. A plasticiser is usually a liquid with a high boiling point, such as large *ester* molecules. The molecules of the plasticiser get between the polymer chains and reduce the intermolecular forces between the chains so that they can slide past each other. Rigid uPVC (unplasticised) is suitable for window frames and guttering. Adding a plasticiser to PVC changes the polymer to a material suitable for squeeze bottles, hose pipes and the insulation on electric cables.

plastics: materials made of long-chain molecules which at some stage can be easily moulded into shape. *Thermoplastics* become *plastic materials* when they are hot and harden on cooling. Some thermoplastics are rigid and brittle at room temperature such as polystyrene and uPVC, others, such as polythene, are flexible. (See also *recycling*.)

platinum (Pt) is a valuable metal which is the most abundant of the group of so-called 'platinum metals' which occur together in sulfide ores. The platinum metals are the *d-block elements* ruthenium and osmium, rhodium and iridium plus palladium and platinum.

Platinum is about as abundant as gold. It is mined in South Africa where the yield of metal is about 30 g from 10 tonnes of rock.

Platinum is an attractive metal and it is very unreactive so it does not tarnish. The main use of platinum is for jewellery. As well as being expensive, it is a *malleable* and *ductile* metal so it can be worked into shape.

Chemists take advantage of the inertness of the metal in electrochemistry. Platinum electrodes are required to measure standard electrode potentials. The conductor in a *standard hydrogen electrode* is platinum covered with a thin layer of finely divided metal (platinum black) deposited by electrolysis.

Platinum is an effective *catalyst*. Its catalytic activity is enhanced by alloying with rhodium. The second main use of platinum is to make *catalytic converters* for cars. Platinum–rhodium gauzes catalyse the oxidation of ammonia in *nitric acid manufacture*. Another use for a platinum–rhodium catalyst is 'platforming' which is **plat**inum catalysed *reforming* in oil refining.

Platinum forms a range of *complex ions* some of which, like *cisplatin*, are used to treat cancer.

polar covalent bonds are covalent bonds formed between atoms of different elements so that the shared electrons are drawn towards the more *electronegative* atom. One end of the bond has a slight excess of negative charge ($\delta-$). The other end of the bond has a slight deficit of electrons so that the charge cloud of electrons does not cancel the positive charge on the nucleus ($\delta+$).

Examples of molecules with polar bonds

$$\begin{array}{c} H \\ | \\ H - \overset{\delta+}{C} - \overset{\delta-}{Br} \\ | \\ H \end{array} \qquad \begin{array}{c} H_3C \diagdown \overset{\delta+}{} \quad \overset{\delta-}{} \\ C = O \\ H_3C \diagup \end{array}$$

Molecules with polar bonds are generally more reactive than non-polar molecules, especially with ionic reagents such as acids, alkalis and oxidising agents. The presence of polar bonds makes molecules open to attack by *electrophiles* and *nucleophiles*.

polar molecules have polar bonds which do not cancel each other out, so that the whole molecule is polar.

$$\begin{array}{c} \overset{2\delta-}{O} \\ \diagup \quad \diagdown \\ \underset{\delta+}{H} \quad \underset{\delta+}{H} \end{array} \qquad\qquad \overset{\delta-}{O} = \overset{2\delta+}{C} = \overset{\delta-}{O}$$

$$\begin{array}{c} H \\ | \\ H - \overset{\delta+}{C} - \overset{\delta-}{Cl} \\ | \\ H \end{array}$$

Overall polar   Overall non-polar

Molecules with polar bonds. Note that in the examples on the left the net effect of all the bonds is a polar molecule. In the examples on the right the overall effect is a non-polar molecule.

Polar molecules are little electrical dipoles (they have a positive pole and a negative pole). Dipoles tend to line up in an electric field. The bigger the dipole the bigger the twisting effect (dipole moment). By making measurements with a polar substance between two electrodes it is possible to calculate dipole moments. The units are Debye units named after the physical chemist Peter Debye (1884–1966).

The *intermolecular forces* are stronger between molecules with permanent dipoles.

Molecule	Dipole moment (in debye units)
HCl	1.08
H_2O	1.94
CH_3Cl	1.86
$CHCl_3$	1.02
CCl_4	0
CO_2	0

Measuring dipole moments can distinguish geometrical isomers.

Polarity of geometric isomers

Overall polar   Overall non-polar

polar solvents are *solvents* made of *polar molecules*. Highly polar solvents such as water are needed to dissolve ionic compounds (see *hydration*). In organic chemistry the use of polar solvents favours *heterolytic bond breaking* and reactions with ionic intermediates.

polarisability: an indication of the extent to which the electron cloud in a molecule or ion is distorted by a nearby electric charge.

Larger molecules with more electrons are generally more polarisable. The *intermolecular forces* which account for weak intermolecular forces between non-polar molecules are stronger between molecules which are more polarisable.

Fajan's rules are a guide to the degree of polarisation of negative ions by neighbouring positive ions.

polarised light: a light beam in which all the waves are vibrating in the same plane.

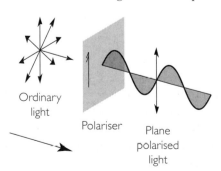

Ordinary light

Polariser

Plane polarised light

Ordinary light and polarised light

Light is plane-polarised after passing through a sheet of Polaroid. *Optical isomers* rotate the plane of polarised light.

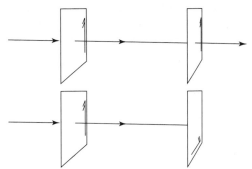

Effect of passing a light beam through two sheets of Polaroid. After the first sheet the light is polarised. The beam passes through the second sheet if it is aligned the same as the first. No light gets through if the second sheet is rotated through 90°.

polarising power: an indication of the extent to which a positive ion is able to distort the electron cloud around a neighbouring negative ion. The larger the charge on a positive ion and the smaller its size the greater its polarising power (see *Fajan's rules*).

pollution: contamination of the *atmosphere, hydrosphere* or *lithosphere* as a result of natural processes or events (such as excretion, the decay of dead organisms and volcanic eruptions) and human activities (such as agriculture, sewage discharges, combustion of fuels, extraction of resources, industrial processes and the disposal of *wastes*).

polyamides are polymers in which the monomer molecules are linked by amide bonds. The first important polyamides were formed by *condensation polymerisation* between diamines and dicarboxylic acids. (See also *Kevlar*.)

Condensation polymerisation to make Nylon-6,6. The numbers in the name indicate the numbers of carbon atoms in the two monomers.

polyesters are *polymers* formed by *condensation polymerisation* between acids with two carboxylic acid groups and alcohols with two or more hydroxyl groups. Units in the polymer chains are linked by a series of *ester* bonds.

The monomers for making polyester fibres are benzene-1,4-dicarboxylic acid and ethan-1,2-diol.

Condensation polymerisation to produce polyester

The traditional names for these reactants are **ter**ephthalic acid and eth**ylene** glycol, hence the commercial name Terylene for one brand of polyester. The alternative name for the polymer is **p**olyethylene **t**erephthalate which gives rise to the name PET when the same polymer is used as a plastic to make bottles for carbonated drinks.

Alkyd *resins*, ingredients of gloss paint, also consist of long chains linked by ester bonds. The polymer chains have *unsaturated hydrocarbon* side-chains. When exposed to air, the *double bonds* in the side-chains of neighbouring polymer molecules react so that the chains become *cross-linked*. As a result the resin hardens to a smooth film as the paint dries and sets.

Some polyesters, such as Biopol, are *biodegradable* because bacteria can break them down by hydrolysing the ester links.

polymer chemistry is the study of the synthesis, structure and properties of *polymers*. It is a branch of chemistry which has developed in the twentieth century. Leo Baekland developed *Bakelite* in 1905 with little help from theory. The first *thermosetting polymers* were discovered before chemists understood the structure of big molecules.

In 1922 the German chemist, Hermann Staudinger published his theory that natural rubber, cellulose and related substances consist of long chain molecules. He had to fight hard to persuade other chemists to accept his theory which is now taken for granted.

One of the people who was convinced by the new theory was the American industrial chemist Wallace Carothers. He wrote an article in 1931 in which he introduced the terms *addition polymerisation* and *condensation polymerisation*. His team at Du Pont invented the synthetic *rubber*, Neoprene. They also discovered the first completely synthetic polymer (nylon) which went into production in 1939.

The 1930s were probably the most important years in the development of the plastics industry. It was the time when the addition polymers polythene, PVC, polystyrene and Perspex were all developed commercially. The high-pressure process for making polythene was discovered and developed in the mid-1930s as an unexpected offshoot of the study of high-pressure gas reactions by Eric Fawcett and Reginald Gibson working for the UK chemical company ICI.

Also in the 1930s, Otto Bayer was pursuing his interests in isocyanates in the laboratories of the IG Farben Industries in Germany. He too faced opposition in his pursuit of new polymeric materials. His breakthrough came in 1941 when he and his team discovered the possibility of producing *polyurethane* foams. It took ten more years of development work before large scale manufacturing could begin.

It was in the early 1950s that Karl Ziegler in Germany discovered how to polymerise ethene at low temperatures using a new kind of catalyst. At the same time Guilio Natta in Italy discovered the benefits of producing *isotactic* polymers (see *Ziegler–Natta catalysts*).

Research and development since the 1950s has produced many new specialised polymer materials including the *polyamide Kevlar* and biodegradable polymers.

polymers are long chain molecules. Natural polymers include *proteins*, polysaccharides (see *carbohydrates*) and *nucleic acids*. Synthetic polymers include *polyesters*, *polyamides* and the many polymers formed by the *addition polymerisation* of compounds with $C = C$ bonds.

The properties of polymers are very varied. Polymeric materials include *plastics*, *elastomers* and *fibres*.

As polymer science has developed, chemists and materials scientists have learnt how to develop new materials with particular properties. Some of the ways of modifying polymeric materials include:

- altering the average length of the polymer chains
- changing the structure of the monomer perhaps by adding side groups which increase *intermolecular forces*
- varying the degree of *cross-linking* between chains
- *copolymerisation*
- controlling the three dimensional shape of the polymer (see *isotactic polymers*)
- changing the alignment of the polymer chains, for example by spinning the polymer into fibres and then stretching the fibres
- adding fillers, pigments and *plasticisers*
- making *composites*.

polymerisation: a process in which many small molecules (*monomers*) join up in long chains (see *addition polymerisation* and *condensation polymerisation*).

polymorphism: the existence of two or more crystalline forms of a substance. Some ionic compounds are polymorphic. Zinc sulfide, for example, can crystallise either with the zinc blende structure or with the alternative wurtzite structure. Polymorphism in solid elements is an example of *allotropy*.

polyols are compounds with more than one hydroxyl group. They include diols, such as ethan-1,2-diol and triols such as propan-1,2,3-triol (better known as glycerine or glycerol). Polyols are used to make *polyurethanes*.

polysaccharides: see *carbohydrates*.

polyurethanes are a varied range of cross-linked polymers made from two liquids – a *polyol* and an isocyanate. Polymerisation is exothermic and happens at room temperature. Adding a chemical to make a gas means that the polymer forms as a plastic foam. By choosing different isocyanates and polyols, manufacturers can vary the properties of the polyurethane. The products range through flexible foam for bedding and upholstery, rigid foam for wall panels and refrigerators, hard wearing but bend-able soles for shoes and ingredients for paints and adhesives.

One of the great advantages of polyurethanes is that they can be made where they are needed without any complex machinery. A furniture manufacturer, for example, can buy the two liquid ingredients and then mix them in a mould so that the poly-mer takes up the required shape for a chair as it forms.

p-orbitals: see *atomic orbitals*.

position of equilibrium: see *Le Chatelier's principle*, the *equilibrium law* and the *temperature effect on equilibria*.

potassium (K): is a very soft, shiny metal which rapidly tarnishes in moist air. It is the third member of *group 1* with the electron configuration [Ar]$4s^1$.

Like other group 1 metals, potassium:

- is stored in oil
- floats on water, melts and reacts violently forming hydrogen which catches fire and KOH which is soluble and strongly alkaline
- forms an ionic, crystalline chloride K^+Cl^-.

Unlike lithium and sodium, potassium produces a *superoxide* (KO_2) when it burns in oxygen. This is an ionic compound with the O_2^- ion. One of the main uses of potassium is to make this oxide for use in emergency breathing apparatus. The oxide removes carbon dioxide from moist breathed-out air and replaces it with oxygen.

$$4KO_2(s) + 4CO_2(g) + 2H_2O(l) \longrightarrow 4KHCO_3(s) + 3O_2(g)$$

Potassium, as potassium ions, is an essential nutrient for plants and an ingredient of NPK *fertilisers*. Large, underground deposits of potassium chloride are mined as the mineral sylvinite just south of Teesside in the UK.

potassium manganate(VII) (KMnO$_4$) consists of greyish-black crystals which dissolve in water to give a deep purple solution. It is used as a powerful *oxidising agent*.

Potassium manganate(VII) is an important reagent in *redox titrations* because it will oxidise many reducing agents in acid conditions, such as iron(II) ions. The reactions go according to their equations which makes them suitable for quantitative work.

$$MnO_4^-(aq) + 8H^+(aq) + 5e^- \longrightarrow Mn^{2+}(aq) + 4H_2O(l)$$

No indicator is required for a manganate(VII) titration. When the solution is added-from a burette the manganate(VII) rapidly changes from purple to colourless (because the colour of the Mn^{2+} ion is so pale). At the end-point it takes only the slightest excess of manganate(VII) to give a permanent red-purple colour.

In organic chemistry potassium manganate(VII) is a reagent used to oxidise the side-chains of *arenes*.

precious metals are expensive metals such as gold and silver. Chemists also use the term to describe other valuable metals which are used as catalysts, including the *platinum* metals.

precipitation: a reaction which produces an insoluble product from soluble chemicals in solution. A common example is the formation of an insoluble salt on mixing solutions of two soluble salts (see *ionic precipitation*).

Ionic precipitation is the basis of many *anion tests* and *cation tests*.

In *organic analysis* the 2,4-dinitrophenylhydrazine reagent for aldehydes and ketones forms a precipitate by an *addition–elimination reaction*. The *Fehling's solution* or *Tollen's reagent* tests, used to distinguish aldehydes from ketones, both make precipitates by redox reactions.

precision of data: data is precise if repeat measurements have values which are close to each other. Precise measurements have a small random *error*.

Precise measurements may or may not be accurate. As a result of a systematic error a series of precise measurements may give values which are almost the same but are not the true value.

pressure is defined as force per unit area. The *SI unit* of pressure is the pascal (Pa) which is a pressure of one newton per square metre (1 N m^{-2}). The pascal is a very small unit so pressures are often quoted in kilopascals (kPa).

When gases are bing studied the standard pressure is *atmospheric pressure* which is:

$$101.3 \times 10^3 \text{ N m}^{-2} = 101.3 \text{ kPa}.$$

In accounts of chemical processes, multiples of atmospheric pressure give an indication of the extent to which gases are compressed (see for example *ammonia manufacture*).

The standard pressure for definitions in thermodynamics is now 1 bar which is:

$$100\ 000 \text{ N m}^{-2} = 100 \text{ kPa}$$

The *partial pressure* of a gas is a measure of its concentration in a mixture of gases.

primary, secondary and tertiary organic compounds are labels which distinguish different chemical situations for functional groups in *alcohols, halogenoalkanes, amines* and *carbocations*.

The labels have the same meaning for alcohols, halogenoalkanes and carbocations but a different meaning for amines.

primary standard: a chemical which can be weighed out accurately to make up a *standard solution* for *volumetric analysis*. A primary standard must:

- be very pure
- not gain or lose mass when exposed to the air (so it must not be *hygroscopic* or *deliquescent*)
- be soluble in water
- have a relatively high molar mass so weighing errors are minimised
- react exactly as described by the chemical equation.

A convenient primary standards for *acid–base titrations* is anhydrous sodium carbonate.

Primary standards for *redox titrations* include: potassium dichromate(VI) and potassium iodate(V).

promoters are chemicals which make catalysts more effective. The iron catalyst in the Haber process for *ammonia manufacture* contains potassium hydroxide as a promoter. The vanadium(V) oxide catalyst used in *sulfuric acid manufacture* has potassium sulfate on a silica support to act as a promoter.

propagation is a step in a *free radical chain reaction*.

propanone (acetone) is the simplest *ketone*, CH_3COCH_3, which is widely used as a solvent. Three to four million tonnes of propanone are manufactured world-wide each year. Propanone is also used to make monomers for the manufacture of *polymers*, including acrylics, polycarbonates and *epoxyresins*. Most propanone is made in the cumene process which also produces *phenol*.

proportionality: a description of any relationship which takes the form:

$y \propto x$ or $y = $ constant $\times x$

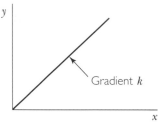

A graph showing that y is proportional to x. A straight line through the origin. $y = kx$. The gradient of the line gives the value of the constant k.

The volume of a fixed amount of an ideal gas, for example, is proportional to the temperature on the *Kelvin* scale at constant pressure. The rate of a *first order reaction* is proportional to the concentration of the reactant in the rate equation.

protecting group: a group introduced into an organic molecule during a synthesis to stop an unwanted reaction. Once the reaction is complete the protecting group is removed.

proteins consist of long chains of *amino acids* joined by *peptide* bonds. Some fibrous proteins are vital to the structure of animals, including the proteins in muscles, ligaments, tendons, skin and hair. Other proteins coil up into a globular shape and dissolve in body fluids where they act as oxygen carriers, *enzymes* and *hormones*.

Biochemists describe the structure at a series of levels:

- **the primary structure** is the sequence of amino acids in the polymer chains
- **the secondary structure** is the way in which the chains are arranged and held in place by *hydrogen bonding* within and between chains – this includes the coiling of chains into an α-helix in proteins such as keratin and the formation of layers of parallel chains as in the β-pleated sheets of silk
- **the tertiary structure** describes the three dimensional folding of protein chains which gives some proteins, such as enzymes, a definite three-dimensional shape held in place by hydrogen bonding, disulfide 'bridges' and interactions between amino acid side-chains with surrounding water molecules
- **the quaternary structure** describes the linking of two or more amino acid chains as in the hormone *insulin* which consists of two chains linked by disulfide bonds, or in antibodies consisting of four chains.

protium (^1_1H) is the abundant isotope of hydrogen. It is sometimes called protium to distinguish it from deuterium, ^2_1H, and tritium, ^3_1H.

proton: one of the two types of particle which make up the nucleus of an atom. The relative mass of a proton is 1 and its charge is +1. The number of protons in the nucleus of an atom is the *atomic number* (or proton number).

A hydrogen atom, H, normally consists of one proton in the nucleus and one electron in the first shell. So a hydrogen ion, H^+, is simply a proton. Chemists often use the terms 'hydrogen ion' and 'proton' interchangeably, especially when describing *acid–base reactions* as *proton transfer* reactions.

proton magnetic resonance spectroscopy: *nuclear magnetic resonance spectroscopy* (nmr) when the nuclei involved are hydrogen nuclei (or protons).

proton number: the number of protons in the nucleus of an atom. An alternative to *atomic number*.

proton transfer describes the movement of a hydrogen ion (or *proton*) from an acid to a base during an *acid–base reaction*. According to the *Brønsted–Lowry theory* an acid is a proton donor and a base is a proton acceptor.

purification: a process used by chemists to remove impurities from the products of reaction. Common methods of purification include:

- *distillation* or *fractional distillation* for liquids
- *steam distillation* to obtain a liquid with a high boiling point which does not mix with water
- *recrystallisation* to separate solids from solid impurities
- *solvent extraction* for liquids or solids.

Some types of *chromatography* are also used not only for analysis but also to separate pure products.

Chemists use a range of techniques to test for purity. A chemical is pure if it:

- melts exactly at its melting point
- boils at its boiling point
- gives only one spot or peak when analysed by chromatography
- gives a print out which matches the spectrum of a known pure sample when analysed by *infra-red spectroscopy*.

qualitative analysis is any method for identifying chemicals in a sample. Examples of qualitative analysis include:

- *gas tests* such as the use of limewater to detect carbon dioxide
- *anion tests* such as the use of silver nitrate solution to detect and distinguish chlorides, bromides and iodides
- the separation and identification of amino acids by *paper chromatography*
- the identifying of *functional groups* in an organic compound using its *infrared spectroscopy*.

quantitative analysis: any method for determining the amount of a chemical in a sample. Examples of quantitative analysis include:

- an *acid–base titration* to determine the concentration of a solution of hydrochloric acid
- the use of a *colorimeter* to determine the concentration of a coloured *complex ion*
- testing the blood-alcohol concentration by *infra-red spectroscopy*
- finding the relative abundances of the *isotopes* of chlorine with a *mass-spectrometer*.

quantum numbers identify the *energy level* or *orbital* occupied by an electron in an atom. Theory shows that four quantum numbers uniquely identify each electron in an atom:

- the principal quantum number indicates the distance of the electron from the nucleus
- the second quantum number identifies the type of *atomic orbital, s, p, d* or *f*
- the third quantum number shows the direction in space of the orbital, p_x, p_y or p_z
- the fourth quantum number states the alignment of the *spin*, spin up or spin down.

The principal quantum number identifies the main *shell* occupied by an electron. The first shell ($n = 1$) can hold 2 electrons, the second shell ($n = 2$) up to eight electrons and the third shell ($n = 3$) up to 18 electrons. Thus the maximum number of electrons in a shell is given by $2n^2$.

quantum theory states that radiation is emitted or absorbed in discrete amounts called energy quanta. Max Planck, the German physicist, put forward the theory in a paper published in 1900. Quanta have energy $E = h\nu$ where h is Planck's constant and ν is the frequency of the radiation.

The Danish physicist, Niels Bohr took up quantum theory in 1913 to explain the lines in hydrogen's *atomic emission spectra* . Bohr's theory could account very well for the frequencies of the lines in the spectrum of the hydrogen atom by making these assumptions:

- electrons in a hydrogen atom can only be at certain definite *energy levels*
- a photon of light is emitted or absorbed when an electron jumps from one energy level to another
- the energy of the photon, *E*, equals the difference between the two energy levels
- the frequency of the radiation emitted or absorbed is related to the energy of the photon by $E = h\nu$.

quartz is a crystalline form of *silica* (silicon dioxide). Quartz has a high melting point because it consists of a covalently bonded *giant structure* in which each silicon atom is linked to four oxygen atoms and each oxygen atom to two silicon atoms.

quaternary ammonium cations are the cations formed when all the hydrogen atoms in an ammonium ion are replaced by *alkyl groups*.

General structure of a quarternary ammonium cation

$$R - \overset{\overset{\displaystyle R'}{|}}{\underset{\underset{\displaystyle R'''}{|}}{N}}\!\!\overset{+}{-} R''$$

Cationic *surfactants* used in fabric and hair conditioners are quaternary ammonium salts such as dodecyltrimethylammonium bromide, $CH_3(CH_2)_{11}N^+(CH_3)_3Br^-$.

quenching: the rapid cooling of a hot metal sample during heat treatment by plunging it into cold oil or water. Quenching a hot metal produces a different structure to slow cooling.

Similarly, chemists 'quench' reactions by sudden cooling. This effectively stops (or 'freezes') the reaction. In this way it is possible to find the composition of an equilibrium mixture or measure the concentration of a reactant or product during a study of rate of reaction.

R

R_f values: ratios which measure how far a particular chemical moves during *paper chromatography* and *thin-layer chromatography* (tlc).

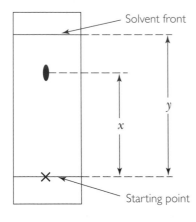

$$R_f = \frac{\text{distance moved by solute}}{\text{distance moved by solvent}}$$

$$= \frac{x}{y}$$

The R_f value is the ratio of the distance moved by the chemical in the mixture to the distance moved by the solvent front

R_f values can help to identify components of mixtures so long as conditions are carefully controlled. The values vary with the type of paper or tlc plate and the nature of the solvent.

racemic mixture: a mixture of equal amounts of the two mirror image forms of a *chiral compound*. The mixture does not rotate *polarised light* because the two *optical isomers* have equal and opposite effects so they cancel each other out.

Lactic acid (2-hydroxypropanoic acid) from muscles is optically active because it consists of the (+) isomer. Lactic acid from sour milk is not optically active because it is a 50:50 racemic mixture of the two mirror image forms. Laboratory synthesis of lactic acid also produces the racemic mixture because there is an equal chance of the two isomers forming. (See diagram on page 303.)

radioactive decay: the decay of a radioactive nucleus to form the nucleus of another element by *alpha decay* or *beta decay*. *Gamma rays* may also be emitted.

radioactive nucleus $\longrightarrow$ daughter nucleus + alpha or beta radiation

Radioactive decay is a *first order* process. The rate of decay is proportional to the number of radioactive atoms. This means that each radionuclide decays with a constant *half-life*.

Radionuclides with a short half-life decay more quickly and tend to give off more intense radiation. The unit of radioactivity is the *becquerel* (symbol Bq).

Half-lives for radioactive decay are unaffected by changes in temperature, pressure or chemical state. This is the principle of the technique used to find the age of the remains of living things or of rocks. Knowing the half-lives it is possible to estimate ages by measuring the proportions of different nuclides in a sample.

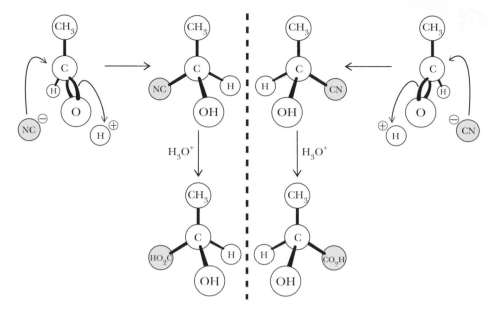

Racemic mixture A normal laboratory synthesis produces a racemic mixture of the mirror image forms of 2-hydroxypropanoic acid (see page 302)

Raoult's law: a law which predicts the *vapour pressure* of mixtures of liquids and solutions. The law states that in a mixture of two liquids A and B:

vapour pressure of
A over the mixture = *mole fraction* of A × vapour pressure of pure A

vapour pressure of
B over the mixture = mole fraction of B × vapour pressure of pure B

The total vapour pressure of the mixture is the sum of the vapour pressures of A and B over the mixture.

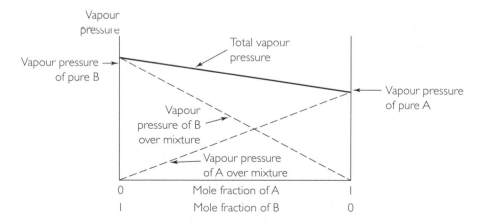

Graph to illustrate the meaning of Raoult's law

Only *ideal mixtures* obey the law. These are mixtures of very similar liquids so that the *intermolecular forces* between molecules of A and B in the mixture are closely similar to the intermolecular forces in pure A and pure B.

If the intermolecular forces between A and B are overall stronger than in the pure liquids, then the mixture is less volatile than Raoult's law predicts and has a lower vapour pressure. This leads to a negative deviation from the law.

If the intermolecular forces between A and B are overall weaker than in the pure liquids, then the mixture is more volatile than Raoult's law predicts and has a higher vapour pressure. This leads to a positive deviation from the law.

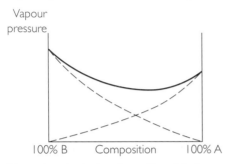

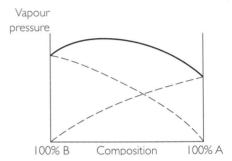

Negative deviation from Raoult's law because of stronger forces between the molecules in the mixture

Positive deviation from Raoult's law because of overall weaker forces between the molecules in the mixture than in the pure liquids

rare earths: the traditional name of the *lanthanide elements.*

rate constants: the constants in *rate equations.* The units of a rate constant depend on the overall order of reaction.

Overall order	Units of the rate constant
zero	$mol\ dm^{-3}\ s^{-1}$
first	s^{-1}
second	$mol^{-1}\ dm^3\ s^{-1}$
third	$mol^{-2}\ dm^6\ s^{-1}$

rate determining step: the slowest step in the mechanism of a reaction which therefore controls the overall rate of reaction. It is generally the molecules or ions involved (directly or indirectly) in the rate determining step which appear in the rate equation for the reaction.

Nucleophilic substitution in halogenoalkanes such as 2-bromo-2-methylpropane involves a two step mechanism. The first step, which is the bond breaking step in this S_N1 mechanism, is the rate determining step. It involves only halogenoalkane molecules. Hydroxide ions are not involved. So the hydroxide ion concentration does not appear in the rate equation which has the form:

rate $= k[C_4H_9Br]$.

The rate determining step in a multi-step reaction is the one with the highest activation energy.

rate equations are equations which show how changes in concentration, temperature and catalysts affect the rate of a reaction.

For the reaction of hydrogen peroxide with hydriodic acid:

$$H_2O_2(aq) + 2I^-(aq) + 2H^+(aq) \longrightarrow 2H_2O(l) + I_2(aq)$$

$$\text{rate} = k[H_2O_2(aq)][I^-(aq)]$$

Note that the rate equation cannot be deduced from the balanced equation; it has to be found by experiment. The equation shows that this reaction is *first order* with respect to each reactant. Overall it is a *second order reaction*.

Methods for finding the order of reaction include:

- the *half-life* method
- the *initial rate method*.

The *rate constant, k*, is constant only for a particular temperature. The value of k varies with temperature. A rise in temperature of 10 K doubles the value of k and so doubles the rate of reaction if the *activation energy* is about 50 kJ mol^{-1}. Adding a *catalyst* lowers the activation energy and so increases the value of k and hence the rate of reaction.

The *Arrhenius equation* is the relationship between the rate constant, the activation energy and the temperature.

rates of reaction, in mol dm^{-3} s^{-1}, are found by measuring the rate of formation of a product or the rate of removal of a reactant. The usual procedure for finding rates is to measure some property of the reaction mixture, such as its volume, and to see how this property varies with time.

$$\text{Rate} = \frac{\text{change in the measured property}}{\text{time}}$$

Rates are measured in practical units such as 'cm per second' and then converted to mol dm^{-3} s^{-1} if necessary.

Methods for studying rates include:

- collecting and measuring the volume of a gas formed
- removing measured samples of the mixture, stopping the reaction and then determining the concentration of one reactant or product by *titration*
- using a *colorimeter* to follow the formation of a coloured product or the removal of a coloured reactant
- using a conductivity cell and meter to measure the changes in electrical conductivity of the reaction mixture and the number or nature of the ions changes.

Factors which affect reaction rates are:

- the *concentration* of reactants in solution – in general the higher the concentration the faster the reaction (see also *collision theory*)
- the *pressure* of gaseous reactants – high pressures compress gases and increase their concentration
- the surface area of solids – breaking a solid into smaller pieces increases the surface area in contact with a liquid or gas

- the temperature – raising the temperature increases the proportion of atoms, molecules or ions with enough energy to react when they collide
- *catalysts* – adding a *homogeneous catalyst* or *heterogeneous catalyst* provides an alternative reaction mechanism with lower *activation energy*
- radiation – *electromagnetic radiation* can initiate *free-radical reactions* and the more intense the radiation the faster the reaction goes.

raw materials: see *chemical industry*.

reacting masses: the masses of elements and compounds which take part in a chemical reaction. Chemists work out *empirical formulae* and *balanced equations* by measuring the masses of reactants and products. They use equations to work out how much of each chemical they need for a chemical process and the theoretical yield of the products (see *yield calculations*).

Worked example:

Carbon monoxide reduces iron(III) oxide to iron in a blast furnace. What mass of carbon monoxide is required to reduce 20 tonnes of the oxide?

Notes on the method
Start by writing the balanced equation for the reaction.

Work out the *molar masses*, M, of the reactants in g mol^{-1}.

The reacting masses are in the same ratio whether measured in grams or in tonnes.

Answer
$Fe_2O_3(s) + 3CO(g) \rightarrow 2Fe(s) + 3CO_2(g)$

1 mol $Fe_2O_3(s)$ reacts with 3 mol $CO(g)$.

$M(Fe_2O_3) = (2 \times 56 \text{ g mol}^{-1}) + (3 \times 16 \text{ g mol}^{-1}) = 160 \text{ g mol}^{-1}$

$M(CO) = (12 + 16) \text{ g mol}^{-1} = 28 \text{ g mol}^{-1}$

So (1 mol × 160 g mol^{-1}) = 160 g iron reacts with (3 mol × 28 g mol^{-1}) = 84 g carbon monoxide,

or 160 tonnes iron reacts with 84 tonnes carbon monoxide

Hence mass of CO needed to react with 20 tonnes iron $= \dfrac{20}{160} \times 84$ tonnes

$= 10.5$ tonnes

---

reacting volumes of gases: see *gas-volume calculations*.

reaction kinetics: the study of the factors that affect the *rates of chemical reactions*. Rates are followed either by measuring the rate of formation of one of the products or the rate of removal of one of the reactants. The results of these studies are summed up in *rate equations* for reactions. *Collision theory* is one model which helps to explain the factors affecting rates. Studying the kinetics for a change can help to work out the *mechanism of a reaction*.

reaction mechanisms: see *mechanism of a reaction*.

reactors are reaction vessels used by the chemical industry for *synthesis*. There are two main types of reactor:

- batch reactors – for *batch processes* in which all the reactants are mixed in a vessel where they are heated or cooled until the process is complete
- continuous reactors – for *continuous processes* in which the reactants are fed continuously into a stirred tank or tube and products drawn out at an equal flow rate.

reagent: any pure chemical, mixture or solution supplied for chemical analysis or synthesis.

'Reagent grade' chemicals match the minimum standards specified for normal laboratory use.

Analytical reagents (AR) are chemicals which are much purer than standard laboratory grade reagents. They are intended for accurate analytical work where impurities in the reagents would affect the results. AR chemicals have to match a British standard. They are labelled AnalaR.

real gases are gases which deviate to a greater or lesser extent from *ideal gas* behaviour as predicted by the ideal gas equation. In practice, many gases at room temperature follow the gas laws sufficiently closely for it to be reasonable to treat them as ideal gases.

The assumptions of *kinetic theory* explain why real gases tend to deviate from ideal behaviour at:

- low temperatures and moderate pressures – when gases are not far above their boiling points so that they are close to condensing because of *intermolecular forces*
- very high pressures – when the volume of the gas is so small that the volume of the molecules cannot be ignored and the gases are less compressible than the gas laws predict.

rearrangement reaction: a reaction which rearranges the atoms in a molecule to convert one *structural isomer* to another.

One example of a rearrangement reaction is the isomerisation of alkanes to make branched compounds to increase the octane number of *petrol*.

$$CH_3 - CH_2 - CH_2 - CH_2 - CH_3 \longrightarrow CH_3 - \overset{\overset{\displaystyle CH_3}{|}}{CH} - CH_2 - CH_3$$

pentane   2-methylbutane

Isomerisation rearranges the atoms in pentane to make 2-methylbutane which has a higher octane number

recrystallisation is a procedure often used to purify solid products of *organic preparations*. The procedure is based on a *solvent* which dissolves the product when hot but not when cold. The choice of solvent is usually made by trial and error. Use of a *Buchner flask and funnel* speeds filtering and makes it easier to recover the purified solid from the filter paper.

The procedure is as follows:

1 warm the impure solid with the hot solvent
2 if the solution is not clear, filter the hot solution though a heated funnel to remove insoluble impurities
3 cool the solution so that the product recrystallises leaving the smaller amounts of soluble impurities in solution
4 filter to recover the purified product
5 wash the solid with small amounts of pure solvent to wash away the solution of impurities still clinging to the solid
6 allow the solvent to evaporate in a stream of air and then in a *desiccator.*

recycling is a method of dealing with *wastes.* The economics of recycling often depend on the energy required to process wastes compared to the energy needed to produce materials directly from raw materials, examples of which follow:

- steel, 32 MJ kg^{-1} (megajoules per kilogram)
- aluminium, 146 MJ kg^{-1}
- plastics such as polythene, 90 MJ kg^{-1}
- cardboard 56 MJ kg^{-1}.

The large amount of energy involved in *aluminium extraction* makes recycling particularly worthwhile.

There are other benefits to recycling such as reducing the environmental impact of extracting and processing raw materials.

Recycling metals – recycling *steel* can be done repeatedly because steel is as good as new after reprocessing. For every tonne of steel recycled there is a saving of 1.5 tonnes of iron ore and half a tonne of coal. There is also a big reduction in the total amount of water needed since large quantities of water are involved in mineral processing.

Worldwide, the steel industry recycles over 430 million tonnes of the metal each year which is a recycling rate of over 50%. The basic oxygen process uses a minimum of 25% scrap steel. Electric arc furnaces make steel plate, beams and bars by remelting nearly 100% scrap steel. Much of the recycled steel is waste from various stages of manufacturing steel objects. Another large source of scrap is from old motor vehicles.

Some steel is recovered from the tin cans in household waste. In the UK the recovery rate is still relatively low but in Germany is is over 80%. The advantage of steel is that it is magnetic so magnets can easily separate cans from other waste materials.

Aluminium scrap from manufacturing processes is always recycled. Less than 1% of the mass of household waste is aluminium, almost all of it in the form of drinks cans. Aluminium is not magnetic but can be separated from a waste stream by a rapidly varying magnetic field which induces eddy currents in the metal of the can. The interaction of the external field and the magnetic effect of the eddy currents leads to a force which can push cans out of the waste stream.

Recycling plastics – plastics are much more difficult to recycle than iron or steel. One problem arises from the low density of plastic waste. It can take as many as 20 000 bottles to make up a tonne of waste. Also, there are many types of plastic which have to be sorted. The industry has introduced codes for labelling plastic products to help consumers separate them for recycling.

Plastics vary greatly in their value. The material used to make sparkling drink bottles, PET, is generally the most valuable. PET can be remelted and spun into fibres for carpets, bedding and clothing. Another plastic worth recycling is PVC, which can be recycled to make sewage pipes, flooring and even the soles of shoes.

An alternative approach to recycling plastics is to use cracking to break the polymer molecules into small molecules which can be added to the feedstock for oil refineries and chemical plants.

redox potential: see *standard electrode potential.*

redox reactions: reactions which involve **red**uction and **ox**idation. In every redox reaction an atom, molecule or ion is reduced while another atom, molecule or ion is oxidised. Reduction and oxidation always go together.

Redox reactions involve the transfer of electrons from the *reducing agent* to the *oxidising agent.* Writing *half-equations* for redox reactions help to show electron transfer.

Oxidation numbers help to identify redox reactions. In any redox reaction the oxidation number of one element becomes more positive (or less negative) while the oxidation number of the other element becomes less positive (or more negative).

Equations for redox reactions, like other balanced equations, show the *amounts* (in moles) of reactants and products involved. Oxidation numbers help to balance redox equations because the total decrease in oxidation number for the element reduced must equal the total increase in oxidation number for the element oxidised. This is illustrated here by the oxidation of iron(II) ions by manganate(VII) ions in acid solutions.

Step 1 –Write down the formulae for the atoms, molecules and ions involved in the reaction:

$$MnO_4^- + H^+ + Fe^{2+} \rightarrow Mn^{2+} + H_2O + Fe^{3+}$$

Step 2 –Identify the elements which change in oxidation number and the extent of change:

change of –5

$$MnO_4^- + H^+ + Fe^{2+} \longrightarrow Mn^{2+} + H_2O + Fe^{3+}$$

change of +1

Step 3 –Balance so that the decrease in oxidation number of one element equals the total increase of the other element.

In this example the decrease of –5 in the oxidation number of manganese is balanced by five iron(II) ions which each increase their oxidation number by +1.

$$MnO_4^- + H^+ + 5Fe^{2+} \rightarrow Mn^{2+} + H_2O + 5Fe^{3+}$$

Step 4 – Balance for oxygen and hydrogen.

In this example the four oxygens from the manganate ion join with eight hydrogen ions to form four water molecules:

$$MnO_4^- + 8H^+ + 5Fe^{2+} \rightarrow Mn^{2+} + 4H_2O + 5Fe^{3+}$$

Step 5 – In an *ionic equation*, check that the positive and negative charges balance and add state symbols.

The net charge on the left is now 17+, which is the same as the net charge on the right.

$$MnO_4^-(aq) + 8H^+(aq) + 5Fe^{2+}(aq) \rightarrow Mn^{2+}(aq) + 4H_2O(l) + 5Fe^{3+}(aq)$$

redox titration: a practical technique used to determine the *concentration of a solution* of an *oxidising agent* or of a *reducing agent*. A *titration* measures the volume of a *standard solution* of oxidising agent or reducing agent needed to react exactly with a measured volume of the unknown solution. The procedure gives accurate results only if the reaction is rapid and is exactly described by the chemical equation. An acidic standard solution of the oxidising dichromate(VI) ions, for example, can be used to estimate the concentration of iron(II) ions. The end-point for a dichromate(VI) titration is detected using a redox indicator – a chemical which gives a clear colour change when oxidised.

The procedure and method of calculating results is similar to those for other titrations for example *iodine–thiosulfate titrations* and *potassium manganate(VII) titrations*.

reducing agents: chemical reagents which can reduce other atoms, molecules or ions by giving electrons to them. Common reducing agents are metals such as zinc, iron and tin, often with acid; other common reducing agents are sulfur dioxide, iron(II) ions and iodide ions.

Iodide ions acting as a reducing agent by giving electrons to iron(III) ions. A reducing agent is itself oxidised when it reacts. Reduction and oxidation always go together in redox reactions.

reduction

$$2Fe^{3+}(aq) + 2e^- \longrightarrow 2Fe^{2+}(aq)$$

$$2I^-(aq) \longrightarrow I_2(aq) + 2e^-$$

oxidation

Some reagents change colour when they are reduced which makes them useful for detecting reducing agents.

Test	Observations	Explanation
Add a solution of potassium manganate(VII) acidified with dilute sulfuric acid	Purple solution turns colourless	Purple MnO_4^- ions are reduced to very pale pink Mn^{2+} ions
Add potassium dichromate(VI) solution acidified with dilute sulfuric acid	Orange solution turns green	Orange $Cr_2O_7^{2-}$ ions are reduced to green Cr^{3+} ions

reducing sugars: *sugars* such as glucose and fructose which give a positive result with *Fehling's solution* or *Benedict's solution*. These sugars reduce the blue copper(II) complex in the test solution to an orange–red precipitate of copper(I) oxide.

All monosaccharides and many disaccharides are reducing sugars, but the disaccharide sucrose is not a reducing sugar.

reductant: an alternative to *reducing agent* which is convenient when describing half-equations for *redox reactions* and *standard electrode potentials*. By convention, when electrode potential values are assigned the half-equation takes the form:

$$\text{oxidant} + n\text{e}^- \rightleftharpoons \text{reductant}$$

Every half-equation involves an oxidant and a reductant.

reduction: originally this meant removal of oxygen or the addition of hydrogen but the term now covers all reactions in which atoms, molecules or ions gain electrons. The definition is further extended to cover molecules, as well as ions, by defining oxidation as a change which makes the *oxidation number* of an element more negative or less positive.

$$\underset{-I}{2Br^-} + 2H^+ + \underset{+6}{H_2SO_4} \longrightarrow Br_2 + \underset{+4}{SO_2} + 2H_2O$$

+6   sulphur reduced   +4

−I   bromine oxidised   0

Bromide ions reducing concentrated sulfuric acid to sulfur dioxide

Oxidation and reduction always go together in *redox reactions*.

Oxidation number rules apply in principle in organic chemistry but it is often easier to use the older definitions. Reduction is either removal of oxygen to a molecule or the addition of hydrogen.

$$\underset{H_3C}{\overset{H_3C}{\diagdown}}C=O \xrightarrow{\text{NaBH}_4\text{(aq)}} H_3C-\underset{OH}{\overset{H}{\underset{|}{\overset{|}{C}}}}-CH_3$$

Reduction of ketone to an alcohol by addition of hydrogen

reference electrodes: electrodes used to measure electrode potentials in place of the standard *hydrogen electrode*. A hydrogen electrode is difficult to use so it is much easier to use a secondary standard such as a silver/silver chloride electrode or a calomel electrode. These electrodes are available commercially and are reliable to use. They have been calibrated against a standard hydrogen electrode. (Calomel is an old-fashioned name for mercury(I) chloride.)

$$Hg_2Cl_2(s) + 2e^- \rightleftharpoons 2Hg(l) + 2Cl^-(aq) \qquad E^\ominus = +0.27 \text{ V}$$

refining: the processes which separate, convert and purify chemicals in *crude oil* in an oil refinery, or the removal of impurities from metals (see *copper refining*).

reflux condenser: a condenser fitted to a chemical apparatus to prevent vapour escaping while heating a liquid. Vapour from a boiling reaction mixture condenses and flows back into the flask.

reforming: a process in oil refining which converts *alkanes* to *arenes* such as benzene and methylbenzene. This helps to raise the octane number of mixtures used in *petrol.*

$$CH_3CH_2CH_2CH_2CH_2CH_2CH_3 \xrightarrow[\text{heat, pressure}]{\text{Pt catalyst}}$$

heptane

methylbenzene

$$+ \quad 4H_2(g)$$

cyclohexane   benzene

$$\xrightarrow[\text{heat, pressure}]{\text{Pt catalyst}} \quad + \quad 3H_2(g)$$

Examples of reforming

refractive index: a *physical property* of chemicals and materials which can help to identify unknown samples.

Forensic scientists use an oil immersion method to find the reactive index of fragments of glass from the scene of an accident or crime. The procedure is to immerse a crushed fragment in silicone oil on a slide. The slide is slowly heated, making the refractive index of the oil change as the temperature rises. The fragments of glass seem to disappear when their refractive index exactly matches that of the oil, then reappear when the oil is heated beyond the point where the refractive indices match. The technique is extremely accurate. Repeat measurements lie within a range of $\pm 0.000\,05$ units.

refractories: materials with very high melting points used to line furnaces and make crucibles. *Ceramics* are refractory materials.

Fireclay is the raw material for making refractory bricks. Some industries need more specialised refractories. Molten *glass* and the *slags* formed during smelting are very corrosive when molten and would quickly destroy ordinary refractory bricks made of fireclay. These industries use refractories such as pure silicon dioxide (an *acidic oxide*) or pure magnesium oxide (a *basic oxide*).

relative atomic mass, A_r: the mean mass of the atoms of an element relative to the mass of atoms of the isotope carbon-12 for which A_r is defined as exactly 12. The values are relative so they do not have units.

Mass spectrometry is the technique for determining accurate relative atomic masses.

Amount of substance in chemistry is defined in such a way that the *molar mass* of an element is numerically equal to its relative atomic mass (but the latter is a ratio and has no units).

Relative atomic masses often do not have whole number values because elements have *isotopes*. Relative atomic masses are average values for the mixture of isotopes found naturally (see *isotopic abundance*).

relative formula mass, M_r is a term used for the relative mass of ionic compounds or ions to avoid the suggestion that their formulae represent molecules. The scale is the same as for relative atomic and relative molecular masses that is $^{12}C = 12$.

The relative formula mass of an ionic compound or ion is the sum of the *relative atomic masses* for the atoms in the formula.

For anhydrous magnesium nitrate, $M_r[Mg(NO_3)_2]$
$= 24 + (2 \times 14) + (6 \times 16) = 148$

$A_r(Mg) \quad A_r(N) \quad A_r(O)$

Amount of substance in chemistry is defined in such a way that the *molar mass* of an ionic compound or an ion is numerically equal to its relative formula mass (but the relative mass has no units).

relative isotopic mass: the mass of the atoms of an isotope of an element relative to the mass of atoms of the isotope carbon-12. Chemists determine relative isotopic masses with a *mass spectrometer*. The values are always very close to the mass numbers of the elements. It is usually accurate enough to assume that the value of the relative atomic mass is the same as the *mass number*.

relative molecular mass, M_r : the relative mass of the molecules of an element or compound on the scale $^{12}C = 12$.

The relative molecular mass of the molecules of an element or compound is the sum of the *relative atomic masses* for the atoms in the *molecular formula*.

For ethanol, $M_r(CH_3CH_2OH) = (2 \times 12) + (6 \times 1) + 16 = 46$

$A_r(C) \quad A_r(H) \quad A_r(O)$

Amount of substance in chemistry is defined in such a way that the *molar mass* of the molecules of an element or compound is numerically equal to its relative molecular mass but as with other relative masses, the relative value has no units.

reserves and resources describes the quantities of *minerals* and *fossil fuels* available in the Earth's crust. Estimates of reserves and resources include some which are proven because they have been identified and evaluated and others which it seems likely to exist by a mixture of inference and guesswork.

The proven reserves of a mineral have been identified and evaluated so that it is known that it is economical to extract and process them at the current market price of the product. The resources of a mineral include all the possible sources, including those that it is not economical to extract at the time.

Judgements about changes in technology, in the demand for materials and in the economy can all shift the balance of reserves and resources. Figures for reserves have to be interpreted with care because commercial companies do not find it economic to establish reserves for more than about 25 years ahead.

residence time: a measure of the time a chemical stays in any part of a system where chemicals are continuously flowing in and out of containers or reservoirs. In environmental systems the reservoirs are parts of the Earth system such as the *atmosphere*, oceans or land. In a *steady state system* the residence time equals the amount of the chemical in the reservoir divided by the rate of addition (or removal) of the chemical. The residence time of calcium ions in the oceans is a million years. The residence time of oxygen in the atmosphere is 7600 years. If the residence time of a pollutant is large it may stay in the environment for a long time before being inactivated.

residues: materials left over at the end of a chemical process. The term sometimes refers to a solid caught in a filter paper. Laboratories which use large quantities of valuable chemicals may have a 'residues bottle' to collect wastes for recovery of the expensive substances. Residues worth collecting include silver compounds, iodine compounds and solvents.

resins: originally these were natural materials from the sap of trees or from insects Examples are the rosin used for violin bows, shellac used in varnishes and myrrh which is added to perfumes. These gummy materials consist of long chain molecules and on warming they soften before they melt.

Now the term resin refers to synthetic materials made by *polymerisation*. *Thermosetting polymers*, for example, are produced in two stages. The first stage produces a polymer resin. Compressing this resin in a hot mould shapes it and at the same time creates *cross-links* between the polymer chains so that the material sets to a hard plastic.

Similarly, one component of an *epoxy resin* adhesive is a polymer resin which sets on mixing with a hardener to start the chemical changes which cross-link the chains.

Synthetic *ion-exchange* resins consist of polymer beads which have been treated chemically so that they can hold cations or anions.

respiration: biochemical processes in cells which provide living organisms with the energy for growth, movement and warmth.

Aerobic respiration is respiration with oxygen. In the process *glucose* is *oxidised* to carbon dioxide and water. Overall:

$$C_6H_{12}O_6(aq) + O_2(g) \longrightarrow 6CO_2(g) + 6H_2O(l) \quad \Delta H = -2816 \text{ kJ mol}^{-1}$$

This is not a one step reaction; it takes place in a complex series of biochemical reactions which harness the energy from the oxidation process to produce *ATP*.

Anaerobic respiration is respiration without *oxygen*. Anaerobic respiration in yeast converts *sugars* to carbon dioxide and ethanol. This is the process of *fermentation* used in baking, brewing and winemaking. Anaerobic respiration in animal cells produces lactate ions (the anions of lactic acid). Anaerobic respiration produces much less ATP than aerobic respiration and so is a much less efficient source of energy for processes.

restricted rotation: see *geometrical isomerism.*

retention times: see *gas-liquid chromatography.*

reversible changes are processes which can be reversed by altering the conditions.

Melting is an example of a reversible change of state. Changing the temperature alters the direction of change. Ice melts above 0°C. Water freezes below 0°C. Water and ice are in equilibrium at 0°C. This is an example of *dynamic equilibrium.*

$$H_2O(s) \rightleftharpoons H_2O(l)$$

Many chemical reactions are reversible, including the reactions used in *ammonia manufacture* and *sulfuric acid manufacture. Acid–base reactions* and *redox reactions* are generally reversible too. There are many important reversible processes in living things. *Haemoglobin,* for example, picks up oxygen from the air in the lungs and then releases the oxygen for *respiration* in body tissues.

The direction and extent of change for a reversible process may vary with the temperature or the concentration of chemicals. *Pressure* is an important variable affecting reactions of gases.

Chemists predict the direction and extent of change for reversible processes when they consider *feasibility.*

reverse osmosis: see *osmotic pressure.*

rhodium is one of the *platinum* metals which is largely used as a catalyst in the platinum–rhodium alloys found in *catalytic converters* and in chemical plants for *nitric acid manufacture.*

risk, in chemistry, is an estimate of the chance that a hazardous substance or process will cause harm.

There are three questions to consider when assessing the risk of a chemical process:

- what are the hazards?
- what is the likelihood that someone will come to harm because of the hazards?
- what can be done to control and reduce the risks?

When a chemical investigation is being planned the possibilities for controlling risks include the following:

- choosing a non-practical approach using secondary sources, models and simulations
- adopting an alternative safer procedure with less hazardous chemicals
- modifying the experimental design with different apparatus or a different method of heating
- creating a barrier between the apparatus and people, using safety screens or a fume cupboard
- wearing personal protection such as eye protection and gloves.

(See also *hazard warning signs* and *COSHH regulations.*)

RNA: a type of *nucleic acid* in which the five carbon sugar is ribose and the four nitrogenous bases are adenine, cytosine, guanine and uracil. Two types of RNA are important in *protein* synthesis:

- messenger RNA (m-RNA) which takes the genetic code from *DNA* molecules in the nucleus to the sites of protein synthesis in the cell
- transfer RNA (t-RNA) which helps to assemble amino acids in the right order during the translation of the genetic code into protein molecules.

In some viruses, such as HIV, the genetic material is RNA rather than DNA.

roasting converts sulfides to oxides during *metal extraction*. Oxides are much easier to reduce to metals than sulfides. Roasting is exothermic so once the reaction has begun it needs little fuel to keep the process going. Roasting produces hot metal oxides which may be fed directly to a furnace for reduction to the metal.

$$2PbS(s) + 3O_2(g) \longrightarrow 2PbO(s) + 2SO_2(g)$$

In the past roasting sulfide ores was a major cause of air pollution. The sulfur dioxide was simply released into the air where it caused acid rain. Now the sulfur dioxide is recovered and used for *sulfuric acid manufacture*. (See also *zinc extraction*.)

rotation about covalent bonds is rotation of one part of a molecule relative to another around a single covalent bond. This allow molecules to take up different *conformations*.

Double bonding stops free rotation and this accounts for the existence of *geometric isomers*.

rubber: an elastic material which is easily stretched or bent but which tends to spring back to its original shape when the force is removed.

Rubbers are **elast**ic **po***ly***mers** often called *elastomers*.

Elastomer molecules tend to coil up in such a way that they can uncoil as they stretch and coil up again when unstretched. The molecules are *cross-linked* by strong covalent bonds so that they cannot permanently slide past each other unless pulled much harder.

Natural rubber comes from the sap of the rubber tree. This white latex coagulates to a very soft material. Vulcanising converts natural rubber into a useful elastomer by cross-linking the polymer chains with sulfur atoms. This was the process discovered by Charles Goodyear in 1839.

The science of *polymer chemistry* has helped to develop a range of alternatives to natural rubber suited for specific purposes. Examples are:

- neoprene – used for rubber dinghies
- styrene-butadiene rubber – used for motor vehicle tyres
- *polyurethanes* – used for furniture foams and stretch fabrics.

Tyres are black because the rubber is mixed with carbon *filler* which makes the material more resistant to wear. Carbon also absorbs *ultraviolet radiation* helping to slow down the rate at which rubber degrades in sunlight.

rusting is the *corrosion* of iron.

S

sacrificial protection: see *cathodic protection.*

salt bridge: the ionic connection between the solutions of the two *half-cells* which make up an *electrochemical cell.* A salt bridge makes an electrical connection between the two halves of the cell by allowing ions to flow while preventing the two solutions from mixing. At its simplest a salt bridge consists of a strip of filter paper soaked in potassium nitrate solution. Potassium salts and nitrates are soluble so the salt bridge does not react with the ions in the half-cells.

In commercial *reference electrodes* the salt bridge is often a small plug of porous glass.

salt hydrates: see *hydration.*

salting out effect: this consists of making a chemical much less soluble in water by adding a high concentration of a salt. Adding common salt (sodium chloride) helps to separate *soap* from water after *hydrolysis* of fats or oils.

salts are ionic compounds formed when an *acid* reacts with a *base.* Salts therefore have two 'parents'. Salts are related to a parent acid and to a parent base.

Acid	Salts
hydrochloric acid, HCl	sodium chloride, $NaCl$
	calcium chloride, $CaCl_2$
	ammonium chloride, NH_4Cl
sulfuric acid, H_2SO_4	sodium sulfate, Na_2SO_4
	calcium sulfate, $CaSO_4$
	ammonium sulfate, $(NH_4)_2SO_4$
ethanoic acid, CH_3CO_2H	sodium ethanoate, CH_3CO_2Na
	calcium ethanoate, $(CH_3CO_2)_2Ca$
	ammonium ethanoate, $CH_3CO_2NH_4$

Base	Salts
sodium hydroxide, $NaOH$	sodium chloride, $NaCl$
	sodium sulfate, Na_2SO_4
	sodium ethanoate, CH_3CO_2Na
calcium oxide, $Ca(OH)_2$	calcium chloride, $CaCl_2$
	calcium sulfate, $CaSO_4$
	calcium ethanoate, $(CH_3CO_2)_2Ca$
ammonia, NH_3	ammonium chloride, NH_4Cl
	ammonium sulfate, $(NH_4)_2SO_4$
	ammonium ethanoate, $CH_3CO_2NH_4$

Neutralisation is not the only way to make a salt. Some metal chlorides, for example, are made by heating metals in a stream of chlorine. This is useful for making anhydrous chlorides, such as *aluminium chloride* or iron(III) chloride, both of which are hydrolysed by water so cannot be dehydrated by heating.

Insoluble salts are conveniently prepared by *ionic precipitation.*

salts of strong and weak acids: see *hydrolysis of salts* and *neutralisation*.

sampling for analysis is a vital first step in any analysis because it is important to make sure that the sample is representative of the whole specimen. This is easy when analysing solutions which are well mixed. It is more difficult when sampling an *ore* made up of lumps of rock each with variable but small amounts of a precious mineral. In these circumstances an analyst has the difficult task of preparing a sample with a mass of about 1 g from a batch of ore with a mass of several tonnes.

Analysts generally prepare multiple samples from the same specimen and work through the analysis with all of them to check on the *uncertainty* of the results.

saturated compounds are compounds containing only single bonds between the atoms in their molecules. Examples of saturated *hydrocarbons* are the *alkanes*.

The term 'saturated' is also used for compounds with saturated hydrocarbon chains such as the saturated *fats* and *fatty acids*, even though these compounds include $C = O$ bonds. Saturated compounds do not undergo *addition reactions*.

saturated solutions are solutions which contain as much of the dissolved substance as possible at a particular temperature. A saturated solution is in equilibrium with undissolved excess of the substance in solution. The concentration of a saturated solution is the *solubility* of the substance at that particular temperature.

saturated vapour pressure: see *vapour pressure*.

s-block elements: the elements in groups 1 and 2 in the *periodic table*. For these elements the last electron added to the atomic structure goes into the *s-orbital* in the outer shell. All the elements in the s-block are reactive metals. (See also *atomic orbitals*.)

scaling up processes: converting a small scale synthesis in a laboratory or pilot plant to a process which can operate successfully and safely on a commercial scale (see Table on page 319).

second law of thermodynamics: see *thermodynamics (laws of)*.

second order reaction: a reaction is second order with respect to a reactant if the rate of reaction is proportional to the concentration of that reactant squared. The concentration term for this reactant is raised to the power two in the *rate equation*. At its simplest the rate equation for a second order reaction is:

$$Rate = k[X]^2$$

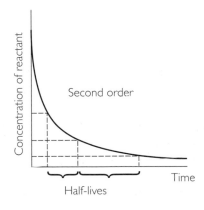

Variation of concentration of a reactant plotted against time for a second order reaction. The half-life for a second order reaction is not a constant. The time for the concentration to fall from c to c/2 is half the time for the concentration to fall from c/2 to c/4. The half-life is inversely proportional to the starting concentration.

Scaling up processes (see page 318)

On a small scale	On a big scale
• Laboratory preparations are normally *batch processes*.	• In industry, manufacturers use *continuous processes* where there is a steady demand for the product.
• A chemist in a laboratory stores chemicals in bottles and transfers them to a reaction vessel with a spatula or by pouring through a funnel.	• In industry, pumps and pipes transfer liquids to reactors. Solids may be handled by the bagful or ingot.
• Manual mixing or shaking is adequate in the laboratory.	• A large reactor typically has a stirrer driven by an electric motor to ensure thorough mixing.
• A laboratory synthesis may happen in a beaker or flask easily heated with a Bunsen flame or cooled by dipping in water.	• It is not to easy to heat or cool large quantities of material. Reactors often have pipes running through them to carry steam for heating or cold water for cooling. Energy from exothermic reactions can raise steam to generate electricity.
• To separate the product, laboratory apparatus can be rearranged for distillation or tipped to pour the products into a filter funnel.	• Solids can be separated as pumps force a liquid through large cloth filters.
• Chemists can pour products from one container to another, perhaps using hand-held tap funnels to extract impurities. *Distillation* or *recrystallisation* help to produce a pure product.	• One large-scale approach to crystallisation and drying is to spray a solution into a stream of hot air.
• The product can be dried in a *desiccator* or small oven.	

The rate of reaction of 1-bromopropane with hydroxide ions is overall second order. It is *first order* with respect to the halogenoalkane and first order with respect to hydroxide ions. The overall order is the sum of the powers in the rate equation.

Rate = $k[CH_3CH_2CH_2Br][OH^-]$

secondary organic compounds: see *primary, secondary and tertiary organic compounds*.

seed crystal: a crystal added to a *supersaturated* solution to encourage it to crystallise. *Recrystallisation* of an organic product sometimes produces a solution which is reluctant to crystallise. Scratching the sides of the container with a glass rod can encourage crystals to form. If this fails, adding a minute crystal of the product may be enough to start rapid crystallisation.

semiconductors are not *electrical insulators* but they do not conduct electricity as well as metals. Examples are elements such as *silicon* and *germanium* and compounds such as gallium arsenide. Semiconductors consist of covalent *giant structures* in which a few of the bonding electrons can break free and become *delocalised* in the structure. The conductivity of semiconductors is enhanced by 'doping' with traces of impurity atoms. Doping silicon (in group 4) with an element such as arsenic (group 5) increases the number of conducting electrons. An arsenic atom has one more electron in its outer shell than is needed for bonding in the giant structure.

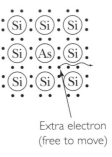

Extra electron
(free to move)

Silicon with an impurity atom from a group 5 element

selectively-permeable membrane: a membrane which allows solvent molecules and perhaps some other small molecules or ions to pass through but is impermeable to larger molecules and other ions. *Osmosis* and *dialysis* both depend on selectively permeable membranes. So does the manufacture of chlorine by *electrolysis of brine* in a membrane cell.

separating funnel: a tap funnel used to separate liquids which do not mix (*immiscible liquids*). Separating funnels are used to:

- separate an organic product by *solvent extraction* with a solvent such as ether (ethoxyethane) – the product is more soluble in the organic solvent but the ionic reagents and ionic by-products remain in the water
- purify an impure organic product by 'washing' it with aqueous reagents (such as dilute acids or alkalis) which dissolve impurities.

Pressure can build up in a separating funnel when a mixture is being shaken if one liquid is very volatile or if a reaction produces a gas. Hence the technique of inverting the funnel while holding in the stopper and releasing excess gas or vapour through the tap.

sequestering agent: a ligand which forms such stable complexes with metal ions that it effectively makes them chemically inactive. Examples of sequestering agents are:

- *edta* used to treat people suffering from metal poisoning
- tripolyphosphates added to detergents to complex with calcium or magnesium ions in hard water so a scale or scum precipitate is not formed.

shapes of complex ions depend on the number of ligands around the central metal ion. There is no simple rule for predicting the shapes of complexes from their formulae. Typically complexes with:

- six smaller ligands, such as H_2O and NH_3, are octahedral, as in $[Mn(H_2O)_6]^{2+}$ and $[Fe(H_2O)_5(SCN)]^{2+}$
- four larger ligands, such as Cl^-, are usually tetrahedral, as in $[CuCl_4]^{2-}$, but may be planar, as in $[Pt(NH_3)_4]^{2+}$
- two ligands are linear, as in $[Ag(NH_3)_2]^+$, $[Ag(S_2O_3)_2]^{3-}$ and $[Ag(CN)_2]^-$.

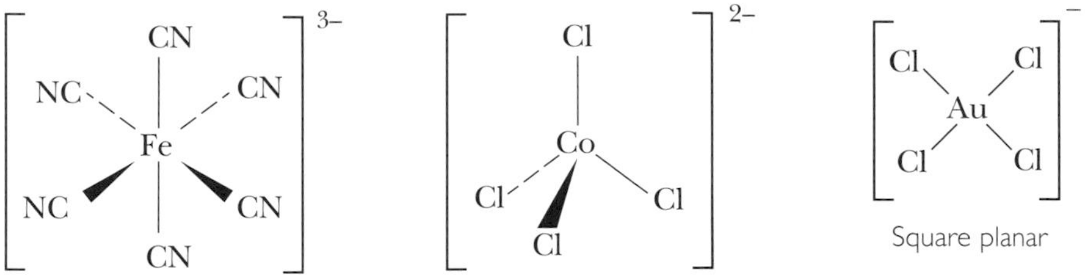

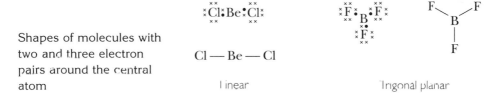

Octahedral   Tetrahedral

Square planar

$$[H_3N\!-\!Ag\!-\!NH_3]^+$$

Linear

Shapes of complex ions

shapes of molecules and ions can be predicted by examining the number of bonding and non-bonding *lone pairs of electrons* in the outer shell of the central atom. The expected shape for a molecule is the one which minimises the repulsion between electron pairs by keeping them as far apart as possible in three dimensions.

Shapes of molecules with two and three electron pairs around the central atom

$$:Cl:Be:Cl:$$

$$Cl\!-\!Be\!-\!Cl$$

Linear

Trigonal planar

Non-bonding lone pairs are held closer to the central atom. The result is that the order of repulsion between electron pairs is:

lone pair–lone pair > lone pair–bond pair > bond pair–bond pair

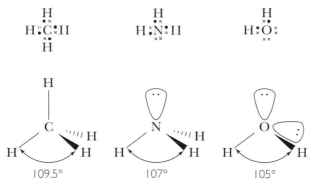

Bond angles in molecules with four electron pairs around the central atom. The greater repulsion between lone pairs, and between lone pairs and bonding pairs, means that the angles between the covalent bonds decrease as the number of lone pairs increases.

the UK is about 2600 μSv per year. The dose for an individual, however, depends on where they live, how they travel and whether or not they have certain medical treatments. One dental X-ray may involve a dose of about 20 μSv. Radiotherapy can involve very large doses such as 40 Sv.

sigma (σ) bond: a single covalent bond formed by a pair of electrons in a *molecular orbital* with the electron density concentrated between the two nuclei. Free rotation is possible about single bonds unlike about *pi bonds*.

Sigma bonds can form by overlap of two *s-orbitals*, an *s*-orbital and a *p-orbital* or two *p*-orbitals.

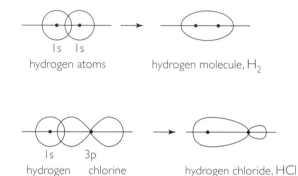

1s 1s
hydrogen atoms

hydrogen molecule, H₂

Examples of sigma bonds
in molecules

1s   3p
hydrogen   chlorine
atom   atom

hydrogen chloride, HCl

significant figures: the number of significant figures quoted for a measurement shows the degree of uncertainty. Measuring instruments vary in their *accuracy* and there is *uncertainty* associated with any measurement.

Putting values into *standard form* provides the clearest way of indicating the number of significant figures. This removes any doubt about any zeros, which sometimes just indicate the position of the decimal point, as in 0.0056 g which is a very small mass quoted to two significant figures. This is clearer when the value is in the form 5.6×10^{-3}. In other values the zeros are included to show the number of significant figures. When writing the distance of 3500 m as 3.50×10^{3} m it become clear that this is a quantity quoted only to three significant figures.

Generally, when multiplying or dividing values the number of significant figures in the result is the same as the measurement with the **least** number of significant figures.

When values are being added or subtracted in a calculation the number of significant figures in the result is determined by the **least** precise measurement included in the calculation. In other words, the measurement with the fewest figures after the decimal point when the measurement is written out in non-standard form.

silica (SiO₂): also known as silicon dioxide, this is the oxide of silicon which is abundant in rocks as quartz. Silica has a very high melting point at 1710°C, turning to a viscous liquid. On cooling the liquid becomes a *glass*.

Amethyst is crystalline quartz coloured purple due to the presence of iron(III) ions. Sandstone and sand consist mainly of silica. Flint is a non-crystalline form of silica.

Giant structure of quartz. Each Si atom is at the centre of a tetrahedron of oxygen atoms. The arrangement of silicon atoms is the same as the arrangement of carbon atoms in diamond but there is an oxygen atom between each silicon atom.

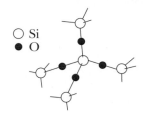

silica gel: a non-crystalline, hydrated form of *silica*. Heating silica gel produces hard granules which absorb water strongly, so silica gel is used as a drying agent. Self-indicating silica gel contains enough of an anhydrous cobalt(II) salt to colour the granules blue. After the gel has absorbed a certain quantity of moisture the cobalt ions become hydrated and turn pink as a warning that the gel is no longer as effective as a drying agent. Heating drives off the water so that the gel can be used again. Silica gel is also used as the stationary phase in *chromatography*.

silicates are the minerals that make up most of the Earth's crust. The basic building block for silicates is a SiO_4^{4-} tetrahedron. Zircon ($ZrSiO_4$) is an example of a simple silicate mineral with metal ions and silicate ions.

Silicate tetrahedra can join up in chains or strands as in asbestos. They can also join up in sheets as in mica, talc or *clay minerals*. In these minerals the negative charges on the silicate part is balanced by positive charges from metal ions such as Na^+, Ca^{2+}, Mg^{2+} or Al^{3+}.

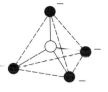

A single SiO_4^{4-} tetrahedron

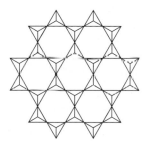

A fragment of a sheet of silicate tetrahedra

In *zeolites* the silicate tetrahedra are built up into three dimensional networks.

A very wide range of silicate minerals is possible because some of the silicon atoms in the tetrahedra can be replaced by aluminium, boron or beryllium atoms.

silicon (Si) is the second most abundant element in the Earth's crust. Combined with oxygen it forms many minerals including *silica* (SiO_2) and *silicates*. Silicon is the second element in *group 4* of the periodic table. Its electron configuration is [Ne]$2s^22p^2$.

Solid silicon has a diamond structure. It is a shiny grey material made by reducing silicon dioxide with carbon in an electric furnace. The silicon formed in this way is not pure enough for use in electronics as a *semiconductor*.

Silicon compounds are typical of the compounds of a *non-metal*:

- the oxide SiO_2 is an *acidic oxide* though relatively unreactive because it has a *giant structure*
- the chloride ($SiCl_4$) is a molecular liquid which is rapidly *hydrolysed* by water to hydrated silica and hydrogen chloride
- the hydrides, such as SiH_4 and Si_2H_6, are molecular gases (silanes).

silicones are *polymers* based on chains of alternating silicon and oxygen atoms. They are made by hydrolysis of compounds such as dimethylchlorosilane.

$$\begin{array}{ccccccc} CH_3 & & CH_3 & & CH_3 & & CH_3 \\ | & & | & & | & & | \\ -Si & -O- & Si & -O- & Si & -O- & Si- \\ | & & | & & | & & | \\ CH_3 & & CH_3 & & CH_3 & & CH_3 \end{array}$$

A silicone polymer

By controlling the chain length and degree of *cross-linking* it is possible to make a range of silicones for use as oils, greases and rubbery materials. Silicones are water repellent and can be more safely used at higher temperatures than polymers based on carbon atoms. They are electrical insulators and other materials do not stick to them. Silicones are colourless, have no smell and are inert.

Silicones are used:

- to waterproof fabrics
- as an ingredient of polishes
- to coat non-stick surfaces such as the paper backing for self-adhesive labels
- as *lubricants*, especially at high temperatures.

silver halides are used in qualitative analysis and in *photography*. The three insoluble silver salts form on mixing solutions of silver nitrate and a soluble chloride, bromide or iodide. Silver fluoride is soluble so there is no precipitate when silver nitrate is added to a solution of fluoride ions.

$$Ag^+(aq) + X^-(aq) \longrightarrow AgX(s) \text{ where } X = Cl, Br \text{ or } I$$

This precipitation reaction is used as a test for halide ions (see *anion tests*). The three silver compounds can be distinguished by their colour and the ease with which they redissolve in ammonia solution. The values for the *solubility product constants* show the trend in solubility. Silver chloride is the most soluble. (See Table on page 327.)

sizes of atoms and ions: see *atomic radius, covalent radius, ionic radius* and *van der Waals radius*.

skeletal formulae: outline formulae for carbon compounds which are a useful shorthand for complex molecules such as many natural products (see page 327). The formulae need careful study because they represent only the hydrocarbon part of the molecule with lines for the bonds between carbon atoms, leaving out the symbols for the carbon and hydrogen atoms. *Functional groups* are included.

Silver halides (see page 326)

Silver halide	Colour	Effect of adding ammonia solution to a precipitate of the compound	Solubility/ mol dm^{-3}	K_{sp}/mol^2 dm^{-6}
silver chloride, AgCl	white	redissolves readily in ammonia solution forming a complex ion	1.4×10^{-5}	2×10^{-10}
silver bromide, AgBr	cream	redissolves but only in concentrated ammonia	5.5×10^{-7}	3×10^{-13}
silver iodide, AgI	yellow	does not redissolve in ammonia solution	2.8×10^{-9}	8×10^{-18}

Skeletal formula for vitamin A compared to its full structural formula

slag: the unwanted waste material from *metal extraction* at high temperature (pyrometallurgy). Slag is tapped from a furnace, such as a *blast furnace*. The slag solidifies as it cools and can be dumped or crushed for use in construction or road building.

smelting is a process of *metal extraction* at high temperature (pyrometallurgy). Smelting involves melting the concentrates from metal ores to remove impurities and to reduce metal compounds to metals. Examples of smelting include the extraction of iron in a *blast furnace* and *aluminium extraction* by electrolysis of a melt.

smog: a smoky fog caused by air pollution. Burning coal in homes and industry created the dense city smogs of the nineteenth and first half of the twentieth centuries. The damaging components of these smogs were sulfur dioxide and soot. Many people affected by these acidic smogs died of lung disease. Smokeless zones and the shift of fuels from coal to oil and natural gas have largely eliminated this type of smog.

Today cities are affected by another type of smog caused by motor traffic. This is *photochemical smog*.

smoke: a *colloid* with specks of solid dispersed in a gas. Smokes are examples of aerosols.

SN1 and SN2 reactions: see *nucleophilic substitution in halogenoalkanes.*

soaps are made by the *hydrolysis* of *fats* or *vegetable oils* with alkali. Soaps are the sodium or potassium salts of *fatty acids.*

glycerol   sodium stearate (soap)

Saponification – the hydrolysis of a fat or vegetable oil (triglyceride) with alkali to make soap

Soaps are *surfactants* which help to remove greasy dirt because they have an ionic (water-loving) head and a long (water-hating) hydrocarbon tail.

Most toilet soaps are made from a mixture of animal fat and coconut palm oil. Soaps from animal fat are less soluble and longer lasting. Soaps from palm oils are more soluble so that they lather quickly but wash away more quickly. A bar of soap also contains a dye and perfume together with an antioxidant to stop the soap and air combining to make irritant chemicals.

sodium (Na) is a soft, shiny metal which rapidly tarnishes in moist air. It is the second member of *group 1* with the electron configuration $[Ne]4s^1$.

Like other group 1 metals, sodium:

- is stored in oil
- floats on water, melts and reacts violently forming hydrogen which catches fire and NaOH which is soluble and strongly alkaline
- forms an ionic, crystalline chloride Na^+Cl^-.

Sodium produces a mixture of the oxide, Na_2O, and peroxide, Na_2O_2, when it burns in air.

Electrolysis of molten sodium chloride is the process used to manufacture sodium. The electrolyte also contains some calcium chloride to lower the melting point. Sodium forms at steel cathodes while chlorine bubbles off the graphite anodes. The cells are designed to keep these two reactive elements apart.

Sodium is a powerful reducing agent used for *titanium extraction* and the extraction of some other metals such as zirconium. Molten sodium is also the fluid which circulates through *heat exchangers* to transfer energy and raise steam in some nuclear power stations and other processes. Sodium is used in street lights.

sodium chloride structure: the cubic crystal structure of the ionic compound sodium chloride, NaCl. Each positive ion is surrounded by six nearest neighbours and each negative ion is surrounded by six positive ions, so the *co-ordination numbers* are 6 and 6.

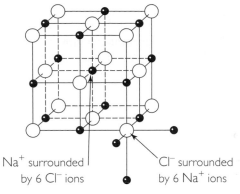

Structure of sodium chloride showing 6:6 co-ordination. The structure consists of a face-centered cubic array of negative ions with all the octahedral holes filled by positive ions.

Na$^+$ surrounded by 6 Cl$^-$ ions

Cl$^-$ surrounded by 6 Na$^+$ ions

Many other compounds have this structure, including: the chlorides, bromides and iodides of Li, Na and K; the oxides and sulfides of Mg, Ca, Sr, Ba as well as the fluoride, chloride and bromide of Ag.

sodium hydroxide is a white, translucent solid supplied as flakes or pellets. It is *deliquescent* . Sodium hydroxide is a *strong base*; it dissolves in water to form a highly alkaline solution. It is fully ionised both in the solid and in solution. Note that the solution is alkaline because of the hydroxide ions (not because of the sodium ions).

Sodium hydroxide is a useful test reagent (see *anion tests, cation tests* and *organic analysis*).

The traditional name for the alkali is caustic soda, which is a reminder that it is highly corrosive. Sodium hydroxide is more hazardous to the skin and eyes than many acids.

Sodium hydroxide is manufactured by the *electrolysis of brine*. UK industry produces over a million tonnes of the alkali each year. Sodium hydroxide is widely used for manufacturing other chemicals including *soaps* and *detergents*, rayon *fibres* as well as *aluminium*, sodium cyanide and sodium peroxide.

sodium tetrahydridoborate(III) (NaBH$_4$) is a *reducing agent* used in organic chemistry. It is a milder reducing agent than *lithium tetrahydridoaluminate(III)* and has the advantage that it can be used in aqueous solution. NaBH$_4$ reduces *aldehydes* and *ketones* to alcohols.

sodium thiosulfate (Na$_2$S$_2$O$_3$) is used in *iodine–thiosulfate titrations*. In *photography* it is the fixer which removes unexposed silver salts after developing the image (see also *thio compounds*).

A milky precipitate of sulfur forms on adding dilute acid to sodium thiosulfate. This is an example of a *disproportionation* reaction.

$$S_2O_3^{2-}(aq) + 2H^+(aq) \longrightarrow SO_2(aq) + S(s) + H_2O(l)$$

solid: one of the *states of matter*. (See also *crystal structures of ionic compounds, crystal structures of non-metals* and *crystal structures of metals*. See also *ceramics, glasses, metals* and *polymers*.)

solubility: the mass (g) or amount (mol) of a substance that will dissolve in 100 g water.

As a general rule, 'like dissolves like'. *Polar solvents*, such as water, dissolve polar or ionic compounds. *Non-polar solvents*, such as hexane, dissolve other hydrocarbons and non-polar elements or compounds. Solids generally become more soluble in water as the temperature rises. Gases become less soluble as the temperature rises. Boiling water, for example, removes gases dissolved from the air. Bubbles of gas appear around the edge of a saucepan before the water boils. The bubbles contain air coming out of solution as the gases become less soluble with the rise in temperature. Gases become more soluble as their pressure rises (see *Henry's law*).

No chemical is completely soluble and none is completely insoluble (see *solubility product constant*). Even so, chemists find it useful to use a rough classification of solubility based on what they see on shaking a little of the solid with water in a test tube:

- **very soluble,** like potassium nitrate, plenty of the solid quickly dissolves
- **soluble,** like copper(II) sulfate; crystals visibly dissolve to a significant extent
- **sparingly or slightly soluble**, like calcium hydroxide; little solid seems to dissolve but the solution becomes quite strongly alkaline
- **insoluble**, like iron(III) oxide; there is no sign that any of the material dissolves.

A similar rough classification applies to gases dissolving in water. Ammonia and hydrogen chloride are very soluble. Sulfur dioxide is soluble. Carbon dioxide is slightly soluble. Nitrogen is insoluble.

Some generalisations about solubilities help to interpret observations during qualitative analysis. The generalisations in the table apply to solutions in water at room temperature. Adding acid or alkali changes the patterns of solubility.

	Soluble in water	Insoluble in water
Acids	All common *acids* are soluble	
Bases	*Alkalis:* the hydroxides of sodium and potassium (calcium hydroxide which is slightly soluble), ammonia, plus the carbonates of sodium and potassium	All other metal oxides, hydroxides and carbonates
Salts	All nitrates	
	All chlorides …	*except* silver and lead chlorides
	All sulfates …	*except* barium sulfate, lead sulfate, and calcium sulfate which is slightly soluble
	All sodium and potassium salts	All other carbonates, chromates, sulfides and phosphates

solubility product constants: equilibrium constants for almost insoluble salts in equilibrium with solutions of their own ions. Even salts which are insoluble for practical purposes do dissolve to a very slight extent. There is an equilibrium between the solid and its ions in solution.

$$AgCl(s) \rightleftharpoons Ag^+(aq) + Cl^-(aq)$$

The *equilibrium law* applies. As with other *heterogeneous equilibria*, the concentration of the solid silver chloride is constant and does not appear in the equilibrium law equation. In this context the equilibrium constant, K_{sp}, is the solubility product constant.

$$K_{sp} = [Ag^+(aq)][Cl^-(aq)] = 2 \times 10^{-10} \text{ mol}^2 \text{ dm}^{-6} \text{ for equilibrium concentrations}$$

K_{sp} values can be used to predict whether or not a precipitate will form on mixing two solutions.

solute: a substance which dissolves in a *solvent* to make a *solution*. In a sugar solution the solvent is water and the solute is sucrose.

solutions are formed when solids, liquids or gases dissolve in a *solvent*. Water is so abundant on Earth that solutions in water (*aqueous* solutions) are particularly important to the natural environment, to life and to chemistry in laboratories and in industry.

Most solutions are solids, liquids or gases dissolved in a liquid but there are also 'solid solutions'. Nickel–copper *alloys* are examples of solid solutions. In a solid solution atoms of one metal replace atoms of the other metal in the crystal lattice.

solvation takes places when solvent molecules bond to ions or molecules as they dissolve. The bonding may be through weak intermolecular forces, attraction between ions and polar molecules or via covalent bonds. *Hydration* describes solvation when the solvent is water.

Solvay process: a process for manufacturing sodium carbonate from salt and limestone.

solvent: a liquid used to dissolve things. Chemists find it helpful to quote the rule 'like dissolves like'. What this means is that *non polar solvents* dissolve non-polar substances while *polar solvents* dissolve ionic and polar compounds.

Water is the commonest solvent. It is a polar solvent and dissolves many ionic compounds. Water molecules also dissolve compounds with which they can form *hydrogen bonds* such as glucose molecules.

White spirit is a non-polar solvent consisting of a mixture of *hydrocarbons*. It dissolves oily and greasy materials, including oil paints.

solvent extraction: a technique for separating and purifying substances with a *solvent* which dissolves the product required but leaves all other compounds dissolved in the original solvent. The solvents must not mix so that the two liquids separate after being shaken up together.

The substance being extracted *partitions* itself between the two solvents until the two solutions are in equilibrium. From the *equilibrium law* it is possible to show that it is more efficient to extract with two or three smaller volumes of solvent than to add all the solvent at once in a single extraction.

Solvent extraction is used in the *perfume* industry to extract fragrant oils from chopped up plant blossom. Solvent extraction is also used to decaffeinate coffee. In both these processes it helps to use a liquefied gas as the solvent. One possibility is carbon dioxide under pressure. After the extraction the solvent is easily removed by lowering the pressure. The solvent turns back to a gas at a low temperature. This allows the use of a non-toxic solvent. It also means that there is no need for heating to distil off the solvent. Heating can easily destroy organic compounds.

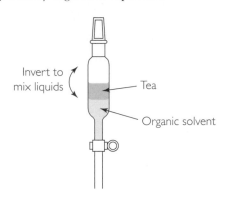

Use of a tap funnel for solvent extraction. Caffeine is more soluble in dichloromethane than in water. The other chemicals in tea are much more soluble in water. After solvent extraction the caffeine is recovered by distilling off the dichloromethane.

s-orbitals: *see atomic orbitals.*

space-filling models: atomic models which show the space taken up by atoms in molecules or crystals. They show the sizes of atoms, molecules or ions and how they pack together. They do not show the bond angles or the numbers of bonds between atoms in molecules as clearly as *ball and stick models.*

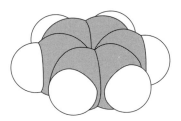

A space-filling model of benzene, C_6H_6

species: a useful collective noun used by chemists to refer generally to the atoms, molecules or ions taking part in a chemical process. A pure chemical species is a collection of identical chemical *entities*.

specific heat capacity: the energy transferred per unit mass when the temperature of a material changes by one degree kelvin. The SI unit is $J\,kg^{-1}\,K^{-1}$. When working on a small scale with a *calorimeter* chemists often work with values in $J\,g^{-1}\,K^{-1}$.

$$\begin{array}{ccccc} \text{energy} & & \text{specific heat} & & \text{mass} \quad \text{temperature change} \\ \text{transfer (J)} & = & \text{capacity }(J\,g^{-1}\,K^{-1}) & \times & \text{(g)} \quad \times \quad \text{(K)} \end{array}$$

In symbols this becomes:

$$q = mc\Delta T$$

spectator ions: ions in a solution during a reaction which do not take part in the chemical change. For clarity, chemists omit spectator ions from *ionic equations*. Adding silver nitrate to a solution of potassium chloride produces a precipitate of insoluble silver chloride. Both silver nitrate and potassium chloride are ionised in solution.

$$Ag^+(aq) + NO_3^-(aq) + K^+(aq) + Cl^-(aq) \longrightarrow AgCl(s) + NO_3^-(aq) + K^+(aq)$$

The potassium and nitrate ions remain in solution unchanged. They are the spectator ions left out of the ionic equation:

$$Ag^+(aq) + Cl^-(aq) \longrightarrow AgCl(s)$$

spectroscopy: a range of practical techniques for studying the composition, structure and bonding of elements and compounds. These instrumental techniques have been developed in the last 75 years and continue to become more powerful. Spectroscopic techniques are now the essential 'eyes' of chemistry.

The instruments used are variously called spectroscopes (emphasising the uses of the techniques for making observations) or spectrometers (emphasising the importance of measurements).

Spectroscopy uses the full range of the electromagnetic spectrum to study atoms, molecules, ions and the bonding between them:

- **radio waves** in *nuclear magnetic resonance spectroscopy*
- **microwaves** to study the rotations of polar molecules
- **infra-red radiation** in *infrared spectroscopy*
- **visible and UV radiation** in *atomic absorption spectroscopy, atomic emission spectroscopy* and *ultraviolet spectroscopy*
- **X-ray** spectroscopy to study electron jumps between the electron shells in atoms.

spin: the property of electrons which accounts for their behaviour in a magnetic field. Electrons behave like tiny magnets. In a magnetic field electrons either line up with the field or against the field.

An *atomic orbital* can hold only two electrons and they must have opposite spins. Arrows pointing up or down represent electrons in *energy level* diagrams.

If all the electrons in molecules or ions are paired with opposite spins the substance is diamagnetic. Elements and compounds with unpaired electrons are *paramagnetic*.

spontaneous reaction: a reaction which tends to go. In thermochemistry, spontaneity has the same meaning as a *feasibility*. So strictly speaking any reaction which naturally tends to happen is spontaneous even if it is very slow.

The control of spontaneous reactions is of vital importance in *metabolism*. The hydrolysis of *ATP* is spontaneous. However, if all the ATP in cells were to react rapidly with water, the energy from respiration would be wasted and life would cease. Hydrolysis happens only with *enzymes* to speed up the reaction. With enzymes the energy from hydrolysis can be harnessed to growth and movement.

In practice chemists sometimes use the word spontaneous in its everyday sense to describe reactions which not only tend to go but also go fast when the reactants are mixed at room temperature. Here is a typical example:

'The hydrides of silicon catch fire spontaneously in air unlike methane which has an ignition temperature of about 500°C.'

The reaction of methane with oxygen is also spontaneous in the thermodynamic sense, even at room temperature. However, the *activation energy* for the reaction is so high that nothing happens until the gas is heated with a flame. A mixture of methane and oxygen at room temperature is kinetically inert (see *inertness of chemicals*).

stability of benzene: the greater stability of *benzene* because of *delocalisation* of electrons. Benzene is more stable than expected for a compound which is often shown with three double bonds as in the *Kekulé formula*.

A measure of the greater stability of benzene comes from a comparison of the experimental enthalpy change on adding three moles of hydrogen to benzene (*hydrogenation*) with three times the enthalpy change on adding a mole hydrogen to cyclohexene.

$$\bighexagon + 3H_2 \longrightarrow \bighexagon \qquad \Delta H^\circ = -208 \text{ kJ mol}^{-1}$$

benzene

$$\bighexagon + H_2 \longrightarrow \bighexagon \qquad \Delta H^\circ = -120 \text{ kJ mol}^{-1}$$

cyclohexene

Enthalpy changes for hydrogenating benzene and cyclohexene

Real benzene is more stable than might be expected by about 150 kJ mol^{-1}.

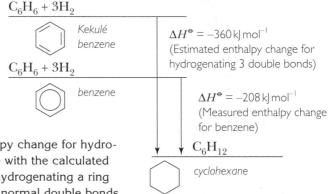

$C_6H_6 + 3H_2$

Kekulé benzene

$C_6H_6 + 3H_2$

benzene

$\Delta H^\circ = -360 \text{ kJ mol}^{-1}$
(Estimated enthalpy change for hydrogenating 3 double bonds)

$\Delta H^\circ = -208 \text{ kJ mol}^{-1}$
(Measured enthalpy change for benzene)

C_6H_{12}

cyclohexane

Comparing the enthalpy change for hydrogenating real benzene with the calculated enthalpy change for hydrogenating a ring compound with three normal double bonds

stability of compounds: compounds are stable if they do not tend to decompose into their elements or into other compounds. A compound which is stable at room temperature and pressure may become more or less stable as conditions change (see *thermal stability*).

Chemists often use standard *enthalpy changes* as an indicator of stability. Strictly they should use standard *free energy change*, ΔG_f°, values but in many cases $\Delta G_f^\circ \approx \Delta H_f^\circ$.

When discussing the stability of a compound it is important to specify the decomposition reaction. At a high enough temperature, for example, *calcium carbonate* becomes unstable relative to decomposition into calcium oxide and carbon dioxide. The *nitrogen oxides*, however, are unstable relative to decomposition into the elements.

A compound such as the gas N_2O is thermodynamically unstable; the compound tends to decompose into its elements. The expressions ΔH_f° (= + 82 kJ mol^{-1}) and

$\Delta G_f^{\ominus}$ (= + 104 kJ mol^{-1}) are both positive. The decomposition reaction is exothermic but the rate is very slow under normal conditions. Chemists sometimes say that N_2O is kinetically 'stable'. It is better to use a different word and to refer to the kinetic inertness of N_2O. (See also *inertness of chemicals*.)

ΔG	Activation energy	Change observed	
positive	high	no reaction	reactants stable relative to products
negative	high	no reaction	reactants unstable relative to products but kinetically inert
positive	low	no reaction	reactants stable relative to products
negative	low	fast reaction	reactants unstable relative to products

stability constants, K_{stab}: equilibrium constants which are a measure of the stability of *complex ions*. The greater the value of the stability constant the more stable the complex. The values for stability constants show that *chelate* complexes formed by bidentate and hexadentate *ligands* are more stable than complexes formed by monodentate ligands.

Complex	$K_{stab}/(\text{mol dm}^{-3})^{-n}$ where n = the number of ligands
$[Ag(NH_3)_2]^+$	1.6×10^7
$[Cu(NH_3)_4]^{2+}$	1×10^{12}
$[Ni\ en_3]^{2+}$ where en is the bidentate ligand 1,2-diaminoethane	2×10^{18}
CuY^{2-} where Y is the hexadentate ligand *edta*	6×10^{18}

Stability constants show the position of equilibrium when a new ligand replaces water molecules in the hydrated ions.

$$[Co(H_2O)_6]^{2+}(aq) + 6NH_3(aq) \rightleftharpoons [Co(NH_3)_6]^{2+} + 6H_2O(aq)$$

For simplicity this is often written as:

$$Co^{2+}(aq) + 6NH_3(aq) \rightleftharpoons [Co(NH_3)_6]^{2+}(aq)$$

The usual rules for writing an equilibrium constant and its units apply.

$$K_{stab} = \frac{[Co(NH_3)_6^{2+}]}{[Co^{2+}][NH_3]^6} = 1 \times 10^5 \text{ mol}^{-6} \text{ dm}^{18}$$

standard conditions: the conditions used in *thermochemistry* to define standard enthalpy and free energy changes. These conditions are the temperature 298 K and a *pressure* of 100 000 N m^{-2} = 10^5 Pa (1 bar).

When standard electrode potentials are being measured the standard conditions are the temperature 298 K, solutions at a concentration of 1.0 mol dm^{-3} and a pressure of 10^5 Pa if gases are involved.

standard electrode potentials are the basis of the *electrochemical series* for predicting the direction of *redox reactions*. They are also used to calculate the emfs of *electrochemical cells*. The standard electrode potential for a *half-cell* is measured relative a standard *hydrogen electrode* under *standard conditions*.

$$Pt[H_2(g)] \mid 2H^+(aq) \parallel Cu^{2+}(aq) \mid Cu(s)$$

The emf of this cell under standard conditions is, by definition, the standard electrode potential of the $Cu^{2+}(aq)|Cu(aq)$ electrode

standard enthalpy changes: see *enthalpy change.*

standard form: the form which mathematicians and scientists use to write very large or very small numbers. Standard form is based on powers of 10.

$10^{-9} = 0.000\,000\,001$	$10^1 = 10$
$10^{-6} = 0.000\,001$	$10^2 = 100$
$10^{-3} = 0.001$	$10^3 = 1\,000$
$10^{-2} = 0.01$	$10^4 = 10\,000$
$10^{-1} = 0.1$	$10^6 = 1\,000\,000$
$10^0 = 1$	$10^9 = 1\,000\,000\,000$

Worked example:

The dissociation constant for ethanoic acid, K_a = 0.000 017 mol dm^{-3}

$$= 1.7 \times 0.000\,01 \text{ mol dm}^{-3}$$
$$= 1.7 \times 10^{-5} \text{ mol dm}^{-3},$$

which is standard form

Worked example:

The *Faraday constant* = 9 6480 C mol^{-1} = $9.648 \times 10\,000$ C mol^{-1}

$$= 9.648 \times 10^4 \text{ C mol}^{-1} \text{ which is standard form.}$$

---

standard free energy changes: see *free energy change.*

standard hydrogen electrode: see *hydrogen electrode.*

standard molar entropy, $S^\ominus$: the *entropy* per mole for a substance under *standard conditions.* Chemists use values for standard molar entropies to calculate entropy changes and so predict the direction and extent of chemical change.

The units for standard molar entropy are joules per kelvin per mole (J K^{-1} mol^{-1}).

solids	$S^\ominus$/J K^{-1} mol^{-1}	liquids	$S^\ominus$/J K^{-1} mol^{-1}	gases	$S^\ominus$/J K^{-1} mol^{-1}
carbon(diamond)	2.4	mercury	76.0	hydrogen	130.6
magnesium oxide	26.9	water	69.9	carbon dioxide	213.6
copper	33.2	ethanol	160.7	propane	269.9

The table shows standard molar entropies at 298 K. Gases have higher standard molar entropies than solids. The number of ways of distributing particles and energy in a system of gas molecules is much higher than in an ordered crystalline solid. The more rigid and regular a crystal, the lower its entropy.

Liquids have intermediate values of standard molar entropies. The more atoms in the molecules the greater the opportunities for vibration and rotation so the higher the entropy values.

standard solution: a solution with an accurately known concentration. The direct method for preparing a standard solution is to dissolve a weighed sample of a *primary standard* in water and to make the solution up to a definite volume in a graduated flask.

Standard solutions are used in *titrations* to determine the concentrations of other solutions. They are also needed for the *calibration* of instruments such as colorimeters.

standard state: the stable state of an element or compound under the *standard conditions* which apply in thermochemistry. Standard *enthalpy changes* and *free energy changes* are defined for substances in their standard states.

When an element such as carbon has two *allotropes*, diamond and graphite, the standard state of carbon is graphite because it is the more *stable* form.

Standard states normally refer to substances in their stable physical states under standard conditions. Carbon dioxide and methane are gases but water is a liquid.

starch is a *carbohydrate* which consists of long chains of glucose units. It is a polysaccharide. Starch is insoluble in cold water. In hot water, starch gelatinises forming a colloidal dispersion. Starch solution gives an intense blue–black colour with iodine. The solution is used as an indicator to detect the end-point in *iodine–thiosulfate titrations*.

state functions: measurable properties which depend only on state of a system, not on how the system got to that state. Examples are *pressure, volume* and *temperature*, as are thermochemical quantities such as enthalpy, *entropy* and *free energy*.

Changes in state functions depend only on the initial and final states of the system and not on the pathway from one state to the other. This is the basis of *Hess's law*.

Equations of state show the mathematical connections between state functions. An important example is the *ideal gas* equation which relates the pressure, volume and temperature of an amount of a gas.

states of matter: the solid, liquid and gaseous states (see *changes of state*) as well as intermediate states such as *liquid crystals* and mixtures, including aqueous solutions and the colloidal state (see *colloids*).

State symbols in equations indicate the states of the chemicals: (s) solid, (l) liquid, (g) gas, (aq) aqueous (dissolved in water).

stationary phase: see *chromatography*.

steady state systems: systems in which the concentrations of chemicals stay constant because they are being supplied or formed as fast as they are removed or destroyed. The temperature is constant too because energy is transferred to the system as fast as it is lost to the surroundings.

The blue flame of the Bunsen burner is a steady state system. There is a constant supply of gas and oxygen which burn to release energy. The products of burning and energy are constantly lost to the surroundings.

Steady state systems are important in *environmental chemistry*. The *atmosphere* approximates to a steady state, with gases such as oxygen being added to the air by photosynthesis but removed by respiration and burning.

Problems arise when human activity on a large scale upsets one or more of the processes in a steady state system. *Ozone*, for example is naturally formed and destroyed in the stratosphere:

rate of formation of ozone – rate of destruction of ozone

The release of chlorine compounds such as *CFCs* has increased the rate of destruction of ozone, upsetting the steady state and lowering the ozone concentration, especially over the poles, hence the 'hole' in the ozone layer.

steam cracking: see *thermal cracking*.

steam distillation is a useful technique for separating oils from plant materials such as rose petals, cloves, lavender, thyme or fennel. It is an important technique in the *perfume* industry. Steam distillation makes it possible to separate compounds which decompose if heated at their boiling points. The technique works only for compounds which do not mix with water.

Steam distillation is also sometimes used to separate products of organic preparations, leaving behind reagents and products which are soluble in water.

Steam distillation works because the *vapour pressure* of a mixture of *immiscible liquids* is the sum of the separate vapour pressures when they are shaken up together. The mixture boils when the total vapour pressure equals atmospheric pressure. So the mixture of steam and oily product distils at a little below the boiling point of water and well below the boiling point of the oil.

steam reforming is the reaction of steam with methane (or other hydrocarbon) in the presence of a nickel oxide *catalyst* at 800°C under pressure.

$$CH_4(g) + H_2O(g) \rightleftharpoons 3H_2(g) + CO(g)$$

It is followed by further reaction with excess steam. This 'shift reaction' converts the CO into CO_2. The reaction happens in the presence of an iron(III) oxide catalyst at 400°C.

$$H_2O(g) + CO(g) \rightleftharpoons H_2(g) + CO_2(g)$$

This mixture of hydrogen and carbon dioxide can be converted directly to *methanol*. Alternatively, absorbing the carbon dioxide in alkali provides hydrogen for *ammonia manufacture*.

steels are *alloys* of iron with carbon and often other metals.

Mild steel contains about 0.2% carbon as iron carbide. Crystals of iron carbide in the metal structure make the steel strong and yet it is still *malleable*. Mild steel is used for car bodies. As the carbon content increases the steel becomes stronger and harder so that it is suitable for rail and tram lines.

Steel is made by the basic oxygen steelmaking (BOS) process which removes impurities from *iron extraction* in a blast furnace. During the process a blast of oxygen converts impurities in the liquid metal, such as carbon, silicon and phosphorus, into their oxides. Carbon dioxide escapes as a gas. The oxides of the non-metals silicon and phosphorus are *acidic oxides*. They are converted to a molten *slag* by adding the *basic oxides* of calcium and magnesium. The slag floats on the surface of the liquid steel and can be poured off separately.

$$SiO_2 + CaO \longrightarrow CaSiO_3(l)$$

Alloy steels consist of iron with small amounts of carbon together with up to 50% of one or more of these metals: aluminium, chromium, cobalt, manganese, molybdenum, nickel, titanium, tungsten, and vanadium. The presence of other metals distinguishes alloy steels from carbon steels. Examples of alloys steels are:

- stainless steels, which include chromium and nickel
- tool steels with tungsten or manganese which make the alloy harder, tougher and keep their properties at higher temperatures so that they are suitable for drill bits and cutting tools.

stereochemistry is the study of molecular shapes and the effect of shape on chemical properties. Stereochemistry is especially concerned with the study of the contrasting properties of *stereoisomers.*

Smell and taste seem to depend on molecular shape. One mirror-image form of limonene smells of oranges while the other form smells of lemons.

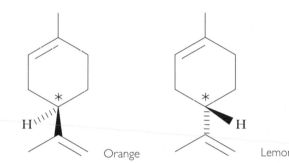

The enantiomers of limonene and their smells. Asterisks mark the chiral centres.

Orange

Lemon

Molecular shape can also subtly alter the physiological effects of drugs as tragically illustrated by the two enantiomers of thalidomide (see *chiral compounds*).

The *antibiotic* penicillin destroys bacteria by disrupting their cell walls. Penicillin molecules resemble the shapes of molecules used to make the membrane around bacterial cells. The resemblance is enough for enzymes to build penicillin molecules into the structure. But penicillin lacks the part of the natural molecules which *cross-links* the structure. The walls are fatally weakened so that the bacterial cells break open and die.

Molecular shape affects the *physical properties* of materials such as polymers. This is illustrated by the differences between poly(propene) as an *isotactic polymer* and as an *atactic polymer.*

Steric factors can also affect the mechanisms of reactions and the rates of chemical change.

stereoisomers are distinct compounds with the same molecular formulae and structural formulae but with different three-dimensional shapes. There are two kinds of stereoisomerism: *geometrical isomerism* and *optical isomerism.*

stereoregular polymer: a polymer with a regular three-dimensional shape. Poly(propene) can be an *isotactic polymer.* This is the useful stereoregular form. Or it can be an *atactic polymer* which is irregular and does not have useful properties.

stereospecific reactions are reactions which involve one of the *optical isomers* of a *chiral* compound but not the other. All the biochemical processes involving *amino acids* are stereospecific. The enzymes involved can act on the L-amino acids but not on the mirror-image D-forms.

Enzymes act only on molecules which fit their *active sites*. They are so selective that they can pick out from a *racemic mixture* those molecules that are either left-handed or right-handed.

Since the thalidomide tragedy, the pharmaceutical industry has become much more aware of the importance of chirality and the need to test mirror image forms of chiral compounds separately. This has encouraged chemists to develop stereospecific reactions to produce particular optical isomers.

steroids are *lipids*. An important steroid is *cholesterol*. Cholesterol can be converted to biologically active steroids such as the sex hormones oestrogen and progesterone.

stiffness is the opposite of flexibility. It is a very important property of materials. Materials scientists seek to develop materials which are both stiff and strong while not being too dense.

J. E. Gordon, the author of the Penguin book *The New Science of Strong Materials*, uses these examples to explain the difference between strength and stiffness:

> 'A biscuit is stiff but weak, steel is stiff and strong, nylon is flexible and strong, raspberry jelly is flexible and weak'.

Stiffness is measured by the Young modulus for the material.

stoichiometry: a stoichiometric equation is the *balanced equation* for a reaction. It shows the amounts, in moles, of the substances involved in a reaction.

A stoichiometric compound has a composition which corresponds exactly to its formula. A stoichiometric reaction is one which uses up reactants and produces products in amounts exactly as predicted by the balanced equation.

The word stoichiometry sounds mysterious but is simply based on Greek words meaning 'element-measure'. Stoichiometry is the basis of *quantitative analysis* where amounts are measured in moles.

storage cells: see *lead–acid cell* and *nickel–cadmium cells*.

stp (standard temperature and pressure): the standard conditions for describing the properties of gases. The standard temperature is 273 K (0°C) and the standard *pressure* is 101.3 kPa (1 atmosphere).

The molar volume of an ideal gas at stp is 22 400 cm^3 (22.4 dm^3).

Note the distinction between these values and the values for defining *standard states* in thermochemistry.

strength is measured by the stress needed to deform and break a material in tension or compression. Stress is the force per unit area of the cross section and is measured in N m^{-2}. Defining stress in this way allows for the fact that it is easier to stretch a thin sample than a thick one.

Metals such as steel have high tensile strength. Ceramics are weak in tension but have high strength in compression.

For many applications engineers look for materials which combine high strength with high *stiffness*.

strong acids are *acids* which are fully ionised when they dissolve in water. Examples are *hydrochloric acid*, *nitric acid* and *sulfuric acid*.

strong bases are *bases* which are fully ionised when they dissolve in water. Examples are the hydroxides of *sodium* and *potassium.*

structural formulae show the arrangements of atoms and *functional groups* in molecules.

Sometimes it is enough to show structure in a condensed form, such as $CH_3CH_2CH_2CO_2H$ for butanoic acid. Often it is clearer to write the full structural formula, showing all the atoms and bonds. This type of formula is also called a *displayed formula* or graphical formula.

propene

butanoic acid

Examples of structural formulae. Drawn like this the formulae do not show the true shape in three dimensions.

cyclohexane

structural isomers: see *isomers.*

sublimation is the change of a solid directly to a gas on heating. Heating iodine crystals makes them sublime to a purple vapour which condenses to shiny crystals on a cold surface. This process is used to purify iodine.

Another substance which sublimes is solid carbon dioxide ('dry ice') because it turns to gas at minus 78°C without melting.

substitution reactions are reactions which replace an atom or group of atoms by another atom or group of atoms. An example is the reaction of butan-1-ol with hydrogen bromide (from sodium bromide and concentrated sulfuric acid).

$$CH_3CH_2CH_2CH_2OH + HBr \longrightarrow CH_3CH_2CH_2CH_2Br + H_2O$$

In organic chemistry, substitution reactions are characteristic of *halogenoalkanes* (see *nucleophilic substitution*) and *arenes* (see *electrophilic substitution*).

The ligand exchange reactions of inorganic complex ions are also substitution reactions (see *ligands*).

substrates, in biochemistry, are the molecules on which *enzymes* act as they catalyse change. An enzyme is specific because it acts only on the substrate which fits into its *active site.*

Literally the word 'substrate' means a lower layer on which something else can form or grow. One way of creating a large surface area for an expensive *catalyst* is to spread it over the surface of an inert substrate. This helps to keep down the cost of *catalytic converters* in car exhausts.

sugars are water soluble *carbohydrates.* Sugars vary in their *sweetness.* Sugar molecules have many — OH groups and can *hydrogen bond* with each other and with water. This means that they are solid at room temperature and very soluble in water. (See also *reducing sugars.*)

Ribose and deoxyribose sugars are important in the formation of nucleic acids such as *DNA* and *RNA.*

sulfates are salts of *sulfuric acid.* Soluble sulfates such as blue copper(II) sulfate and green iron(II) sulfate are familiar laboratory reagents.

Some insoluble sulfates are natural minerals, including calcium sulfate which exists in a hydrated form as gypsum and in an anhydrous form called anhydrite. Heating gypsum produces *plaster of Paris.* Barytes is barium sulfate, called 'heavy spar' because of its density. (See also *solubilities of salts* and *anion tests.*)

sulfides of metals are salts of hydrogen sulfide, H_2S. Some valuable metals occur as sulfide ores, including copper, silver, mercury and iron (see also *roasting*). Sodium sulfide is soluble in water but most metal sulfides are insoluble.

sulfites are salts of sulfurous acid which form when sulfur dioxide dissolves in water. Sulfurous acid is unstable and exists only in solution.

$$SO_2(g) + H_2O(l) \longrightarrow H_2SO_3(aq)$$

The most important sulfite is sodium sulfite (Na_2SO_3), made by dissolving sulfur dioxide in sodium hydroxide solution.

Sulfur dioxide and the sulfite ion are *reducing agents.* They reduce chlorine, iron(III) ions, dichromate(VI) ions and manganate(VII) ions.

Sulfur dioxide and sulfites are used as preservatives They are antioxidants used in foods such as lemon juice. They inhibit the growth of bacteria. (See also *anion tests.*)

sulfonamide drugs: see *chemotherapy.*

sulfonation of benzene is an *electrophilic substitution* reaction which takes place in the presence of fuming sulfuric acid – a solution of sulfur trioxide in concentrated sulfuric acid. The product is benzene sulfonic acid.

The *electrophile* is sulfur trioxide. A sulfur atom attached to three more *electronegative* oxygen atoms is electrophilic.

benzene sulfonic acid

Sulfonation of benzene

Sulfonation of arenes is important because it helps to produce a range of useful products including *surfactants, ion-exchange resins, dyes* and sulfonamide drugs.

sulfur (S) is a yellow crystalline solid which normally consists of S_8 molecules. Sulfur is a *non-metal,* coming below oxygen in group 6 of the periodic table. Its electron configuration is $[Ne]3s^23p^4$.

Sulfur has two solid *allotropes*. Under atmospheric pressure, rhombic sulfur is the stable form of the element at room temperature and up to 95.5°C. Above this temperature the stable form is monoclinic sulfur. At the transition temperature the two allotropes are in equilibrium. Both forms consist of S_8 molecules. Heating sulfur crystals produces a runny, pale-yellow liquid which darkens as the temperature rises, producing a highly *viscous liquid*. The S_8 rings break and form long, tangled chains of sulfur atoms. Further heating below the boiling point (445°C) makes the liquid more fluid because the chains start to break into shorter lengths. Pouring this dark red liquid into cold water produces an elastic, non-crystalline mass of plastic sulfur. In time plastic sulfur hardens as the sulfur chains gradually reform S_8 rings and produce rhombic sulfur again.

Sulfur is a reactive element but it is a less powerful *oxidising agent* than oxygen. It combines with most other elements with the exception of nitrogen, iodine, the noble gases and some of the less reactive metals such as gold.

Hydrogen sulfide (H_2S) is a toxic, foul-smelling compound. Despite having a higher relative molecular mass than water, hydrogen sulfide is a gas, because sulfur is not sufficiently electronegative for *hydrogen bonding* between H_2S molecules.

Sulfur forms two *acidic oxides* SO_2 and SO_3, which are both important in the process of *sulfuric acid manufacture*. Sulfur dioxide is a *reducing agent*, as are sulfurous acid and its salts, the *sulfites*.

Bonding in the oxides of sulfur.
Note that the rules for predicting
the *shapes of molecules* apply.

The reactions of *halide ion*s with concentrated sulfuric acid illustrate the range of oxidation states of sulfur.

sulfur extraction: processes which obtain sulfur from underground deposits of the element or from crude oil and *natural gas*.

The Frasch process is an ingenious process for extracting sulfur from underground deposits of the element. Sulfur is a non-metal with a molecular structure. It melts at 113°C. Superheated steam at 165°C passes down a pipe into the sulfur deposits, where the steam melts the sulfur. Compressed air pumped down a second pipe forces the liquid up a third pipe to the surface where the sulfur cools and solidifies.

Natural gas and crude oil often contain unwanted sulfur compounds which must be removed before the gas and oil are used as fuels or in the chemical industry. The hydrocarbons are mixed with hydrogen under pressure and passed over a *catalyst*. This converts the sulfur to hydrogen sulfide. The hydrogen sulfide is separated from the hydrocarbons and converted to a mixture of sulfur and sulfur dioxide.

Ninety per cent of sulfur is used to make sulfuric acid.

sulfuric acid (H₂SO₄) is a highly corrosive but important chemical reagent because it can act as a *strong acid*, a *dehydrating agent*, an *oxidising agent* and as a sulfonating agent. Pure sulfuric acid is a colourless *viscous liquid*. The molecules of this *oxoacid* are attracted to each other by *hydrogen bonding*.

343

Some 'instant hand warmers' consist of a supersaturated solution of sodium ethanoate in a plastic bag. Flexing a metal disc in the side of the bag starts crystallisation which is exothermic and heats the bag to about 60°C. The hand warmer can be reused after heating in boiling water to redissolve the salt and then allowing it to cool to room temperature so that the solution is once more supersaturated.

surface tension: the tendency of surfaces to contract to the minimum surface area. Surface tension accounts for the spherical shape of soap bubbles.

Surface tension arises because of *intermolecular forces*. Molecules inside the liquid are pulled in all directions by surrounding molecules. Molecules near the surface experience a net pull inwards. This means that as many molecules as possible leave the surface which tends to shrink.

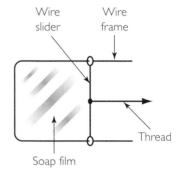

The surface tension of the soap film makes the slider move to the left unless balanced by a pull on the thread

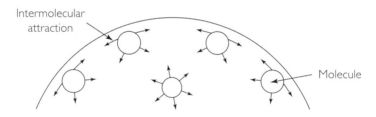

Intermolecular forces acting on molecules inside the liquid and at the surface

The surface tension of many organic liquids is in the range 15 to 30 mN m^{-1}. The surface tension of water is higher because of strong *hydrogen bonding*. For water the value is about 73 mN m^{-1}. *Surfactants* lower the surface tension of water so that it can wet greasy surfaces (see *wetting*).

surfactants are surface-active agents. They are chemicals with molecules which seek out the boundary surface between two liquids or between a liquid and a gas. One of the important effects of surfactants is that they change the *surface tension* of liquids.

A surfactant molecule has an ionic or polar group attached to a hydrocarbon chain. The polar group is water-loving (hydrophilic). The hydrocarbon chain is water-hating (hydrophobic).

The hydrophobic chains of surfactant molecules tend to escape from water by moving to the surface or by forming *micelles*. Surfactant molecules at the surface lower the surface tension of the water so that it wets surfaces more effectively. Surfactant molecules adsorbed at the boundary between air and water stabilise bubbles of foam by stopping all the water draining away.

Structure of a surfactant molecule — Non-polar; hydrophobic 'tail' — Ionic; hydrophilic 'head'

Surfactants help to separate greasy dirt from surfaces. They keep dirt dispersed in water so that it rinses away. They also help to prevent dirt re-attaching itself to the surface of fabrics.

Surfactants come in three kinds based on the type of hydrophilic group:

- anionic surfactants such as *soap*, $CH_3(CH_2)_{16}CO_2^-Na^+$
- non-ionic surfactants made from *epoxyethane*, $CH_3(CH_2)_{11}(OCH_2CH_2)_6OH$
- cationic surfactants such as *quaternary ammonium salts*, $CH_3(CH_2)_{11}N^+(CH_3)_3Br^-$

Synthetic anionic surfactants are widely used. They are included in bath foam, shampoos and washing up liquids. They make stable foams with water.

Non-ionic surfactants are used in many household cleaners because they allow smooth drainage and leave no deposit even when they are not fully rinsed away. Non-ionic surfactants also make less stable foams so they are included in washing powders for dish washers and washing machines.

Cationic surfactants are used in fabric softeners and hair conditioners.

surroundings: see *system*.

suspension: particles of a solid suspended by shaking or stirring in a liquid or gas. In time the particles of a suspension settle out, unlike the smaller particles in *colloids*.

sweetness: a taste sensation on the tip of the tongue produced by *sugars* and a number of synthetic chemicals such as saccharin, cyclamate and aspartame. Sugars vary in their sweetness. Fructose is sweeter than sucrose, which is sweeter than glucose.

Most synthetic sweeteners have been discovered when chemists have accidentally tasted chemicals made for other purposes. Synthetic sweeteners are taken in small amounts especially by people trying to cut the quantity of energy foods in their diet.

Chemical formulae of two sweeteners. Saccharin is 300 times sweeter than sucrose when equal quantities are compared.

saccharin

cyclamate

Cyclamates are banned in the US and UK because some people metabolise the sweetener to another chemical which causes bladder cancer if fed in large doses to rats.

Aspartame is an artificial sweetener which is about 200 times sweeter than *sucrose*. It consists of the methyl *ester* of a dipeptide consisting of two *amino acids* linked by a *peptide* bond.

Structure of aspartame   aspartame (Nutrasweet®)

Aspartame has no aftertaste but cannot be used in cooking because *hydrolysis* on heating destroys its sweetness. There is a warning on the labels of soft drinks and other foods sweetened with aspartame that they contain a source of phenylalanine. This can harm some people with a genetic disorder. Hydrolysis of aspartame produces phenylalanine.

synthesis: a process of making compounds from simpler starting materials. Synthesis puts things together. It is the opposite of analysis which takes things apart to see what they are made of.

The Haber process is an example of synthesis: two elements (nitrogen and hydrogen) combine to make a compound (ammonia). This is the method of *ammonia manufacture* on a large scale in industry.

Synthesis of more complex molecules often takes several reaction steps.

synthetic routes: see *organic routes.*

system: a term used in *thermochemistry* to describe the material or mixture of chemicals being studied. Everything around 'the system' is 'the surroundings' which includes the apparatus with maybe a waterbath, the air in the laboratory – in theory everything else in the universe.

An open system can exchange energy and matter with its surrounding. Most chemical reactions in laboratories take place in open systems.

A closed system cannot exchange matter with its surroundings. Energy can transfer in or out of a closed system but not the reactants or products.

An isolated system cannot exchange energy or matter with its surroundings. A mixture of chemicals in a vacuum flask with a stopper comes close to being an isolated system.

temperature effect on equilibria: the shift in the position of equilibrium which happens when the temperature changes. The effect depends on the enthalpy change for the reaction. *Le Chatelier's principle* predicts that raising the temperature makes the equilibrium shift in the direction which is endothermic. In *sulfuric acid manufacture*, for example, raising the temperature lowers the percentage of sulfur dioxide at equilibrium.

$$2SO_2(g) + O_2(g) \rightleftharpoons 2SO_3(g) \qquad \Delta H = -98 \text{ kJ mol}^{-1}$$

The equilibrium shifts to the left as the temperature rises because this is the direction in which the reaction is endothermic.

These shifts happen because equilibrium constants vary with temperature.

$$K_p = \frac{p_{SO_3}{}^2}{p_{SO_2}{}^2 p_{O_2}}$$

At 500 K, $K_p = 2.5 \times 10^{10} \text{ atm}^{-1}$, but at 700 K, $K_p = 3.0 \times 10^4 \text{ atm}^{-1}$. The value of the equilibrium constant falls as the temperature rises. With a smaller value of K_p the proportion of $SO_3(g)$ falls while the proportions of $SO_2(g)$ and $O_2(g)$ rise at equilibrium.

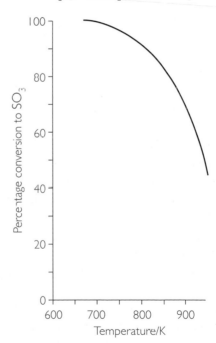

Effect of rising temperature on the equilibrium between SO_2, O_2 and SO_3

temperature effects on reaction rates: see *activation energy* and the *Arrhenius equation*.

teratogens: substances which may cause harm to an unborn child if inhaled, swallowed or absorbed through the skin of the mother.

termination: a step in a *free-radical chain reaction* which tends to stop the process by removing free radicals. In a termination step two free radicals combine to form a molecule.

tertiary organic compounds: see *primary, secondary and tertiary organic compounds.*

theoretical yield: see *yield calculations.*

thermal cracking breaks up bigger *hydrocarbon* molecules into smaller molecules by heating them with steam at a high temperature. The process is especially useful for converting the *naphtha* fraction from crude oil distillation into starting materials for synthesis in the petrochemical industry.

Thermal cracking with steam converts ethane to *ethene* which is a very important starting compound for chemical synthesis in industry. Ethane comes from the natural *gas* piped out of North Sea gas fields.

A mixture of the hydrocarbon vapour and steam passes through tubes in a furnace where they are heated to about 1000°C. Thermal cracking involves *free radical chain reactions,* unlike *catalytic cracking* which has an ionic mechanism. The first step in the thermal cracking of ethane is for the molecules to split into methyl free radicals.

thermal decomposition is a reaction in which a compound decomposes on heating. An important example for industry and agriculture is the thermal decomposition of *calcium carbonate* (limestone):

$$CaCO_3(s) \longrightarrow CaO(s) + CO_2(g)$$

Sometimes heating causes decomposition because a compound is stable at room temperature but becomes unstable at a higher temperature. This is the case with the calcium carbonate and the other carbonates of *group 2* metals.

Sometimes heating causes decomposition of a compound which is unstable at room temperature but does not decompose because the rate of reaction is so slow. This is true of the *nitrogen oxides* which, at room temperature, are examples of unstable but *inert chemicals.* They all tend to decompose into nitrogen and oxygen but they do so only on heating.

thermal dissociation: a reversible process which splits a compound into fragments as the temperature rises, but re-forms the starting material on cooling.

The brown gas which forms on heating some metal *nitrates* contains an equilibrium mixture of N_2O_4 and NO_2.

$$N_2O_4 \rightleftharpoons 2NO_2$$
colourless   brown

The mixture darkens on heating as more colourless N_2O_4 dissociates into NO_2 which is brown. The colour fades on cooling as N_2O_4 reforms.

thermal stability of carbonates: the ease with which the carbonates of metals decompose on heating.

Chemists relate differences in the properties of the compounds to two factors:

- the charges on the metal ions
- the sizes of the metal ions.

Group 2 carbonates and nitrates are generally less stable than the corresponding *group 1* compounds. This suggests that the larger the charge on the metal ion the less stable the compounds. Down either group 1 or group 2 the carbonates become more stable. This suggests that the larger the metal ion the more stable the compounds.

Chemists explain the trend in thermal stability by analysing the energy changes. Two of the energy quantities they take into account are:

- the energy needed to break up the carbonate ion into an oxide ion and carbon dioxide
- the *lattice energy* given out as the positive and negative charges get closer together when the larger carbonate ions break up into smaller oxide ions and carbon dioxide.

It turns out that all the carbonates are thermally stable at room temperature but become less stable as the temperature rises. The key factor is the energy released as the ions get closer together. This is greater when the metal ion is small than when the metal ion is large.

Detailed analysis of the energy changes is complex, so chemists find it convenient to correlate the stability of compounds such as carbonates and nitrates with the *polarising power* of the metal ions. Generally, the greater the polarising power of the metal ion the less stable the carbonates and the more easily they decompose to the oxide.

thermit reaction: a highly *exothermic reaction* use for welding rails and in some incendiary devices. The reaction is similar to the process for chromium extraction but operates on a smaller scale. A magnesium fuse heats and sets off the reaction in a mixture of iron(III) oxide and aluminium. The aluminium reduces the oxide to a glowing mass of molten iron.

$$Fe_2O_3(s) + 2Al(s) \longrightarrow 2Fe(s) + Al_2O_3(s)$$

thermochemistry is the study of energy changes during chemical reactions, including *enthalpy changes, free energy changes* and *entropy* changes. It is a major part of chemical thermodynamics. Thermochemistry is important for theory because it helps chemists to explain the *stability of compounds* and to predict the likely direction of chemical change. With the help of thermochemistry, chemists can decide on the *feasibility* of reactions.

Practically, thermochemistry is important because of the significance of:

- calculating the energy from burning fuels
- keeping large scale reactions under control in the *chemical industry*
- estimating the impact of energy changes in the environment.

Thermochemistry was developed mainly in the nineteenth and early twentieth centuries.

thermodynamic control operates where there is a choice of possible products and the main product is the one that is most *stable* (according to thermodynamics) This contrasts with *kinetic control* which operates when the main product is the one that forms fastest.

thermodynamic stability: see *stability of compounds.*

thermodynamics: see *thermodynamics (laws of)* and *thermochemistry.*

Oxygen and sulfur are in the same group of the periodic table with the same number of electrons in the outer shells of their atoms. They therefore form compounds with similar formulae and structures.

tin (Sn) is a shiny metal with a long history. It was mined in Cornwall from the Roman times until the last mine closed in 1998. Tin was valued as an ingredient of a range of *alloys* including pewter (with antimony), solder (with lead) and *bronze* (with copper). The main use of tin today is for coating the steel for tin cans. The layer of tin stops the iron corroding. Tin's Sn^{2+} ions are not toxic. Small traces of dissolved tin contribute to the characteristic taste of some tinned foods such as canned fruit and tomatoes.

At room temperature tin has a metallic structure but below the *transition temperature* 13.2°C the stable form is grey tin which has the diamond structure.

$$\text{grey tin} \quad \underset{\text{below 13.2°C}}{\overset{\text{above 13.2°C}}{\rightleftharpoons}} \quad \text{metallic tin}$$

It takes a long time for atoms in a solid to rearrange themselves. So metallic tin has to stay cold for a long time before it shows the symptoms of 'tin plague' as it gradually alters to the crumbly, brittle, grey form.

Tin, with the *electron configuration* $[Kr]4d^{10}5s^25p^2$, comes below germanium but above lead in *group 4* of the periodic table.

titanium (Ti) is a very strong metal which is much less dense than steel. It melts at the very high temperature of 1675°C . It does not corrode because, like *aluminium*, it is protected by a thin layer of oxide on the surface of the metal. The difficulties of titanium extraction mean that the metal is not as widely used as might be expected considering its properties.

Titanium is a *d-block* element with the electron configuration $[Ar]3d^24s^2$. Titanium forms compounds in the +2, +3 and +4 states but only the +4 state is common. Titanium(IV) chloride is a colourless liquid formed as an intermediate in *titanium extraction* and in the manufacture of *titanium(IV) oxide*.

Smoke grenades produce dense clouds of titanium(IV) oxide by the rapid *hydrolysis* of titanium(IV) chloride.

titanium extraction: a process for extracting the metal from its ores which are rutile (TiO_2) and ilmenite ($FeTiO_3$). Titanium is the fourth most abundant metal in the Earth's crust and it would be more widely used if the methods of extraction were less difficult and expensive. In theory it should be possible to extract titanium from its oxide with carbon but in practice some of the titanium reacts with carbon forming carbides which make the metal brittle.

After the ore has been purified it is heated with carbon in a stream of chlorine gas at about 1100 K.

$$2TiO_2 + 2Cl_2 + 2C \longrightarrow TiCl_4 + 2CO_2$$

The titanium(IV) chloride condenses as a liquid which can be purified by *fractional distillation*.

In the UK, sodium is the *reducing agent* for producing titanium from its chloride. In most other countries the preferred reducing agent is magnesium. Either way the production is a *batch process*.

In the UK method, titanium chloride passes into a reactor containing molten sodium at 800 K in an inert argon atmosphere. Exactly the right amount of the chloride is added to react with all the sodium. The reaction is exothermic and so the temperature rises.

$$TiCl_4 + 4Na \longrightarrow Ti + 4NaCl$$

The reactor is kept hot for about two days then it is removed from the furnace and allowed to cool. The solid product is crushed and *leached* with dilute hydrochloric acid which dissolves the sodium chloride, leaving the titanium metal which is then washed and dried.

Titanium is used mainly to make aircraft engines and airframes. Other major uses are the production components of chemical plants such as *heat exchangers*.

titanium(IV) oxide is a brilliant white *pigment*. In has two crystalline forms: anatase and rutile. Millions of tonnes of titanium(IV) oxide are made each year. The main use of the pigment is in *paint* but it is also a surface coating for paper, a filler in plastics and an ingredient of cosmetics and toothpaste. The oxide makes a good white pigment because it has a very high refractive index but absorbs almost no light in the visible part of the spectrum. When ground to a fine powder it is both intensely white and very opaque so as a pigment it has excellent covering power. The oxide is also inert and non-toxic, which is why it has replaced lead oxide in many applications.

There are two processes for producing titanium(IV) oxide from ilmenite: the long established sulfate process and the newer chloride process, both of which are operated in the UK.

titration: a *volumetric analysis* technique for finding the concentrations of solutions and to investigate the amounts of chemicals involved in reactions. Titrations are widely used because they are quick, convenient, accurate and easy to automate.

The procedure only gives accurate results if the reaction is rapid and is exactly as described by the chemical equation. So long as these conditions apply, titrations can be used to study *acid–base, redox, precipitation* and *complex*-forming reactions.

The measured volume of the unknown is transferred to a flask with a *pipette*, then the standard solution is added carefully from a *burette*. The *end-point* is determined by adding a few drops of an indicator or by using an instrument such as a pH meter, a *colorimeter* or a conductivity meter.

Sometimes no indicator is needed because the excess reagent itself produces a permanent colour change at the end-point. This happens in *potassium manganate(VII)* titrations.

The diagram on page 356 shows the apparatus for a titration involving a solution A which reacts with solution B. Suppose the equation for the reaction takes the form:

$$n_A A + n_B B \longrightarrow products$$

which means that n_A moles of A reacts with n_B moles of B.

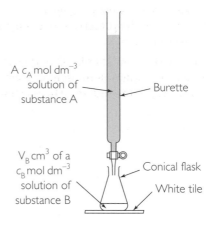

A c_A mol dm^{-3}
solution of
substance A

Burette

V_B cm^3 of a
c_B mol dm^{-3}
solution of
substance B

Conical flask

White tile

Apparatus for a titration

In the laboratory volumes of solutions are normally measured in cm^3 but they should be converted to dm^3 in calculations so that they are consistent with the units used to measure concentrations. (1 dm^3 = 1000 cm^3)

The concentration of B in the flask is c_B mol dm^{-3} and its volume is V_B dm^3.

The concentration of A in the burette is c_A mol dm^{-3}. V_A dm^3 of the solution A are added until the indicator shows that the end-point has been reached.

The amount of B in the flask at the start = $V_B \times c_B$ mol

The amount of A added from the burette = $V_A \times c_A$ mol

The ratio of these amounts must be the same as the ratio of the amounts shown in the equation:

$$\frac{V_A \times c_A}{V_B \times c_B} = \frac{n_A}{n_B}$$

In any titration, all but one of the values in this formula are known. The one unknown is determined from the results.

(See *acid–base titrations* and *iodine–thiosulfate titrations* and *complex-forming titrations* for worked examples.)

Tollen's reagent is a test reagent used to distinguish *aldehydes* from *ketones*. Warming Tollens' reagent with an aldehyde produces a precipitate of silver which coats clean glass with a shiny layer of silver so that it acts like a mirror.

The reagent consist of an alkaline solution of diamminesilver(I) ions, $[Ag(NH_3)_2]^+$. Aldehydes reduce the silver ions to metallic silver. Ketones do not react.

toughness: a property of materials which measures how much energy is needed to break them. Tough materials are hard to break. Metals and polymers are tough. Ceramics and glass are not tough – they are brittle.

toxic substances are labelled with the skull and cross-bones *hazard warning sign* because they can lead to serious, acute or chronic health risks or even death. Toxic substances can cause harm if inhaled, swallowed or absorbed through skin.

Toxic substances include poisons, *carcinogens*, *mutagens* and *teratogens*.

trace elements: the micronutrients which plants need in small amounts from the soil. Examples are cobalt (Co^{2+}), copper (Cu^{2+}), iron (Fe^{2+} or Fe^{3+}), manganese (Mn^{2+}) and zinc (Zn^{2+}). Plants need these nutrients in much smaller amounts than the nitrogen, phosphorus and potassium supplied by NPK *fertilisers*.

tracers: chemicals used to keep track of chemicals in the environment, in living organisms, in a chemical process or in many other circumstances where chemicals are on the move. The tracer is chosen so that it does not interfere with the changes to be studied. For this reason radioactive isotopes are often chosen as tracers because they are chemically identical with the substances being tracked but easily detected from the alpha, beta or gamma particles emitted during radioactive decay. Examples of radioactive tracers include the use of:

- iodine-131 to study the behaviour of the thyroid gland
- hydrogen-3 (*tritium*) to investigate the metabolism of drugs in the human body
- phosphorus-32 to explore the uptake of phosphate fertilisers from the soil by plants.

transition metals are *d-block elements* which have partially filled *d* energy levels in one or more of their *oxidation states*. In the *d*-block series from scandium to zinc this definitions includes all of the metals except for the first and the last. Scandium, $[Ar]3d^14s^2$, is often excluded because it only forms compounds in the +3 state and it loses all its three outer electrons when it forms a 3+ ion. *Zinc*, $[Ar]3d^{10}4s^2$, is excluded because all its compounds are in the +2 state. Losing two electrons gives an ion Zn^{2+} with the electron configuration $[Ar]3d^{10}$ in which all the *d*-energy levels are full.

Transition metals share a number of common features. They:

- are *metals* with useful mechanical properties and high melting points
- form compounds in more than one *oxidation state*
- form *coloured compounds*
- form a variety of *complex ions*
- act as *catalysts* either as metals or as compounds.

Many transition metal salts are *paramagnetic* because of the presence of unpaired electrons in the partially filled inner *d* energy levels.

Sc	Ti	V	Cr	Mn	Fe	Co	Ni	Cu	Zn
				+7					
			+6						
		+5							
	+4	+4		+4					
+3	+3	+3	+3		+3	+3			
					+2	+2	+2	+2	+2
								+1	

Main oxidation states of the elements Sc to Zn. Note that scandium and zinc form ions only in one state.

The chemistries of *chromium, cobalt, copper, iron, manganese, nickel, titanium* and *vanadium* provide many examples of these features.

transition state: the state of the reacting atoms, molecules or ions when they are at the top of the activation energy barrier for a reaction step. Transition states exist for such a brief moment that they cannot be detected or isolated, unlike the *intermediates in reactions* formed between two steps.

The combination of reacting atoms, molecules or ions in a transition state is sometimes called an *activated complex.*

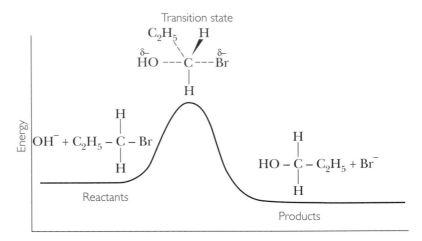

Energy changes during a reaction with a one-step mechanism showing the transition state

transition temperature: the temperature at which two *allotropes* of an element are in equilibrium. One form is stable below this temperature, the other is stable above the transition temperature. (See *sulfur* and *tin.*)

transmittance measures the extent to which a sample in a spectrometer absorbs radiation at a particular wavelength. Printouts from some spectrometers show transmittance on one axis.

Transmittance, T, is defined as the ratio of the intensity of the radiation leaving the sample, I, to the intensity of the radiation entering the sample, I_0.

$$T = \frac{I}{I_o}$$

For a particular wavelength of radiation, transmittance is low when a sample absorbs strongly. It is 100% if the sample does not absorb at all.

trend: a term often used by chemists to describe the way in which a property increases or decreases along a series of elements or compounds. In the periodic table the term can describe the variations of a property down a *group* or across a *period.* In organic chemistry the term may describe the changes in a property from one member of an *homologous series* to the next.

triglycerides are *esters* of *fatty acids* with glycerol (propan-1,2,3-triol) which is an alcohol with three —OH groups. Animal *fats* and *vegetable oils* are examples of triglycerides. Triglycerides, like other *lipids*, are soluble in organic solvents.

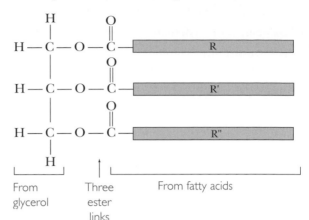

General structure of a triglyceride. In natural fats and vegetable oils, R, R′ and R″ may all be the same or they may be different.

From glycerol

Three ester links

From fatty acids

triiodide ion: the ion that forms when *iodine* dissolves in aqueous potassium iodide.

$$I_2(s) + I^-(aq) \rightleftharpoons I_3^-(aq)$$

Iodine is only very slightly soluble in water. It dissolves in potassium iodide solution because it forms $I_3^-(aq)$. A reagent labelled 'iodine solution' is normally $I_2(s)$ in KI(aq).

The $I_3^-(aq)$ ion is a yellow-brown colour which explains why aqueous iodine looks quite different from a solution of iodine in a *non-polar solvent* such as hexane.

triiodomethane reaction: a reaction which produces triiodomethane (iodoform) from a compound containing either a CH_3C— group (a β-keto group) or a CH_3C— group.

$$\overset{|}{\underset{OH}{}}$$

The reaction conditions are to warm a drop of the compound with a solution of iodine in sodium hydroxide.

The triiodomethane reaction is used as a test for the presence of a methyl group next to a hydroxyl group, or a carbonyl group, in an organic molecule. If the result of the test is positive, triiodomethane appears as a pale yellow precipitate.

The reaction takes place in three steps for an alcohol or two steps for a carbonyl compound. The first step, oxidation to a carbonyl group, is not necessary for a carbonyl compound.

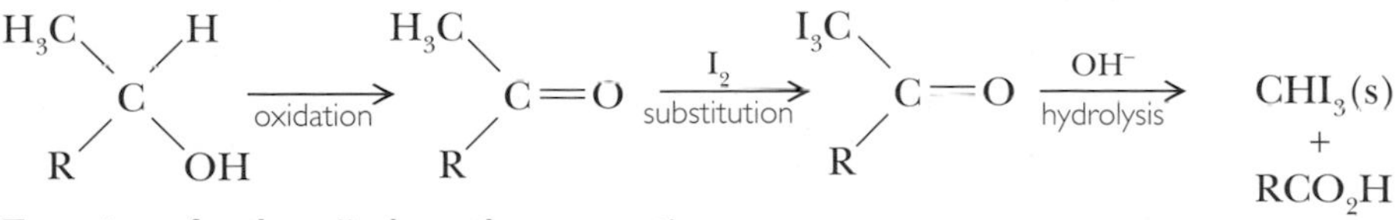

Equations for the triiodomethane reaction

Pure triiodomethane is a lemon-yellow powder which slowly decomposes in the presence of moisture to release iodine. It was once used as a *antiseptic* in wound dressings.

Now its only medical use is as the antiseptic in a paste with bismuth and paraffin for packing abscesses.

Chlorine and bromine in alkali react in a similar way to produce trichloromethane (chloroform) and tribromomethane (bromoform).

trioxygen is the systematic name for *ozone.*

triple bond: three covalent bonds between two atoms as in *nitrogen, alkynes* and the cyanide ion. With three electron pairs involved in the bonding there is a region of high electron density between two atoms joined by a triple bond.

$$\overset{\times}{\underset{\circ}{\text{N}}}\overset{\circ}{\underset{\times}{\text{N}}}\times \qquad\qquad \text{H}\overset{\times}{\underset{\circ}{\text{C}}}\overset{\circ}{\underset{\times}{\text{C}}}\text{H}$$

Examples of molecules with triple bonds $\qquad$ N$\equiv$N $\qquad$ H$-$C$\equiv$C$-$H

The molecular orbital model for a triple bond shows that one of the bonds is a normal σ-*bond* while the other two bonds are π-*bonds.*

triple point: the unique combination of temperature and pressure at which the solid, liquid and gaseous forms of a substance are at equilibrium. The triple point of water is at 611 N m^{-2} (0.006 atmospheres) and 273.16 K (0.01°C).

Three lines meet at the triple point on a *phase* diagram for a pure substance.

tritium is the *isotope* of hydrogen, $^{3}_{1}$H. It is radioactive and emits beta particles with a *half-life* of 12.3 years.

ultraviolet radiation, UV, is *electromagnetic radiation* with shorter *wavelengths* (and so higher frequencies) than the violet end of the visible spectrum. The energy of UV photons is high enough to excite the electrons in covalent bonds to higher energy levels. Excited molecules may then split into atoms (dissociate). Thus UV light can start chemical reactions by forming *free radicals*.

Ozone forms in the upper *atmosphere* (the stratosphere) where UV light from the Sun shines on oxygen molecules, splitting them into oxygen atoms which then combine with other oxygen molecules to make ozone. In the laboratory, UV light can start the reaction of chlorine with alkanes by splitting chlorine molecules into chlorine atoms. This is an example of a *free-radical chain reaction*.

ultraviolet (UV) spectroscopy is particularly useful for studying colourless organic molecules with unsaturated *functional groups* such as $C = C$ and $C = O$. The molecules absorb UV radiation at frequencies which excite shared electrons in double bonds. A UV spectrometer records the extent to which samples absorb UV radiation across a range of *wavelengths*.

In organic molecules with *conjugated systems* the UV absorption peak moves to longer wavelengths as the number of alternating double and single bonds increases. The maximum absorption by ethene is at 185 nm; this shifts to 220 nm for buta-1,3-diene. These are both colourless compounds.

The longer the conjugated chain the stronger the absorption. Beta-carotene with eleven conjugated carbon–carbon double bonds absorbs strongly with a peak shifted so far that it is not in the UV region but lies in the blue region of the visible spectrum (450 nm) so carotene is bright orange (see *coloured compounds*).

UV spectrometers make it possible to extend the techniques of *colorimetry* to colourless compounds. In the *pharmaceutical industry*, for example, scientists use UV spectroscopy to check that medicines contain the correct amounts of drugs and that the products do not deteriorate in storage.

uncertainty of measurement refers to variations in analytical results due to factors which the analyst cannot control. Uncertainty in *titrations* can arise from:

- features of the measuring devices – such as the difficulty of reading a burette if the meniscus lies between two of the graduations
- aspects of the procedure – such as changes in the laboratory temperature which affect the volume of volumetric glassware which is calibrated at a particular temperature
- the behaviour of the analyst – such as the detection of the end point by observing the colour change of an indicator which is a matter of judgement.

Examples of measurement uncertainty for laboratory glassware:

- 50 cm³ grade B burette; ±0.08 cm³ when used to deliver 25 cm³
- 25 cm³ grade B pipette; ±0.04 cm³
- 50 cm³ measuring cylinder; ±0.6 cm³.

Measuring the percentage uncertainty makes it possible to compare the uncertainties due to each measuring device used in an analytical procedure. This makes it possible to identify the measurement that contributes most to the overall uncertainty in the calculated value.

$$\text{percentage uncertainty} = \frac{\text{uncertainty in the result}}{\text{result}} \times 100\%$$

unit cell: the smallest unit of a crystal structure which, when piled up in three dimensions, gives a whole crystal.

The unit cell of a cubic crystal is a minute cube. An atom or ion inside the unit cell entirely belongs to that cell. An atom or ion at a corner is shared between the eight unit cells which meet in three dimensions at a corner.

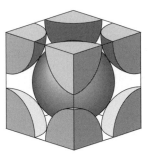

Unit cell for the caesium chloride structure

The unit cell for caesium chloride consists of one caesium ion and $8 \times \frac{1}{8}$ of a chloride ion. So the composition of the unit cell still corresponds to the empirical formula CsCl. (See *caesium chloride structure*.)

universal indicator is a mixture of several *acid–base indicators* which changes colour through the colours of the spectrum from red to indigo as the *pH* rises from 1 to 11. Narrower range indicators are also available which cover a smaller range of pH.

unsaturated compounds are compounds containing one or more double or triple bond between atoms in their molecules. The term is often applied to the *hydrocarbons alkenes* and *alkynes* which typically undergo addition reactions. The term is also commonly used to describe unsaturated *fats* and *fatty acids* which have double $C = C$ bonds in their hydrocarbon side chains.

urea is a white crystalline solid. It is an end product of the metabolism of proteins, and excreted as a waste product in urine.

Structure of urea. Urea is a diamide of carbonic acid so it has the alternative name carbamide.

$$\text{H}_2\text{N} - \overset{\displaystyle \text{O}}{\underset{\displaystyle \text{NH}_2}{\text{C}}}$$

Urea slowly hydrolyses, releasing ammonia, making it a useful nitrogen *fertiliser*. It is manufactured by the reverse process of heating ammonia and carbon dioxide under pressure.

Urea and methanal (formaldehyde) are used to manufacture a range of *thermosetting polymers* which are used as adhesives in chip board, for producing electric sockets and plugs.

vacuum distillation: distillation under reduced pressure which makes it possible to distil liquids with very high boiling points below the temperatures at which they start to decompose. Lowering the pressure lowers the temperature at which liquids boil.

The process is used in oil refineries to separate the *hydrocarbons* in the residue from fractional distillation at atmospheric pressure. Vacuum distillation separates lubricating oils and waxes from other hydrocarbons fed to the plant for *catalytic cracking*.

Vacuum distillation is also used on a small scale to separate products of synthesis which would decompose if heated to their boiling points at atmospheric pressure.

vacuum filtration: see *Buchner flask and funnel.*

valency theory covers all the theories of chemical bonding. The outer electrons of an atom are its valency electrons which take part in chemical bonding. Electron transfer leads to *ionic bonding* or electrovalency. Electron sharing leads to *covalent bonding.*

Chemists use a variety of ways of describing or predicting the number of bonds formed by the atoms of an element. These include:

- the pattern of charges on the ions in a group of element – *group* 2 metals, for example, form 2+ ions
- the number of covalent bonds which atoms normally form in molecules – carbon, 4; hydrogen, 1; oxygen, 2; nitrogen, 3; and the halogens, 1
- the *co-ordination numbers* in crystals and *complex ions* – giving the numbers of nearest neighbours bonded to an atom
- *oxidation numbers* – which provide a formal code for keeping track of the numbers of electrons taking part in bonding.

None of these are rigid guidelines and there are exceptions to all the rules.

Some writers refer to the VSEPR theory which is short for the 'valence shell electron pair repulsion theory'. This is the theory which helps to predict the *shapes of molecules* from the number of bonding pairs of electrons and lone pairs of electrons in the outer (valence) shell of the central atoms in a molecule.

van der Waals forces are weak *intermolecular forces.* Chemists disagree about the definition of the term 'van der Waals forces'. Some use the term for all intermolecular forces; others exclude *hydrogen bonding*, while another group uses the term only for the weakest attraction between temporary and induced dipoles.

Johannes van der Waals (1837–1923) was a Dutch scientist who studied *real gases* and their deviations from *ideal gas* behaviour. He devised a modified gas equation for real gases. His equation includes two correction factors based on *kinetic theory* – one to allow for the existence of intermolecular forces and the other to allow for the fact that gas molecules have a definite volume.

van der Waals radius: the effective radius of atom when held in contact with another atom by weak *intermolecular forces.*

Atoms do not have a definite size. The apparent size of an atom depends on the way it is bonded to a neighbouring atom, so the *ionic radius* and the *covalent radius* of an atom are not the same. Generally, the stronger the bonding the smaller the effective radius. Intermolecular forces are much weaker than covalent bonds, so van der Waals radii are relatively large.

van der Waals radii determine the effective size of a molecule as it bumps into other molecules in a liquid or gas, and when it packs together with other molecules in a solid.

Covalent radius = 0.114 nm

van der Waals radius = 0.190 nm

Covalent and van der
Waals radii for bromine

vanadium (V) is a *d-block* metal with the electron configuration $[Ar]3d^3 4s^1$. The metal is used to make alloy *steels* which are strong and *tough* making them suitable for machine tools and parts of engines. Vanadium is a typical *transition metal*: it forms coloured compounds in several *oxidation states*, it forms *complex ions* and has compounds which can be used as *catalysts* such as vanadium(V) oxide in *sulfuric acid manufacture.*

In solution, vanadium forms ions in the +2, +3, +4 and +5 oxidation states. The +5 state is available as the yellow solid ammonium vanadate(V).

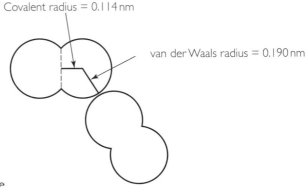

+5	VO_2^+ (aq) yellow
+4	VO^{2+} (aq) blue
+3	V^{3+} (aq) green
+2	V^{2+} (aq) violet
+1	
0	V

Oxidation states of vanadium showing
the colours of the ions

Standard *electrode potentials* help to identify reducing agents to reduce vanadium(V) to the succession of lower states. (See *electrochemical series.*). Iodide ions reduce vanadium(V) to vanadium(VI).

$$\overset{\ominus}{E}/V$$

$$I_2(aq) + 2e^- \rightleftharpoons 2I^- \qquad\qquad +0.54$$

$$VO_2^+(aq) + 2H^+(aq) + e^- \rightleftharpoons VO^{2+}(aq) + H_2O(l) \qquad +1.00$$

Half equations and electrode potentials for the reduction of vanadium(V) to vanadium(IV)

Copper metal reduces vanadium(V) to vanadium(III).

Zinc in acid will reduce vanadium(V) all the way to vanadium(II). The $V^{2+}(aq)$ ion is a strong *reducing agent*.

$$Zn^{2+}(aq) + 2e^- \rightleftharpoons Zn(s) \qquad E^{\ominus} = -0.76\,V$$

$$V^{3+}(aq) + e^- \rightleftharpoons V^{2+}(aq) \qquad E^{\ominus} = -0.26\,V$$

Half equations and electrode potentials for the reduction of vanadium(III) to vanadium(II)

vapour pressure is the pressure of a vapour in a closed container with its own liquid. The full term is either:

- **equilibrium vapour pressure** – a reminder that in a closed container a liquid and vapour reach a state of dynamic equilibrium
- **saturated vapour pressure** – a reminder that the atmosphere above the liquid holds as much vapour as it can at equilibrium at a particular temperature.

The vapour pressure of a liquid rises as the temperature rises. A liquid starts *boiling* when its vapour pressure equals the external pressure.

A solution of an involatile substance in a liquid has a lower vapour pressure than the pure solvent. This follows from *Raoult's law*. Because of this effect the solution has to be hotter before it starts to boil at a given pressure. Dissolving an involatile salt in water raises its boiling point slightly.

The lowering of vapour pressure depends on the mole fraction of dissolved particles. It does not depend on the chemical nature of the particles. (So the lowering of vapour pressure is an example of a *colligative property*.)

vapours are gases formed by evaporation of substances which are usually liquids or solids at room temperature. So chemists talk about oxygen gas but water vapour.

Vapours are easily condensed by cooling or increasing the pressure because of relatively strong *intermolecular forces*. Vapours therefore tend to deviate markedly from *ideal gas* behaviour.

Physicists sometimes broaden the definition of a vapour to includes gases such as butane, ammonia and carbon dioxide which can be liquefied at room temperature by increasing the pressure. (These are gases below their *critical temperature.*)

vat dyes: dyes such as indigo which are insoluble in water but can be converted to a soluble form to dye cloth. Chemical reduction converts indigo to an almost colourless, water-soluble dye which soaks into cloth such as denim for jeans. Oxidation converts indigo back to the insoluble blue form which precipitates in the fibres.

vegetable oils are liquids extracted from plants which are esters of fatty acids with propan-1,2,3-triol. They are *tri-glycerides* which belong to the broad class of *lipids.* The *fatty acids* in vegetable oils contain a higher proportion of *unsaturated* fatty acids than animal *fats.* Examples of vegetable oils are olive oil, sunflower oil and palm oil.

Triglycerides with unsaturated fatty acids have a less regular structure than saturated fats. The molecules do not pack together so easily to make solids, so they have lower melting points and have to be cooler before they solidify.

vibration of bonds: the spring-like vibration of covalent bonds as they bend and stretch. A vibrating *polar covalent bond* is an oscillating *dipole* which can interact with infra-red radiation. Like other energy changes involving atoms and electrons, energy is gained or lost in fixed amounts (quanta). According to *quantum theory,* $\Delta E = h\nu$. It turns out that the sizes of the energy jumps for vibrating bonds correspond to frequencies in the infra-red region of the spectrum. Each type of bond has a particular value for the energy jumps so it absorbs at a characteristic frequency. This is the basis of *infra-red spectroscopy.*

viscous liquids are thick and sticky liquids such as syrups and treacle. Some liquids are viscous because they consist of a tangled mass of long chain molecules. Lubricants from crude oil consist of *hydrocarbons* with chains of over 25 carbon atoms in the molecules.

Other liquids are viscous because of extensive *hydrogen bonding.* Propan-1,2,3-triol (glycerol) is a sticky liquid for this reason. It flows much more slowly than propan-1-ol. Hydrogen bonding also contributes to the high viscosity of concentrated solutions of sugars in water.

On an atomic scale toffee and *glass* have disordered structures like liquids. They are, however, so viscous and flow so slowly that they are effectively solids.

visible radiation is *electromagnetic radiation* with *wavelengths* between 400 nm and 700 nm. This band of radiation is visible because it can bring about reversible chemical changes in cells of the retina. These chemical changes lead to electrical impulses in the nerve cells of the eye interpreted in the brain as colours.

Coloured compounds absorb radiation in the visible region of the spectrum. Chemists study these compounds using visible spectroscopy or *colorimetry.*

vitamins are a group of chemically unrelated organic compounds which are needed in very small amounts in the diet for healthy growth and body functions. The B vitamins and vitamin C are soluble in water. The other vitamins are fat soluble.

The B vitamins are essential for the activity of some *enzymes.* Lack of vitamin B_{12} leads to a form of anaemia. One of the triumphs of twentieth century organic chemistry has been the analysis and total synthesis of vitamin B_{12}. Vitamin C is *ascorbic acid.*

volatile substances are solids or liquids which evaporate easily. A volatile substance easily turns into a *vapour*. *Iodine* is an example of a volatile solid. Most familiar volatile compounds are molecular liquids at room temperature. Examples are water, hexane, *ethanol* and *propanone*.

The equilibrium *vapour pressure* of a substance is a measure of its volatility at a particular temperature.

Makers of *perfumes* take advantage of differences in volatility.

voltmeters measure the potential difference between two points in a circuit in volts. Chemists use voltmeters to measure the *emfs* of electrochemical cells.

The value needed when determining *standard electrode potentials* is the emf when no current is flowing between the electrodes. These values are measured with an accurate voltmeter with a very high internal resistance.

volume is the amount of space taken up by a sample. The *SI unit* of volume is the cubic metre (m^3). Gas volumes are converted to m^3 before substituting values into the *ideal gas* equation.

Chemists generally measure volumes in cubic decimetres (dm^3, formerly in *litres*) or cubic centimetres (cm^3).

Measuring volumes of gases and solutions is an essential part of quantitative chemistry. (See *gas volume calculations, molar volume, reacting volumes of gases* and *volumetric analysis*.)

volumetric analysis: a very important aspect of quantitative analysis because volumetric methods can be very *accurate*. An analyst can deliver a volume of liquid from a 50 cm^3 grade A burette with *uncertainty of measurement* of ±0.06 cm^3. If the total volume is 25 cm^3, this represents an uncertainty of about 1 part in 400. About 80% of the methods recommended by the British Pharmacopoeia for measuring amounts of drugs and medicines are based on volumetric techniques (see *titrations*).

washing powders: see *detergents*.

wastes are unwanted solids, liquids and gases discarded from homes, commerce, industry, agriculture and the public services. They can help to conserve resources and limit the amount of waste dumped in *landfill* by:

- **reduction** – the chemical industry has an extensive programme of research and development to introduce new processes which give higher yields and create less waste so that the disposal problem is cut
- **reuse** – this depends on solvents and detergents which help to clean products or containers so that they can be used again
- *recycling* – a variety of physical and chemical processes recover metals, polymers and other materials so that they can be added to the raw materials used to make new products; in some industries, such as the steel industry, recycling is well established
- **recovery** – this includes the recovery of the energy resources tied up in materials; *incineration* not only disposes of the waste but provides energy to generate electricity. However, incineration can be hazardous if not carried out at a high enough temperature to ensure that harmful chemicals do not escape into the air.

water (H$_2$O) is a familiar liquid with remarkable properties. Despite the small size of its molecules, water is a liquid at room temperature because of *hydrogen bonding*. The O—H bonds in water molecules are highly polar because oxygen is so *electronegative*. Water molecules too are polar because they are not linear thanks to the two lone pairs of electrons which help to determine the *shape of the molecule*.

As a *polar solvent*, water can hydrate ions and dissolve salts. Water can also dissolve some organic compounds because it can form hydrogen bonds with hydroxyl groups in alcohols, sugars or carboxylic acids and with —NH$_2$ groups in amines.

Water plays an important part in many reactions:

- **hydration reactions** – water forms aquo *complex ions* with metal ions
- **acid–base reactions** – water acts as both an acid and as a base because it is an *amphoteric compound*
- **redox reactions** – reactive metals such as *group 1* elements reduce the hydrogen in water to hydrogen gas
- **hydrolysis reactions** – this includes the *hydrolysis of non-metal halides* and of *organic compounds* such as esters (*acid catalysis* or *base catalysis*).

water cycle: the cycling of water in the environment between the oceans, the atmosphere and the crust of the Earth.

The specific latent heat of vaporisation of water is relatively large so huge amounts of energy are transferred by the water cycle. Energy from the Sun evaporates water in the tropics. Winds carry the warm moist air to higher latitudes where the energy is released as the water condenses and falls as rain.

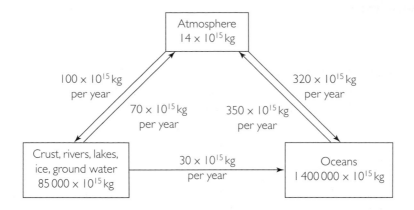

The water cycle, showing the amounts of water in each of the three main reservoirs and the flows between them

water of crystallisation: water molecules which make up part of the crystal structure of a compound. There are five molecules of water of crystallisation for each $CuSO_4$ unit in blue copper(II) sulfate crystals, $CuSO_4.5H_2O$. The full systematic name for copper sulfate is tetraaquocopper(II) tetraoxosulfate(VI)-1-water which shows that four of the water molecules *hydrate* the copper ions; the fifth water molecule forms *hydrogen bonds* with sulfate ions.

Heating hydrated crystals drives off the water as steam, leaving the anhydrous salt. Stronger heating may then lead to other changes (see *nitrates* for example). Other salts with water of crystallisation include:

- hydrated sodium carbonate (washing soda crystals) – $Na_2CO_3.10H_2O$
- hydrated magnesium sulfate (Epsom salts) – $MgSO_4.7H_2O$

water treatment provides water for drinking, water for industry and water for medical uses.

Some of the chemical processes used to provide water for homes include:

- **coagulation** – the addition of aluminium(III) or iron(III) salts to add positive ions which help to coagulate negatively charged *colloid* particles so that finely divided solids clump together and settle out
- **adsorption** – removing organic chemicals on the surface of activated *charcoal*
- **disinfection** – use of chemicals to kill micro-organisms which might otherwise cause disease (see *chlorine water treatment*). Increasingly *ozone* or *ultraviolet radiation* are being used in place of chlorine because they are less hazardous and do not react with organic impurities to produce chlorinated hydrocarbons which may be harmful
- **water softening** – removing calcium and magnesium ions from *hard water* by precipitating them as insoluble carbonates.

Ion exchange can remove all the ions from water to give very pure water needed industrially and commercially.

wavelength: the distance between the peaks (or troughs) of a wave. Wavelenghs of *electromagnetic radiation* vary from about 1000 m for radiowaves down to about 10^{-3} nm for gamma rays.

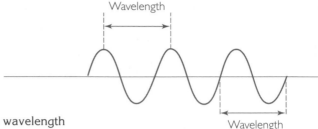

A waveform showing the wavelength

Light in the *visible region* of the spectrum has wavelengths from about 400 nm at the blue end to 700 nm at the red end.

All electromagnetic radiation travels at the same speed, c, in a vacuum. The *frequency*, v, wavelength, λ, and speed are simply related by $c = v\lambda$.

waxes are materials which can be moulded when warm but are hard and brittle when cold. They are insoluble in water and water repellent. Waxes are ingredients of polishes and are used to waterproof cloth and leather. Natural waxes such as beeswax from honeycombs and lanolin from wool are esters of *fatty acids* with alcohols with only one hydroxyl group. This distinguishes them from *fats* and *vegetable oils*. The formula of beeswax is $C_{15}H_{31}CO_2C_{30}H_{61}$. Carnauba wax, an ingredient of polishes, varnishes and lipstick, comes from the leaves of a palm tree which grows in Brazil.

Hydrocarbon wax, such as the paraffin wax used for candles, is one of the products from the *vacuum distillation* of the residue from the *fractional distillation* of crude oil.

The chemical industry makes synthetic waxes by polymerising *epoxyethane*.

weak acids are only slightly ionised when they dissolve in water. In a 0.1 mol dm^{-3} solution of ethanoic acid, for example, only about one in a hundred molecules reacts with water to form *oxonium ions*. The more dilute the solution the greater the degree of ionisation. The *acid dissociation constant*, K_a measures the strength of an *acid*.

Note the important distinction between strength and concentration. Strength is the extent of ionisation. *Concentration* is the amount of acid present in mol dm^{-3}. Note that it takes as much sodium hydroxide to neutralise 25 cm^3 of 0.1 mol dm^{-3} of a weak acid (such as ethanoic acid) as it does to neutralise 25 cm^3 of 0.1 mol dm^{-3} of a strong acid such a hydrochloric acid.

Measuring the *pH* of a solution of acid is not enough to show whether or not the acid is strong or weak. A solution of an acid with pH 4 might be a very dilute solution of a strong acid or a concentrated solution of a weak acid.

The salts made from weak acids and strong bases are alkaline in solution (see *hydrolysis of salts*).

weak bases are only slightly ionised when they dissolve in water. In a 0.1 mol dm^{-3} solution of *ammonia*, for example, only about one in a hundred molecules reacts with water to form ammonium ions. The *base dissociation constant*, K_b measures the strength of a *base*.

As with *weak acids*, it is important to distinguish between strength and concentration.

The salts made from weak bases and strong acids are acidic in solution (see *hydrolysis of salts*).

weathering is a process which breaks down rocks physically and chemically. Physical weathering cracks rocks into smaller fragments increasing the surface area exposed to chemical attack. During chemical weathering rocks react with water, oxygen and carbon dioxide. The types of reaction involved are *hydrolysis* and *oxidation*. Weathering releases soluble ions which plants need for growth. It also forms insoluble *clay minerals*.

weight, measured in newtons (N), is the pull of the Earth's gravity on an object. A laboratory balance responds to weight but is calibrated to give readings in grams or kilograms which measure the mass of the sample. On the surface of the Earth the pull of the Earth is proportional to mass.

$$\text{weight} = \text{mass} \times g \quad \text{where the constant } g = 9.8 \text{ N kg}^{-1}$$

So chemists use 'weighing' to determine the mass of a sample. This accounts for the continuing use of concentration units such as weight/volume per cent (w/v) which should strictly be the mass/volume per cent.

wetting happens when water spreads out and covers the surface of a solid. Water wets clean glassware but breaks up into separate droplets on greasy glassware. This makes it easy to see if pipettes and burettes are clean.

Wetting is an important step in getting materials clean but water does not wet fabrics which are contaminated with greasy dirt. One of the functions of *surfactants* in *detergents* is to lower the *surface tension* of water so that it will wet dirty surfaces.

Surfactants are added to pesticides so that when sprayed onto plants the solution spreads over the leaves and is absorbed. With no surfactant the solution would break up into tiny droplets on the surface of the leaves.

word equation: an equation which describes a chemical change in words instead of symbols. Writing word equations identifies the reactants and products, so it is a useful first step towards *balanced equations* with symbols. For example:

$$\text{sodium} + \text{water} \longrightarrow \text{sodium hydroxide} + \text{hydrogen}$$

Word equations are useful as ways of summarising general patters of behaviour. For example:

$$\text{acidic oxide} + \text{water} \longrightarrow \text{oxoacid}$$

$$\text{carboxylic acid} + \text{alcohol} \longrightarrow \text{ester} + \text{water}$$

X-ray crystallography is a technique used to determine *crystal structures* by interpreting the diffraction patterns formed when *X-rays* are scattered by the electrons of atoms in crystalline solids. Interference between the scattered X-rays produces patterns recorded on photographic film. Crystallographers have worked out techniques for interpreting the patterns so that they can determine:

- the arrangements of atoms, molecules or ions in crystals
- the sizes of atoms and ions
- the *shapes of molecules.*

Since X-rays are scattered by electrons it is possible to draw up *electron density maps* from the diffraction patterns showing the positions of atoms. Small atoms such as hydrogen with few electrons do not show up well, so their position can be hard to determine.

The use of X-ray crystallography to determine the structures of crystals was pioneered by Lawrence Bragg (1890–1971) who developed the theory, and his father William Bragg (1862–1942) who designed the instruments.

Among the great successes of X-ray crystallography were the determination of the structures of:

- the protein α-helix by Linus Pauling in 1951
- myoglobin and haemoglobin and John Kendrew and Max Perutz in the 1950s
- DNA by Rosalind Franklin, Maurice Wilkins, Francis Crick and James Watson in 1953
- vitamin B_{12} by Dorothy Hodgkin in 1956.

X-rays are electromagnetic radiation with very short wavelengths in the region 1×10^{-9} m (1 nm). X-rays have wavelengths of the same order of magnitude as the lengths of atomic bonds. As a result it is possible to use X-ray diffraction to investigate *crystal structures* and the *shapes of molecules* including long chain molecules such as *proteins* and *nucleic acids.*

xenon (Xe) is a noble gas which makes up less than 0.1% of the atmosphere. Xenon is separated from liquid air by *fractional distillation.* For a long time chemists thought that the noble gas elements of *group 8* were completely unreactive so they called them the 'inert gases'. This changed in 1962 when the British-born Canadian chemist Neil Bartlett was working on the properties of platinum fluorides. Based on the accidental discovery that platinum(VI) hexafluoride could combine with oxygen molecules, he used values for *ionisation energies* to predict the formation of a compound between platinum(VI) hexafluoride and xenon and produced the yellow solid $Xe^+PtF_6^-$. Since then chemists have produced a range of xenon compounds but only with the most reactive elements oxygen and fluorine. Examples are XeF_2, XeF_4 and XeF_6.

yield calculations are used to assess the efficiency of a chemical synthesis. A perfectly efficient reaction would convert all of the starting material to the desired product. This would give a 100% yield.

Few reactions are completely efficient and most reactions, especially organic reactions, give lower yields. There are several reasons why the overall yield may be low:

- the reaction may be incomplete (perhaps because it is slow or because it reaches an equilibrium state) so that a proportion of the starting chemicals fails to react
- there may be side-reactions producing by-products instead of the required chemical
- recovery of all the product from the reaction mixture is usually impossible
- some of the product is usually lost during transfer of the chemicals from one container to another when the product is separated and purified.

The 'theoretical yield' is the mass of product assuming that the reaction goes according to the chemical equation and the synthesis is 100% efficient.

The 'actual yield' is the mass of product obtained.

The 'percentage yield' is given by this relationship:

$$\text{percentage yield} = \frac{\text{actual yield}}{\text{theoretical yield}} \times 100\%$$

Worked example:

What is the theoretical yield of aspirin when 15.5 g 2-hydroxybenzoic acid reacts with excess ethanoic anhydride? What is the percentage yield if the actual yield of aspirin is 7.25 g?

Notes on the method

Start by writing the equation for the reaction. This need not be the full balanced equation so long as the equation includes the *limiting reactant* and the product.

Since the ethanoic anhydride is in excess the limiting reactant is the 2-hydroxybenzoic acid. This means that the ethanoic anhydride can be ignored during the calculation.

Answer

The equation:

The molar mass of 2-hydroxybenzoic acid, $C_7H_6O_3$ = 138 g mol^{-1}

The amount of 2-hydroxybenzoic acid at the start of the synthesis

$$= \frac{15.5 \text{ g}}{138 \text{ g mol}^{-1}} = 0.112 \text{ mol}$$

1 mol of the acid produces 1 mol of aspirin.

The molar mass of aspirin, $C_9H_8O_4 = 180$ g mol^{-1}

The theoretical yield of aspirin $= 0.112$ mol $\times 180$ g mol^{-1} $= 20.2$ g

Percentage yield $= \dfrac{7.25 \text{ g}}{20.2 \text{ g}} \times 100\% = 35.9\%$

Z

zeolites are sodium aluminium *silicates* in which the three-dimensional structures of the silicon and oxygen atoms form tunnels and cavities into which ions and small molecules and ions can fit.

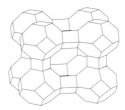

Model of a zeolite crystal structure

Synthetic zeolites can make excellent *catalysts*. They can be developed with active sites to favour the required reactions by acting on molecules with particular shapes and sizes. They are used for *catalytic cracking*.

Natural and synthetic zeolites are *ion exchangers*. Permutit is a synthetic zeolite used to soften water by swapping the calcium ions in *hard water* with sodium ions. For this reason many washing powders contain up to 25% of zeolites.

Zeolites can be very good drying agents; they can absorb water molecules selectively because the water molecules fit into the holes in their crystal structure.

Similarly, zeolites can acts as 'molecular sieves', sorting out molecules by size. Some zeolites will absorb nitrogen from the air while letting oxygen pass through. This is one of the methods used to separate *oxygen* from the *air* on a large scale. Once the zeolite is saturated with nitrogen it can easily be regenerated by lowering the pressure and releasing the waste nitrogen back into the air.

zero order reaction: a reaction is zero order with respect to a reactant if the rate of reaction is unaffected by changes in the concentration of that reactant. The concentration term for this reactant is raised to the power zero in the *rate equation*. (See diagram on page 376.)

$$\text{Rate} = k[X]^0 = k \text{ (a constant)} \quad \text{since } [X]^0 = 1$$

(Any term raised to the power zero equals 1.)

The rate of reaction of iodine with propanone in the presence of acid is zero order with respect to iodine.

$$\text{Rate} = k[\text{propanone}]^1[\text{hydrogen ion}]^1[\text{iodine}]^0$$

Ziegler–Natta catalysts speed up the polymerisation of *alkenes* such as ethene and propene making it possible to produce *addition polymers* at relatively low temperatures and pressures. Ziegler–Natta catalysts are made from titanium(IV) chloride and aluminium alkyls (such as triethyl aluminium) in a hydrocarbon solvent.

There is little chain branching in the polymers, so the poly(ethene) chains formed by this method can pack closely together, forming the high density form of poly(ethene). Poly(propene) made with a Ziegler–Natta catalyst has a regular

APPENDIX I
REVISION LISTS

This appendix helps you with study and revision. Pick the topic you want to revise, then look up the terms listed in alphabetical order under each heading. Cross-references in the text will help you to build on these starting points. Note that some important terms appear more than once in the lists.

Appendix 4 shows you which topics you need to revise for each module of the main AS and A level Chemistry courses. Note that the lists include a few key terms which may not be required for your course or which may be excluded from the examinations. Studying these extra key terms will help your overall understanding of the topics. Check with the specification for your course to find which key terms you have to know and understand in detail.

I Atoms

Atom
Atomic number
Atomic theory
Fundamental particles
Isotopes
Mass spectrometer
Mass number
Relative atomic mass
Relative isotopic mass
Relative molecular mass

2 Electrons in atoms

Atomic orbitals
Atomic spectrum
Aufbau principle
Electron configuration
Energy levels
Hydrogen spectrum
Ionisation energies
Quantum numbers
Shielding

3 Chemical amounts and formulae

Amount of substance
Avogadro constant
Empirical formula
Molar mass
Mole
Molecular formula
Percentage composition

Relative atomic mass
Relative formula mass
Relative molecular mass

4 Chemical amounts and equations

Acid–base titrations
Amount of substance
Balanced equations
Chemical equations
Concentrations of solutions
Formula unit
Gas volume calculations
Limiting reactant
Molar mass
Molar volume
Reacting masses
Titrations

5 States of matter

Boiling
Changes of state
Ideal gases
Giant structures
Kinetic theory of gases
Liquids
Molecule
Melting
Real gases
Solids
Vapours

6 Gases

Ideal gas equation
Kinetic theory of gases
Gas laws
Molar mass
Molar volume
Real gases
SI units
Vapours

7 Structure and properties

Allotropes
Body-centered cubic structure
Close-packed structures
Crystal structures of ionic compounds.
Crystal structures of metals
Crystal structures of non-metals
Giant structures
Ice
Metals
Non-metals
Physical properties
Plastics
Sodium chloride structure

8 Bonding

Atomic radius
Co-ordinate bond
Covalent bonds
Covalent radius
Dative covalent bond
Dot and cross diagrams
Ionic bonding
Ionic radius
Intermolecular forces
Metallic bonds
Octet rule

9 Bonding in ionic compounds

Complex ions
Co-ordination compounds
Electrolysis
Hydration
Ionic bonding
Fajan's rules
Polarisability
Polarising power

10 Bonding in molecules

Bond angles
Bonding molecular orbitals
Covalent bonding
Dative covalent bond
Delocalised electrons
Electronegativity
Intermediate bonding
Multiple bonds
Polar covalent bonds
Shapes of molecules

11 Bonding between molecules

Dipole–dipole interactions
Hydrogen bonding
Ice
Intermolecular forces
Lone pair of electrons
Polar molecules
Van der Waals forces

12 Energetics

Bond enthalpies
Calorimeter
Endothermic reaction
Energy (enthalpy) level diagrams
Enthalpy changes
Enthalpy change of combustion
Enthalpy change of formation
Enthalpy change of neutralisation
Enthalpy change of reaction
Exothermic reaction
Hess's law
Stability of compounds
Spontaneous reaction
System

13 Redox reactions

Electron transfer
Half-equation
Oxidising agents
Oxidation
Oxidation numbers
Oxidation states
Redox reactions
Reducing agents
Reduction

revision lists

Substitution reactions
Yield calculations

21 Hydrocarbons

Addition polymerisation
Alkanes
Alkenes
Carbocations
Electrophilic addition
Epoxyethane
Free-radical chain reactions
Hydrogenation
Markovnikov's rule
Saturated compounds
Unsaturated compounds

22 Organic halogen compounds

Addition polymerisation
CFCs
Elimination reaction
Fluorocarbons
Halogenoalkanes
Hydrolysis
Leaving group
Nitriles
Nucleophilic substitution in
   halogenoalkanes
Polar covalent bonds
Polar molecules

23 Alcohols

Alcohols
Aldehydes
Dehydration
Diols
Elimination reaction
Esters
Ethanol
Fermentation
Ketones
Methanol
Oxidation

24 Petrochemicals

Arenes
Catalytic cracking
Crude oil

Fractional distillation of oil
Isomerisation
Refining
Reforming
Thermal cracking
Zeolite

25 Fuels

Acid rain
Biofuels
Catalytic converter
Combustion
Fuels
Greenhouse effect
Knocking
Petrol
Photochemical smog

26 Waste

Acid rain
Biogas
Dioxins
Incineration
Landfill
Pollution
Recycling
Toxic substances
Wastes

27 Experimental skills

Acid–base titration
Anion tests
Calorimeter
Cation tests
Distillation
Errors of measurement
Flame tests
Gas tests
Hazard warning signs
Organic analysis
Recrystallisation
Titration
Uncertainty

28 Kinetics

Arrhenius equation
First order reaction

revision lists

Half-life (chemical)
Initial rate method
Rate determining step
Rate equations
Rates of reaction
Reaction kinetics
Second order reaction
Transition state
Zero order reaction

29 Reaction mechanisms

Carbocation
Curly arrows
Electrophilic addition
Electrophilic substitution
Free-radical chain reactions
Markovnikov's rule
Mechanism of a reaction
Nitration of benzene
Nucleophilic addition reaction
Nucleophilic substitution in derivatives
    of carboxylic acids
Nucleophilic substitution of
    halogenoalkanes

30 The equilibrium law

Concentration
Equilibrium law
K_c
K_c
Heterogeneous equilibrium
Homogeneous equilibrium
Mole fraction
Partial pressures
Phase
Temperature effect on equilibria

31 Acid–base equilibria

Acid dissociation constants
Acid–base indicators
Brønsted–Lowry theory
Buffer solution
Enthalpy of neutralisation
Henderson–Hasselbalch equation
Ionic product of water, K_w
pH changes during acid–base titrations
pH scale

Proton
Weak acids

32 Redox equilibria

Oxidation numbers
Disproportionation
Electrochemical cells
Electrochemical series
Emf
Half-equations
Hydrogen electrode
Redox reactions
Reference electrodes
Standard electrode potentials

33 Thermochemistry of ionic compounds

Born–Haber cycle
Electron affinity
Enthalpy change of atomisation
Enthalpy change of formation
Enthalpy change of hydration
Enthalpy change of solution
Lattice energy (enthalpy)

34 Energetics and the direction of change

Entropy
Equilibrium law
Feasibility
Spontaneous reaction
Standard molar entropy
Thermochemistry

35 Periodicity

Acidic oxide
Aluminium(III) ions
Amphoteric oxides
Basic oxide
Hydrolysis of non-metal halides
Hydrolysis of salts
Metals
Non-metals
Oxides
Periodicity
Periodicity of chemical properties

revision lists

APPENDIX 2
HINTS FOR EXAM SUCCESS

Your revision should be systematic. Use a copy of the specification to plan your revision.

Make your revision active. For each revision topic, try writing the title of a topic in the centre of a sheet of paper. Then use the definitions and explanations in the main part of this book to help you build up a spider diagram, or concept map ,to show how the ideas in the topic link together.

Suppose you are revising the chemistry of group 7. Have a pile of scrap paper to hand and a pencil. Now, as your read, make jottings, small lists, summary phrases, write equations, sketch diagrams and practise labelling them. Now tear up the paper, close the texts and notes, and write out those lists, equations, diagrams and so on. Then check to see whether you have remembered correctly.

When it comes to learning the reactions of a family of organic compounds such as the alcohols, consider using index cards. Write the equation for each alcohol reaction you have to know on one side of the card. Write the names of the reactants and the conditions for the reaction on the other side. Now you can use the cards for revision. Look at one side of the card and try to recall what is on the other side.

Practice calculations with the help of worked examples. Find questions from a book with the answers in the back so that you can check that you have worked through to the answer correctly.

Look at past papers. The same subject matter appears, year after year, with only very minor differences in emphasis and phrasing of questions. It is because so many questions appear to be the same year after year, that you should read the question with the greatest of care to note the particular emphasis, and precise nature of a question.

Towards the end of your revision programme you will want to look at the more searching 'synoptic' questions, which test your understanding of the inter-relationships in chemistry (see appendix 3).

When answering questions in an exam use the mark scheme to guide you when deciding how much to write. Three marks for part of a question probably means that the examiner is expecting three good points to be made. If asked for a chemical test, for example, one mark might be for the reagent to be used, one for the conditions and the third for describing the observations.

Some questions expect you to write at greater length, drawing together and organising relevant material. No marks are allocated for the plan you sketch out before writing your answer, but making a plan helps you to write a well organised answer with examples and so leads to better marks.

Three don'ts:
- Don't simply rephrase a question in different words.
- Don't use correcting fluid. In an exam it is all too easy to forget to go back and put in the intended correction once the fluid has dried.
- Don't use red or green ink – that's for the examiners. It can be tempting, for instance, to draw curly arrows for a reaction mechanism in a different colour, but you will simply cause aggravation to the examiner if you use red or green.

APPENDIX 3
SYNOPTIC ASSESSMENT

In the second year of an advanced course you are expected to bring together your knowledge and understanding of different areas of chemistry and apply them in contexts that may be new to you. The term 'synoptic assessment' describes questions that you can only answer by drawing on knowledge and skills from a range of topics.

Some of the key term entries in this book are designed to give you an overview of topics so that you can make connections, link ideas and prepare for synoptic assessment.

You can use key terms such as *inorganic chemistry, organic chemistry* and *physical chemistry* to start creating your own overview of the subject. Use the cross references to build up charts showing how ideas link together.

Following up the cross-references from these other key terms will also help you gain a synoptic overview of parts of your course: *atomic theory, bonding, chemical industry, environmental chemistry, group, mechanism of a reaction, organic analysis, organic preparations, organic routes, periodicity, physical properties, pollution, polymer chemistry, qualitative analysis, quantitative analysis, solid, spectroscopy, thermochemistry, valency theory.*

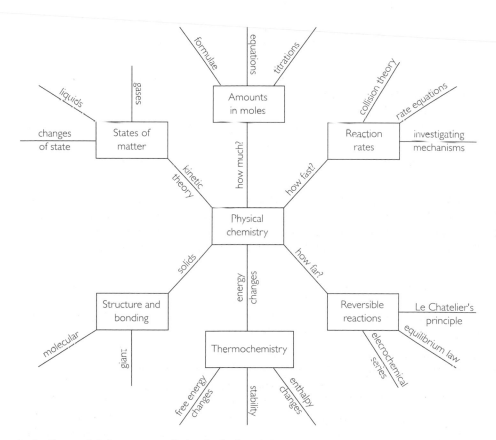

A starting point for a map of physical chemistry

APPENDIX 4
COVERAGE OF SPECIFICATIONS

Use this appendix to plan your studies and preparation for unit or module tests. The tables below cover the ideas covered in the written examinations.

Note that *A–Z Chemistry* also includes key terms to help you with your laboratory work, whether this is assessed as coursework or during a practical examination. Follow the cross-references from *qualitative analysis* and *quantitative analysis* to find the key terms you need.

AQA Chemistry specification

Module with main revision topics (giving the numbers of the topics in Appendix 1)	Other key terms covered by the module	
1 Foundation Chemistry 1 1, 2, 3, 4, 5, 6, 7, 8, 9, 10, 11, 14, 15		
2 Foundation Chemistry 2 12, 13, 17, 18	aluminium extraction chlorine water treatment iodine-thiosulfate titrations iron extraction	metal extraction recycling steels titanium extraction
3 Foundation Chemistry 3 19, 20, 21, 22, 23, 24, 25, 26, 29		
4 Further Chemistry 1 28, 30, 31, 37, 38, 39, 40, 41, 42, 43		
5 Further Chemistry 2 32, 33, 34, 35, 36	autocatalysis amphoteric oxides bond dissociation enthalpy bond enthalpies chromium cisplatin cobalt co-ordination number electroplating	free energy change haemoglobin hydrogen peroxide hydrolysis of salts Lewis acid/base theory photography Tollen's reagent vanadium
6 Further Chemistry 3	Synoptic assessment – see Appendix 3	

Edexcel Chemistry specification A

Unit with main revision terms (giving the numbers of the topics in Appendix 1)	Other key terms covered by the unit	
1 Structure, bonding and main group chemistry 1, 2, 3, 4, 5, 7, 8, 9, 10, 11, 13, 14, 15, 16	group 1	
2 Introductory organic chemistry, energetics, kinetics and equilibrium and applications 12, 17, 18, 19, 20, 21, 22, 23, 25	bauxite bleaching aluminium extraction ammonia manufacture chlorine water treatment	electrolysis of brine fertilisers nitric acid manufacture recycling sulfuric acid manufacture
3 Laboratory chemistry 1 27	errors of measurement risk	
4 Periodicity, quantitative equilibria and functional group chemistry 30, 31, 33, 35, 37, 39, 40	Grignard reagents group 4 inert pair effect	
5 Transition metals, quantitative kinetics and applied organic chemistry 28, 29, 32, 36, 38, 42, 43	cation tests chromium copper drug fats iron margarine	medicine nickel redox titrations soap urea vanadium
6 Laboratory chemistry 2	Synoptic assessment – see Appendix 3	

Edexcel Chemistry specification B (Nuffield)

Unit with main revision terms (giving the numbers of the topics in Appendix 1)	Other key terms covered by the unit	
1 Introductory chemistry 1, 2, 3, 4, 14, 15, 19, 23	acids atomic radius bases calorimetry electron density maps electron transfer endothermic reaction enthalpy changes enthalpy change of formation enthalpy change of reaction exothermic reaction Hess's law	ionic bonding ionic radius ionic precipitation metals metallic bonds redox reactions salts strong acid thermal decomposition weak acid yield calculations
2 Bonding and reactions 7, 10, 11, 13, 16, 17, 19, 20, 21, 22	bond enthalpies bond lengths dipole moments dot-and-cross diagrams dynamic equilibrium enthalpy change of atomisation	enthalpy change of combustion iodine-thiosulfate titrations Le Chatelier's principle reversible changes
4 Energy and reactions 28, 29, 30, 31, 38, 39	aspirin entropy feasibility infra-red spectroscopy	spontaneous reaction standard molar entropy X-ray crystallography
5 The Born–Haber cycle, structure and bonding 32, 33, 34, 35, 36, 37, 40, 41, 42, 43	chromatography combustion analysis Fajan's rules intermediate bonding	polarisation of ions polarising power R_f values stability constants

Special Studies	Key terms which provide starting points for revision of the Special Study	
Biochemistry	active site	nucleic acids
	amino acids	nucleotides
	antibiotics	peptides
	biochemistry	pesticides
	DNA	proteins
	enzymes	RNA
	immobilised enzymes	substrate
	inhibitor	
Chemical engineering	batch process	heat exchangers
	chemical industry	reactors
	continuous process	scaling up processes
	fractional distillation	
Food science	bacteria	fermentation
	carbohydrates	food additives
	enzymes	proteins
	fats	triglycerides
	fatty acids	vitamins
Materials science	amorphous	phase diagram
	annealing	physical properties
	ceramics	plastic materials
	composites	polymers
	co ordination number	polymer chemistry
	corrosion of a metal	recycling
	crystal structures of metals	strength
	density	unit cell
	eutectic metals	waste
Mineral process chemistry	atmosphere	lithosphere
	chemical industry	metal extraction
	density	minerals
	environmental chemistry	ore
	environmental issues	recycling
	leaching	reserves and resources
	froth flotation	wastes
	hydrosphere	zinc extraction

Module with main revision terms (giving the numbers of the topics in Appendix 1)	Other key terms covered by the module	
1 The elements of life; Developing fuels 1, 2, 3, 4, 8, 12, 14, 15, 19, 24, 25	alpha decay alpha particles beta particles catalysts entropy heterogeneous catalyst gamma radiation	nuclear fusion radioactivity shapes of molecules standard molar entropy skeletal formula tracers
2 From minerals to elements; The atmosphere; The polymer revolution; What's in a medicine? 11, 13, 16, 17, 18, 20, 22, 39, 41	acid-base titration acids acid strength of organic hydroxy compounds air alkenes atmosphere bromine extraction Buchner flask and funnel carbon cycle concentration of solutions COSHH regulations crystallinity of polymers crystal structures of ionic compounds crystal structures of non-metals diamond drug drug development electrolysis of brine electromagnetic radiation	electron configuration electrophilic addition geometric isomerism giant structures global warming greenhouse effect infrared spectroscopy ionic precipitation mass spectrometry medicine metal extraction ozone parts per million phenol proton transfer quantum theory risk silica sunscreen thin layer chromatography titration
4 Designer polymers; Engineering proteins; The steel story 28, 30, 32, 36, 37, 40, 41	cation tests corrosion of a metal DNA edta enzymes	recycling stability constants steels wastes

continued

Module with main revision terms (giving the numbers of the topics in Appendix 1)	Other key terms covered by the module	
5 Aspects of agriculture; Colour by design; The oceans; Medicines by design; Visiting the chemical industry 20, 31, 33, 34, 35, 38, 42, 43	active site aldehydes ammonia manufacture atomic emission spectroscopy batch process chemical industry chromophore clay minerals coloured compounds continuous process drug development fats fibre reactive dyes gas-liquid chromatography ice ion exchange ketones	K_p medicine nitrogen nitrogen fixation nitrogen oxides partial pressures partition pesticides pigments scaling up processes silicates solubility solubility product constants triglycerides vegetable oils wastes water

APPENDIX 5
EXAMINERS' TERMS

Every year too many well prepared candidates fail to score as many marks as they should because they do not answer the question set by the examiners.

Examiners try very hard to set questions which are clear to the candidates. Even so, under examination conditions it is all too easy to rush into writing an answer before checking carefully the meaning of the question.

A useful first step is to highlight the words in the question which give instructions. Words such as 'calculate', 'describe' and 'explain'. Here is an A–Z commentary on some of the instructions often used by chemistry examiners.

calculate means that your answers will include a number (and usually units). Always show your working, then you will get almost all the marks even if you make a silly slip while calculating. Most of the marks are given for showing how you arrive at the answer.

With a calculator it may be tempting to work through several steps and then just write down the number you get. You must resist this temptation. Showing the working helps you to give an answer to the correct number of *significant figures* and with the right units. Do not be tricked by your calculator into quoting an answer to more significant figures than are justified by the data.

Some quantities do not have units. If so, it is worth saying so in your answer to show the examiner that you have considered the point. For example, the equilibrium constant, K_p, for the reaction of hydrogen with iodine to form hydrogen is a number with no units. Most equilibrium constants, however, do have units.

comment is an invitation to interpret information given in the question or to relate the information in the question to your knowledge and understanding of the topic.

compare asks you to say something definite about each of the things to be compared. If asked to compare the action of bromine with cyclohexene, and with cyclohexane, you could start by stating that cyclohexene, the unsaturated compound, rapidly decolourises a solution of bromine. You should go on to show that you also know that there is no immediate action with the cyclohexane because it is a saturated hydrocarbon. Setting out your answer in a table is a way of making sure that you say something about each of the chemicals or processes to be compared.

deduce implies that you will work out your answer with the help of information given in the question but using your understanding of the principles involved. Your answer should show the steps in your thinking as well as your conclusions.

define requires a precise statement. Using a simple example often helps to make a definition clearer. Many chemical terms have very specific definitions. This is especially true in thermochemistry. When defining standard *enthalpy change of formation*, it is not good enough to write 'the heat given out when a compound is formed from its elements'. You must specify the conditions of temperature and pressure, the standard states of the elements and the amount of compound formed.

describe asks you to outline in words, and diagrams if helpful, the main points of a topic. If asked to describe and explain the shape of an ammonia molecule you will want to draw both and dot-and-cross diagram to show the bonding electrons and lone pairs as well as a 3D diagram of the molecule with bond angles marked in. Then you will explain the *shape of the molecule* with the help of electron-pair repulsion theory.

'Describe how you would prepare … in the laboratory', asks for words and diagrams outlining the apparatus, the chemicals used and the procedure. A good answer will also include the key observations during the preparation and especially what the product is like.

determine tells you to work out a quantity, formula or equation from the information given in the question.

discuss expects you to give an argument reflecting on the key points in the topic of the question. This might be a review of the benefits and risks of using chemicals such as pesticides. Make sure your answer refers to both sides of the argument where there can be differences of opinion.

explain is one of the most important words on an exam paper. Frequently you will be asked to write down some information and then give an explanation for your answer. Your explanation is likely to be based on a chemical theory. Always give reasons to support your explanation.

estimate probably means that you will have to calculate but that you can make simplifying assumptions to arrive at an order of magnitude answer to the question.

how would you … ? means that you should describe a procedure which you could do. Imagine yourself standing at a bench in the laboratory. Now, what would *you* do? Thinking in this way will help you to avoid describing the Haber process, when asked, 'How would *you* prepare a sample of ammonia?'.

You might be asked 'How would you test for … ?', if so, take care both to write down the procedure for the test and then to describe what you would observe if the test were positive. Remember that observations include smells and temperature changes as well as things which could can see, such as colour changes.

list asks for a list of points – probably just one or two words each. If asked for a certain number of examples in the list do not give more.

name is usually a request for the recommended IUPAC name. You are quite likely to be asked to give the name of an organic structure (see *names of organic compounds*).

outline tells you that the answer can be short with only the key points. Even so, you must include essential details. For instance, even in 'outline', when mentioning sulfuric acid you must state whether it is dilute or concentrated acid.

predict means that you cannot simply recall the answer. You must use information given in the question, or your answers to earlier parts of a question, together with your understanding of chemistry to describe what will happen under specified conditions.

sketch a graph means that what you draw should be the right general shape in the right proportions but the position of the line or curve need only be qualitatively correct. The axes should be labelled but need not have a numbered scale. Sometimes it is important to identify particular features to show that the line passes through the origin or cuts one of the axes at a particular value.

Take care. In an answer about rates of reaction it can be useful to sketch the distribution of molecular kinetic energies in a gas at a given temperature, and superimpose upon it, the distribution when the temperature is ten degrees higher (see *activation energy*). It is all too easy to exaggerate the picture to make a point, and then when the activation energy is added it appears the reaction will be wildly explosive.

sketch a diagram means that a simple outline drawing will be acceptable to the examiners. Make sure that the proportions are about right and that key features are fully labelled and explained.

If asked to, 'Draw/sketch the apparatus used to …', simply take sufficient care to depict the apparatus to be used. There are very seldom any marks allocated for artistic quality, but a good labelled drawing can save a lot of writing. Your drawing should be good enough for you not to have to label the apparatus, only the substances present in the apparatus. Small points are important. Do not draw a stopper at the top of a reflux condenser – it would be explosive.

state or give both ask for short and simple, factual statements.

suggest tells you that you cannot simply recall a right answer. If you are asked for a suggestion it may be that there is more than one possible answer and that you will be given marks for any sensible suggestions. You may also be asked to suggest answers if you are given some new information which you have not met before. The examiners want you to apply your knowledge of chemistry to answer the question in an unfamiliar situation.

why? means 'explain' or 'give reasons'. Be careful not to reword the question without giving an explanation. If asked 'Why does the first *ionisation energy* of the elements in *group 2* decrease with increase in atomic number ?' you have to say more than 'Because there is less attraction on the electrons by the nucleus?'. This is a true statement, but not enough. The examiner expects you to show how *shielding* by inner full shells affects the pull of the nucleus on the outer electrons of atoms such as lithium and caesium.

write, or construct, an equation means that you should describe a reaction with a *balanced equation* complete with chemical formulae. It is not usually good enough to write a 'word equation'.

Sometimes it is very important to include *state symbols*. This is especially true in thermochemistry and electrochemistry. The first ionisation energy of sodium, for example, refers to gaseous atoms so the symbol $Na(g)$ must be used to avoid confusion with the solid element. In general it is better to include state symbols whenever possible especially when aqueous reagents react to form a precipitate or a gas.